Lecture Notes in Economics and Mathematical Systems   567

Founding Editors:

M. Beckmann
H. P. Künzi

Managing Editors:

Prof. Dr. G. Fandel
Fachbereich Wirtschaftswissenschaften
Fernuniversität Hagen
Feithstr. 140/AVZ II, 58084 Hagen, Germany

Prof. Dr. W. Trockel
Institut für Mathematische Wirtschaftsforschung (IMW)
Universität Bielefeld
Universitätsstr. 25, 33615 Bielefeld, Germany

Editorial Board:

A. Basile, A. Drexl, H. Dawid, K. Inderfurth, W. Kürsten, U. Schittko

T0181207

Akira Namatame
Taisei Kaizouji
Yuuji Aruka (Eds.)

The Complex Networks of Economic Interactions

Essays in Agent-Based Economics
and Econophysics

 Springer

Editors

Professor Akira Namatame
Department of Computer Science
National Defense Academy
1-10-20, Hashirimizu, Yokosuka, 239-8686
Japan
e-mail: nama@nda.ac.jp

Taisei Kaizouji
Division of Social Sciences
International Christian University (ICU)
Osawa, Mitaka, Tokyo, 181-8585
Japan
e-mail: kaizoji@icu.ac.jp

Professor Yuuji Aruka
Department of Commerce
Chuo University
3-11, Tama, Tokyo, 180-8585
Japan
e-mail: aruka@tamacc.chuo-u.ac.jp

ISSN 0075-8442
ISBN-10 3-540-28726-4 Springer Berlin Heidelberg New York
ISBN-13 978-3-540-28726-1 Springer Berlin Heidelberg New York

This work is subject to copyright. All rights are reserved, whether the whole or part of the material is concerned, specifically the rights of translation, reprinting, re-use of illustrations, recitation, broadcasting, reproduction on microfilms or in any other way, and storage in data banks. Duplication of this publication or parts thereof is permitted only under the provisions of the German Copyright Law of September 9, 1965, in its current version, and permission for use must always be obtained from Springer-Verlag. Violations are liable for prosecution under the German Copyright Law.

Springer is a part of Springer Science+Business Media

springer.com

© Springer-Verlag Berlin Heidelberg 2006
Printed in Germany

The use of general descriptive names, registered names, trademarks, etc. in this publication does not imply, even in the absence of a specific statement, that such names are exempt from the relevant protective laws and regulations and therefore free for general use.

Typesetting: Camera ready by author
Cover design: *Erich Kirchner*, Heidelberg

Printed on acid-free paper   42/3153DK   5 4 3 2 1 0

Preface

Understanding the mechanism of a socio-economic system requires more than an understanding of the individuals that comprise the system. It also requires understanding how individuals interact with each other, and how the aggregated outcome can be more than the sum of individual behaviors. This book contains the papers fostering the formation of an active multi-disciplinary community on socio-economic systems with the exciting new fields of agent-based modeling and econophysics.

We especially intend to increase the awareness of researchers in many fields with sharing the common view many economic and social activities as collectives of a large-scale heterogeneous and interacting agents.

Economists seek to understand not only how individuals behave but also how the interaction of many individuals leads to complex outcomes. Agent-based modeling is a method for studying socio-economic systems exhibiting the following two properties: (1) the system is composed of interacting agents, and (2) the system exhibits emergent properties, that is, properties arising from the interactions of the agents that cannot be deduced simply by aggregating the properties of the system's components. When the interaction of the agents is contingent on past experience, and especially when the agents continually adapt to that experience, mathematical analysis is typically very limited in its ability to derive the outcome.

Many physicists have contributed to a better understanding of large-scale properties of socio-economic systems, and they open the new research field, "econophysics". An international scientific development has started to gain new insight into the dynamics of socio-economic systems by using methods originally developed in statistical physics and complex theory. This book also covers the current achievements in this rapidly changing field.

This book contains selected papers presented at the 9-th International Workshop on Heterogeneous Interacting Agents (WEHIA), which was held at Kyoto University, Japan, from May 27 to 29, 2004. From the broad spectrum of activities, leading experts presented important papers and numerous practical problems appear throughout this book. We also encouraged papers dealing with applications of agent-based modeling.

WEHIA was initiated as a result of the growing recognition of the importance of agent-based modeling to study large-scale socio-economic systems at University of Ancona, Italy in 1996. The annual series of WEHIA serve for sharing the most recent theoretical applications and methodological advances on agent-based approaches throughout economists, physicists, computer scientists, and other scientists in professionals. The main goals of WEHIA have been to promote interactions and cross-fertilization among different approaches to understanding complex and emergent behaviors and to mange large-scale socio-economic systems.

WEHIA confers especially to encourage papers at the cutting-edge of other approaches that are relevant socio-economic systems. By bringing together three different emerging fields, economics, echonophysics and computer science under the same umbrella, WEHIA stresses the expanding importance of importance close communication and cooperation of the three areas for the future scientific and technological development. The genuinely interdisciplinary approach will enable researchers and students to expand their socio-economic knowledge and to draw up concepts for future interdisciplinary academic achievement.

Based on the success of WEHIA for many years, the new association, "The society for Economic Science with Heterogeneous Interacting Agents" (ES-HIA) (www.es-hia.org) will be launched in 2006. The official society journal, "Journal of Economic Interaction and Coordination" (JEIC) will be published from Springer in 2006. The new society, ESHIA especially features in-depth coverage of important areas and aims to contribute scientific ally in three directions: (1) To examine theoretical and methodological issues of agent-based modeling. (2) To discuss multi-agents based simulations and demonstrate applicability in order to study complex economic behaviors. (3) To contribute to develop methodological tools of agent-based modeling and apply them to complex economic and social problems.

We could solicit many high quality papers that reflect the result of the growing recognition of the importance of the areas. All papers have received a careful and supportive review, and we selected 22 papers out of 94. The contributions were submitted as a full paper and reviewed by senior researchers from the program committee. All authors revised their earlier versions presented at the workshop with reflecting criticisms and comments received at the workshop. The editors would like to thank the program committee for the careful review of the papers and the sponsors and volunteers for their valuable contribution. We hope that as a result of reading the book you will share with us the intellectual excitement and interest in this emerging discipline.

We are grateful to the many people who have made this symposium possible. First and foremost, we thank the authors for providing manuscripts on time and in a standard format. We also thank the many referees who generously contributed time and Dr. Hiroshi Sato to ensure the quality of the finished product.

Finally, we would like to acknowledge the support and encouragement of many peoples in helping us getting this book to be published. Especially the publication of this book and the 9th WEHIA are financially supported by the grant from the Commerative Organization for the Japan World Exposition ('70), Hayasibara Foundation, Kozo Keikaku Engineering Inc. We would like also thank for the grant-in-aid for Scientific Research (C) No.15201038, Japan Society for the promotion of Science (JSPS).

October 2005   *Akira Namatame*
                                                                 Taisei Kaizoji
                                                                 Yuji Aruka

Contents

Part I Econophysics

Five Years of Continuous-time Random Walks in Econophysics
Enrico Scalas.. 3

Why Macroeconomic Price Indices are Sluggish in Large
Economies ?
Masanao Aoki, Hiroshi Yoshikawa.................................. 17

Growth Volatility Indices
Davide Fiaschi, Andrea Mario Lavezzi 37

Financial Fragility and Scaling Distributions in the Laboratory
Giovanna Devetag, Edoardo Gaffeo, Mauro Gallegati, Gianfranco
Giulioni 61

Part II Complex Economic Network

Heterogeneous Economic Networks
Wataru Souma, Yoshi Fujiwara, Hideaki Aoyama.................... 79

The Emergence of Paradigm Setters Through Firms'
Interaction and Network Formation
Rainer Andergassen, Franco Nardini, Massimo Ricottilli 93

Part III Economic Dynamics

Statistical Properties of a Heterogeneous Asset Pricing Model
with Time-varying Second Moment
Carl Chiarella, Xue-Zhong He, Duo Wang 109

Deflationary Recessions in a General Equilibrium Framework
Luca Colombo, Gerd Weinrich 125

Concepts of Thermodynamics in Economic Growth
Jürgen Mimkes 139

Firm Dynamics Simulation Using Game-theoretic Stochastic
Agents
Yuichi Ikeda, Osamu Kubo, Yasuhiro Kobayashi 153

Part IV Agent-based Modeling

A Functional Modularity Approach to Agent-based Modeling
of the Evolution of Technology
Shu-Heng Chen, Bin-Tzong Chie 165

Herding Without Following the Herd: The Dynamics of
Case-Based Decisions with Local Interactions
Andreas Krause 179

Cultural Evolution in a Population of Heterogeneous Agents
Gábor Fáth, Miklos Sarvary 193

Part V Auction and Two-sided Matching

Simulating Auctions
Konrad Richter 209

Counterclockwise Behavior Around the Beveridge Curve
Koji Yokota 225

The Waiting-Time Distribution of Trading Activity in a
Double Auction Artificial Financial Market
Silvano Cincotti, Sergio M. Focardi, Linda Ponta, Marco Raberto,
Enrico Scalas 239

Part VI Minority Games and Collectie Intelligence

Theoretical Analysis of Local Information Transmission in
Competitive Populations
Sehyo Charley Choe, Sean Gourley, Neil F. Johnson, Pak Ming Hui 251

Analysis of Complexity and Time Restriction in Resources
Allocation Problems
Kiyoshi Izumi, Tomohisa Yamashita, Koichi Kurumatani 265

How Does Collective Intelligence Emerge in the Standard
Minority Game?
*Satoshi Kurihara, Kensuke Fukuda, Toshio Hirotsu, Osamu Akashi,
Shinya Sato, Toshiharu Sugawara* 279

Part VII Game-theoretic Approach

What Information Theory Says About Bounded Rational
Best Response
David H. Wolpert 293

Evolution of Reciprocal Cooperation in the Avatamsaka
Game
Eizo Akiyama, Yuji Aruka 307

Game Representation - *Code Form*
Maria Cristina Peixoto Matos, Manuel Alberto M. Ferreira 321

Effect of Mutual Choice Metanorm in Group Dynamics for
Solving Social Dilemmas
Tomohisa Yamashita, Kiyoshi Izumi, Koichi Kurumatani 335

Part I

Econophysics

Five Years of Continuous-time Random Walks in Econophysics

Enrico Scalas

DISTA, Università del Piemonte Orientale, Via Bellini 25/G, 15100 Alessandria, Italy and INFM Unità di Genova, Via Dodecaneso 33, 16146 Genova. Italy
scalas@unipmn.it

Summary. This paper is a short review on the application of continuos-time random walks to Econophysics in the last five years.

1 Introduction

Recently, there has been an increasing interest on the statistical properties of high-frequency financial data related to market microstructural properties [1, 2, 3, 4, 5, 6]. High-frequency econometrics is now well established after research on autoregressive conditional duration models [7, 8, 9, 10].

In high-frequency financial data not only returns but also waiting times between consecutive trades are random variables [11]. This remark is present in a paper by Lo and McKinlay [12], but it can be traced at least to papers on the application of compound Poisson processes [13] and subordinated stochastic processes [14] to finance. Models of tick-by-tick financial data based on compound Poisson processes can also be found in the following references: [15, 16, 17].

Compound Poisson processes are an instance of continuous-time random walks (CTRWs) [18]. The application of CTRW to economical problems dates back, at least, to the 1980s. In 1984, Rudolf Hilfer discussed the application of stochastic processes to operational planning, and used CTRWs as tools for sale forecasts [19]. The revisited and augmented CTRW formalism has been applied to high-frequency price dynamics in financial markets by our research group since 2000, in a series of three papers [20, 21, 22]. Other scholars have recently used this formalism [23, 24, 25]. However, already in 1903, the PhD thesis of Filip Lundberg presented a model for ruin theory of insurance companies, which was later developed by Cramér [26, 27]. The underlying stochastic process of the Lundberg-Cramér model is another example of compound Poisson process and thus also of CTRW.

Among other issues, we have studied the independence between log-returns and waiting times for stocks traded at the New York Stock Exchange in October 1999. For instance, according to a contingency-table analysis performed on General Electric (GE) prices, the null hypothesis of independence can be rejected with a significance level of 1 % [28]. We have also discussed the anomalous non-exponential behaviour of the unconditional waiting-time distribution between tick-by-tick trades both for future markets [21] and for stock markets [28, 29]. Different waiting-time scales have been investigated in different markets by various authors. All these empirical analyses corroborate the waiting-time anomalous behaviour. A study on the waiting times in a contemporary FOREX exchange and in the XIXth century Irish stock market was presented by Sabatelli *et al.* [30]. They were able to fit the Irish data by means of a Mittag-Leffler function as we did before in a paper on the waiting-time marginal distribution in the German-bund future market [21]. Kyungsik Kim and Seong-Min Yoon studied the tick dynamical behavior of the bond futures in Korean Futures Exchange (KOFEX) market and found that the survival probability displays a stretched-exponential form [31]. Finally, Ivanov *et al.* [33] confirmed that a stretched exponential fits well the survival distribution for NYSE stocks as we suggested in [28]. Moreover, just to stress the relevance of non-exponential waiting times, a power-law distribution has been recently detected by T. Kaizoji and M. Kaizoji in analyzing the calm time interval of price changes in the Japanese market [32]. We have offered a possible explanation of the anomalous waiting-time behaviour in terms of daily variable activity [29].

The aforementioned empirical results are important as market microstructural models should be able to reproduce such a non-exponential behaviour of waiting-time distributions in order to be realistic. However, the rest of this paper focuses on the theory and is divided as follows: in Section 2, CTRW theory is presented as applied to finance. Finally, in Sec. 3, a summary of main results is presented together with a discussion on the direction of future research.

2 Theory

Random walks have been used in finance since the seminal thesis of Bachelier [34], a work completed at the end of the XIXth century, more than a hundred years ago. After a rather long period in which the ideas of Bachelier were neglected, they were further developed until recent times [35, 36].

Our approach to random walks in finance is related to that of Clark [14] and to the introductory part of Parkinson's paper [37]. It is a purely phenomenological approach. There is no assumption on the rationality or the behaviour of trading agents and it is not necessary to assume the validity of the efficient market hypothesis [38, 39]. However, as briefly discussed above, even in the absence of a *microfoundation*, a phenomenological model can still be useful to corroborate or falsify the consequences of behavioural or other assumptions on markets. The model itself can be corroborated or falsified by empirical data.

In order to model tick-by-tick data, we use the so-called continuous-time random walk (CTRW), where time intervals between successive steps are random variables, as discussed by Montroll and Weiss [18]. In physics, CTRWs have been introduced as models of diffusion with instantaneous jumps from one position to the next. For this reason they can be used as models of price dynamics as well.

Let $S(t)$ denote the price of an asset at time t. In a real market with a double-auction mechanism, prices are fixed when buy orders are matched with sell orders and a transaction (trade) occurs. It is more convenient to refer to returns rather than prices. For this reason, we shall take into account the variable $x(t) = \log S(t)$: the logarithm of the price. For a small price variation $\Delta S = S(t_{i+1}) - S(t_i)$, the return $r = \Delta S/S(t_i)$ and the logarithmic return $r_{log} = log[S(t_{i+1})/S(t_i)]$ virtually coincide.

CTRWs are essentially point processes with reward [40]. The point process is characterized by a sequence of independent identically distributed (i.i.d.) positive random variables τ_i, which can be interpreted as waiting times between two consecutive events:

$$t_n = t_0 + \sum_{i=1}^{n} \tau_i; \quad t_n - t_{n-1} = \tau_n; \quad n = 1, 2, 3, \ldots; \quad t_0 = 0. \tag{1}$$

The rewards are (i.i.d.) not necessarily positive random variables: ξ_i. In the usual physical intepretation, the ξ_is represent the jumps of a diffusing particle, and they can be n-dimensional vectors. Here, only the 1-dimensional case is studied, but the extension of many results to the n-dimensional case is straightforward. The position x of the particle at time t is given by the following random sum (with $N(t) = \max\{n : t_n \leq t\}$ and $x(0) = 0$):

$$x(t) = \sum_{i=1}^{N(t)} \xi_i. \tag{2}$$

In the financial interpretation outlined above, the ξ_i's have the meaning of log-returns, whereas the *positions* or rewards $x(t)$ represent log-prices at time t. Indeed, the time series $\{x(t_i)\}$ is characterised by $\varphi(\xi, \tau)$, the *joint probability density* of log-returns $\xi_i = x(t_{i+1}) - x(t_i)$ and of waiting times $\tau_i = t_{i+1} - t_i$. The joint density satisfies the normalization condition $\int \int d\xi d\tau \varphi(\xi, \tau) = 1$. It must be again remarked that both ξ_i and τ_i are assumed to be independent and identically distributed (i.i.d.) random variables. This strong assumption is useful to derive limit theorems for the stochastic processes described by CTRWs. However, in financial time series, the presence of volatility clustering, as well as correlations between waiting times do falsify the i.i.d hypothesis. The reader interested in a review on correlated random variables in finance is referred to chapter 8 in McCauley's recent book [41].

In general, log-returns and waiting times are not independent from each other [28]. By probabilistic arguments (see [18, 21, 42]), one can derive the following integral equation that gives the probability density, $p(x,t)$, for the particle of being in position x at time t, conditioned by the fact that it was in position $x = 0$ at time $t = 0$:

$$p(x,t) = \delta(x)\,\Psi(t) + \int_0^t \int_{-\infty}^{+\infty} \varphi(x - x', t - t')\,p(x',t')\,dt'\,dx', \qquad (3)$$

where $\delta(x)$ is Dirac's delta function and $\Psi(\tau)$ is the so-called survival function. $\Psi(\tau)$ is related to the marginal waiting-time probability density $\psi(\tau)$. The two marginal densities $\psi(\tau)$ and $\lambda(\xi)$ are:

$$\psi(\tau) = \int_{-\infty}^{+\infty} \varphi(\xi,\tau)\,d\xi$$

$$\lambda(\xi) = \int_0^\infty \varphi(\xi,\tau)\,d\tau, \qquad (4)$$

and the survival function $\Psi(\tau)$ is:

$$\Psi(\tau) = 1 - \int_0^\tau \psi(\tau')\,d\tau' = \int_\tau^\infty \psi(\tau')\,d\tau'. \qquad (5)$$

Both the two marginal densities and the survival function can be empirically derived from tick-by-tick financial data in a direct way.

The integral equation, eq. (3), can be solved in the Laplace-Fourier domain. The Laplace transform, $\tilde{g}(s)$ of a (generalized) function $g(t)$ is defined as:

$$\tilde{g}(s) = \int_0^{+\infty} dt\,e^{-st}\,g(t), \qquad (6)$$

whereas the Fourier transform of a (generalized) function $f(x)$ is defined as:

$$\hat{f}(\kappa) = \int_{-\infty}^{+\infty} dx\,e^{i\kappa x}\,f(x). \qquad (7)$$

A generalized function is a distribution (like Dirac's δ) in the sense of S. L. Sobolev and L. Schwartz [43].

One gets:

$$\hat{\tilde{p}}(\kappa, s) = \tilde{\Psi}(s)\,\frac{1}{1 - \hat{\tilde{\varphi}}(\kappa, s)}, \qquad (8)$$

or, in terms of the density $\psi(\tau)$:

$$\hat{\tilde{p}}(\kappa, s) = \frac{1 - \tilde{\psi}(s)}{s}\,\frac{1}{1 - \hat{\tilde{\varphi}}(\kappa, s)}, \qquad (9)$$

as, from eq. (5), one has:

$$\Psi(s) = \frac{1 - \widetilde{\psi}(s)}{s}. \tag{10}$$

In order to obtain $p(x,t)$, it is then necessary to invert its Laplace-Fourier transform $\widetilde{\widehat{p}}(\kappa, s)$. As we shall see in the next subsection, for log-returns independent from waiting times, it is possible to derive a series solution to the integral equation (3).

2.1 Limit Theorems: The Uncoupled Case

In a recent paper, Gorenflo, Mainardi and the present author have discussed the case in which log-returns and waiting times are independent [42]. It is the so-called uncoupled case, when it is possible to write the joint probability density of log-returns and waiting times as the product of the two marginal densities:

$$\varphi(\xi, \tau) = \lambda(\xi)\psi(\tau) \tag{11}$$

with the normalization conditions $\int d\xi \lambda(\xi) = 1$ and $\int d\tau \psi(\tau) = 1$.

In this case the integral master equation for $p(x,t)$ becomes:

$$p(x,t) = \delta(x)\,\Psi(t) + \int_0^t \psi(t - t') \left[\int_{-\infty}^{+\infty} \lambda(x - x')\, p(x', t')\, dx' \right] dt' \tag{12}$$

This equation has a well known general explicit solution in terms of $P(n,t)$, the probability of n jumps occurring up to time t, and of the n-fold convolution of the jump density, $\lambda_n(x)$:

$$\lambda_n(x) = \int_{-\infty}^{+\infty} \cdots \int_{-\infty}^{+\infty} d\xi_{n-1} \ldots d\xi_1 \lambda(x - \xi_{n-1}) \ldots \lambda(\xi_1). \tag{13}$$

Indeed, $P(n,t)$ is given by:

$$P(n,t) = \int_0^t \psi_n(t - \tau)\Psi(\tau)\, d\tau \tag{14}$$

where $\psi_n(\tau)$ is the n-fold convolution of the waiting-time density:

$$\psi_n(\tau) = \int_0^\tau \cdots \int_0^{\tau_1} d\tau_{n-1} \ldots d\tau_1 \psi(\tau - \tau_{n-1}) \ldots \psi(\tau_1). \tag{15}$$

The n-fold convolutions defined above are probability density functions for the sum of n independent variables.

Using the Laplace-Fourier method and recalling the properties of Laplace-Fourier transforms of convolutions, one gets the following solution of the integral equation [44, 42, 45, 46]:

$$p(x,t) = \sum_{n=0}^{\infty} P(n,t)\lambda_n(x) \tag{16}$$

Eq. (16) can also be used as the starting point to derive eq. (12) via the transforms of Fourier and Laplace, as it describes a jump process subordinated to a renewal process [14, 47].

Let us now consider the following pseudodifferential equation, giving rise to anomalous relaxation and power-law tails in the waiting-time probability:

$$\frac{d^\beta}{d\tau^\beta}\Psi(\tau) = -\Psi(\tau), \quad \tau > 0, \quad 0 < \beta \le 1; \quad \Psi(0^+) = 1, \tag{17}$$

where the operator d^β/dt^β is the Caputo fractional derivative, related to the Riemann–Liouville fractional derivative. For a sufficiently well-behaved function $f(t)$, the Caputo derivative is defined by the following equation, for $0 < \beta < 1$:

$$\frac{d^\beta}{dt^\beta}f(t) = \frac{1}{\Gamma(1-\beta)}\frac{d}{dt}\int_0^t \frac{f(\tau)}{(t-\tau)^\beta}\,d\tau - \frac{t^{-\beta}}{\Gamma(1-\beta)}f(0^+), \tag{18}$$

and reduces to the ordinary first derivative for $\beta = 1$. The Laplace transform of the Caputo derivative of a function $f(t)$ is:

$$\mathcal{L}\left(\frac{d^\beta}{dt^\beta}f(t); s\right) = s^\beta \widetilde{f}(s) - s^{\beta-1}f(0^+). \tag{19}$$

If eq. (19) is applied to the Cauchy problem of eq. (17), one gets:

$$\widetilde{\Psi}(s) = \frac{s^{\beta-1}}{1+s^\beta}. \tag{20}$$

Eq. (20) can be inverted, giving the solution of eq. (17) in terms of the Mittag-Leffler function of parameter β [48, 49]:

$$\Psi(\tau) = E_\beta(-\tau^\beta), \tag{21}$$

defined by the following power series in the complex plane:

$$E_\beta(z) := \sum_{n=0}^{\infty} \frac{z^n}{\Gamma(\beta n + 1)}. \tag{22}$$

The Mittag-Leffler function is a possible model for a fat-tailed survival function. For $\beta = 1$, the Mittag-Leffler function coincides with the ordinary exponential function. For small τ, the Mittag-Leffler survival function coincides with the stretched exponential:

$$\Psi(\tau) = E_\beta(-\tau^\beta) \simeq 1 - \frac{\tau^\beta}{\Gamma(\beta+1)} \simeq \exp\{-\tau^\beta/\Gamma(\beta+1)\}, \; 0 \le \tau \ll 1, \tag{23}$$

whereas for large τ, it has the asymptotic representation:

$$\Psi(\tau) \sim \frac{\sin(\beta\pi)}{\pi} \frac{\Gamma(\beta)}{\tau^\beta}, \ 0 < \beta < 1, \ \tau \to \infty. \tag{24}$$

Accordingly, for small τ, the probability density function of waiting times $\psi(\tau) = -d\Psi(\tau)/d\tau$ behaves as:

$$\psi(\tau) = -\frac{d}{d\tau} E_\beta(-\tau^\beta) \simeq \frac{\tau^{-(1-\beta)}}{\Gamma(\beta)}, \ 0 \le \tau \ll 1, \tag{25}$$

and the asymptotic representation is:

$$\psi(\tau) \sim \frac{\sin(\beta\pi)}{\pi} \frac{\Gamma(\beta+1)}{\tau^{\beta+1}}, \ 0 < \beta < 1, \ \tau \to \infty. \tag{26}$$

The Mittag-Leffler function is important as, without passage to the diffusion limit, it leads to a time-fractional master equation, just by insertion into the CTRW integral equation. This fact was discovered and made explicit for the first time in 1995 by Hilfer and Anton [50]. Therefore, this special type of waiting-time law (with its particular properties of being singular at zero, completely monotonic and long-tailed) may be best suited for approximate CTRW Monte Carlo simulations of fractional diffusion.

For processes with survival function given by the Mittag-Leffler function, the solution of the master equation can be explicitly written:

$$p(x,t) = \sum_{n=0}^{\infty} \frac{t^{\beta n}}{n!} E_\beta^{(n)}(-t^\beta)\lambda_n(x), \tag{27}$$

where:

$$E_\beta^{(n)}(z) := \frac{d^n}{dz^n} E_\beta(z).$$

The Fourier transform of eq. (27) is the characteristic function of $p(x,t)$ and is given by:

$$\hat{p}(\kappa, t) = E_\beta[t^\beta(\hat{\lambda}(\kappa) - 1)]. \tag{28}$$

If log-returns and waiting times are scaled according to:

$$x_n(h) = h\xi_1 + h\xi_2 + \ldots + h\xi_n, \tag{29}$$

and:

$$t_n(r) = r\tau_1 + r\tau_2 + \ldots + r\tau_n, \tag{30}$$

the scaled characteristic function becomes:

$$\hat{p}_{h,r}(\kappa, t) = E_\beta \left[\frac{t^\beta}{r^\beta} (\hat{\lambda}(h\kappa) - 1) \right]. \tag{31}$$

Now, if we assume the following asymptotic behaviours for vanishing h and r:

$$\hat{\lambda}(h\kappa) \sim 1 - h^\alpha |\kappa|^\alpha; \ \ 0 < \alpha \le 2, \tag{32}$$

and

$$\lim_{h,r \to 0} \frac{h^\alpha}{r^\beta} = 1, \tag{33}$$

we get that:

$$\lim_{h,r \to 0} \hat{p}_{h,r}(\kappa, t) = \hat{u}(\kappa, t) = E_\beta[-t^\beta |\kappa|^\alpha]. \tag{34}$$

The Laplace transform of eq. (34) is:

$$\widetilde{\hat{u}}(\kappa, s) = \frac{s^{\beta-1}}{|\kappa|^\alpha + s^\beta}. \tag{35}$$

Therefore, the well-scaled limit of the CTRW characteristic function coincides with the Green function of the following pseudodifferential *fractional* diffusion equation:

$$|\kappa|^\alpha \widetilde{\hat{u}}(\kappa, s) + s^\beta \widetilde{\hat{u}}(\kappa, s) = s^{\beta-1}, \tag{36}$$

with $u(x, t)$ given by:

$$u(x, t) = \frac{1}{t^{\beta/\alpha}} W_{\alpha,\beta} \left(\frac{x}{t^{\beta/\alpha}} \right), \tag{37}$$

where $W_{\alpha,\beta}(u)$ is given by:

$$W_{\alpha,\beta}(u) = \frac{1}{2\pi} \int_{-\infty}^{+\infty} d\kappa \, e^{-i\kappa u} E_\beta(-|\kappa|^\alpha), \tag{38}$$

the inverse Fourier transform of a Mittag-Leffler function [51, 52, 53, 54, 55].

For $\beta = 1$ and $\alpha = 2$, the fractional diffusion equation reduces to the ordinary diffusion equation and the function $W_{2,1}(u)$ becomes the Gaussian probability density function evolving in time with a variance $\sigma^2 = 2t$. In the general case ($0 < \beta < 1$ and $0 < \alpha < 2$), the function $W_{\alpha,\beta}(u)$ is still a probability density evolving in time, and it belongs to the class of Fox H-type functions that can be expressed in terms of a Mellin-Barnes integral as shown in details in ref. [52].

The scaling equation, eq. (33), can be written in the following form:

$$h \simeq r^{\beta/\alpha}. \tag{39}$$

If $\beta = 1$ and $\alpha = 2$, one recognizes the scaling relation typical of Brownian motion (or the Wiener process).

In the passage to the limit outlined above, $\widetilde{\hat{p}}_{r,h}(\kappa, s)$ and $\widetilde{\hat{u}}(\kappa, s)$ are asymptotically equivalent in the Laplace-Fourier domain. Then, the asymptotic equivalence in the space-time domain between the master equation and the fractional diffusion equation is due to the continuity theorem for sequences of characteristic functions, after the application of the analogous theorem for sequences of Laplace transforms [56]. Therefore, there is convergence in law or weak convergence for the corresponding probability distributions and densities. Here, weak convergence means that the Laplace transform and/or Fourier transform (characteristic function) of the probability density function are pointwise convergent (see ref. [56]).

2.2 Limit Theorems: The Coupled Case

The diffusive limit in the coupled case is discussed by Meerschaert *et al.* [57]. The coupled case is relevant as, in general, log-returns and waiting times are not independent [28]. Based on the results summarized in [42] and discussed in [58, 59], it is possible to prove the following theorem for the coupled case:

Theorem

Let $\varphi(\xi, \tau)$ be the (coupled) joint probability density of a CTRW. If, under the scaling $\xi \to h\xi$ and $\tau \to r\tau$, the Fourier-Laplace transform of $\varphi(\xi, \tau)$ behaves as follows:

$$\widetilde{\widehat{\varphi}}_{h,r}(\kappa, s) = \widetilde{\widehat{\varphi}}(h\kappa, rs) \tag{40}$$

and if, for $h \to 0$ and $r \to 0$, the asymptotic relation holds:

$$\widetilde{\widehat{\varphi}}_{h,r}(\kappa, s) = \widetilde{\widehat{\varphi}}(h\kappa, rs) \sim 1 - \mu|h\kappa|^\alpha - \nu(rs)^\beta, \tag{41}$$

with $0 < \alpha \leq 2$ and $0 < \beta \leq 1$. Then, under the scaling relation $\mu h^\alpha = \nu r^\beta$, the solution of the (scaled) coupled CTRW master (integral) equation, eq. (3), $p_{h,r}(x, t)$, weakly converges to the Green function of the fractional diffusion equation, $u(x, t)$, for $h \to 0$ and $r \to 0$.

Proof

The Fourier-Laplace transform of the scaled conditional probability density $p_{h,r}(x, t)$ is given by:

$$\widetilde{\widehat{p}}_{h,r}(\kappa, s) = \frac{1 - \widetilde{\psi}(rs)}{s} \frac{1}{1 - \widetilde{\widehat{\varphi}}(h\kappa, rs)}. \tag{42}$$

Replacing eq. (41) in eq. (42) and observing that $\widetilde{\psi}(s) = \widetilde{\widehat{\varphi}}(0, s)$, one asymptotically gets for small h and r:

$$\widetilde{\widehat{p}}_{h,r}(\kappa, s) \sim \frac{\nu r^\beta s^{\beta-1}}{\nu r^\beta s^\beta + \mu h^\alpha |\kappa|^\alpha}, \tag{43}$$

which for vanishing h and r, under the hypotheses of the theorem, converges to:

$$\widetilde{\widehat{p}}_{0,0}(\kappa, s) = \widetilde{\widehat{u}}(\kappa, s) = \frac{s^{\beta-1}}{s^\beta + |\kappa|^\alpha}, \tag{44}$$

where $\widetilde{\widehat{u}}(\kappa, s)$ is the Fourier-Laplace transform of the Green function of the fractional diffusion equation (see eq. (36)). The asymptotic equivalence in the space-time domain, between $p_{0,0}(x, t)$ and $u(x, t)$, the inverse Fourier-Laplace transform of $\widetilde{\widehat{u}}(\kappa, s)$, is again ensured by the continuity theorem for sequences of characteristic functions, after the application of the analogous theorem for sequences of Laplace transforms [56]. There is convergence in law or weak convergence for the corresponding probability distributions and densities.

An important consequence of the above theorem is the following corollary showing that in the case of marginal densities with finite first moment of waiting times and finite second moment of log-returns, the limiting density $u(x, t)$ is the solution of the ordinary diffusion equation (and thus the limiting process is the Wiener process). The corollary can be used to justify the popular Geometric Brownian Motion model of stock prices, here with expected return set to zero. Again, in order to derive this result, no reference is necessary to the Efficient Market Hypothesis [38, 39].

Corollary

If the Fourier-Laplace transform of $\varphi(\xi, \tau)$ is regular for $\kappa = 0$ and $s = 0$, and, moreover, the marginal waiting-time density, $\psi(\tau)$, has finite first moment τ_0 and the marginal jump density, $\lambda(\xi)$, is symmetric with finite second moment σ^2, then the limiting solution of the master (integral) equation for the coupled CTRW is the Green function of the ordinary diffusion equation.

Proof

Due to the hypothesis of regularity in the origin and to the properties of Fourier and Laplace transforms, we have that:

$$\widetilde{\widehat{\varphi}}_{h,r}(\kappa, s) = \widetilde{\widehat{\varphi}}(h\kappa, rs) \sim \widetilde{\widehat{\varphi}}(0, 0) +$$

$$+ \frac{1}{2} \left(\frac{\partial^2 \widetilde{\widehat{\varphi}}}{\partial \kappa^2} \right)_{(0,0)} h^2 \kappa^2 + \left(\frac{\partial \widetilde{\widehat{\varphi}}}{\partial s} \right)_{(0,0)} rs =$$

$$= 1 - \frac{\sigma^2}{2} h^2 \kappa^2 - \tau_0 rs, \tag{45}$$

and as a consequence of the theorem, under the scaling $h^2 \sigma^2 / 2 = \tau_0 r$, one gets, for vanishing h and r:

$$\widetilde{\widehat{p}}_{0,0}(k, s) = \widetilde{\widehat{u}}(k, s) = \frac{1}{s + k^2}, \tag{46}$$

corresponding to the Green function (36) for $\alpha = 2$ and $\beta = 1$, that is the solution of the Cauchy problem for the ordinary diffusion equation.

3 Summary and Outlook

In this paper, a discussion of continuous-time random walks (CTRWs) has been presented as phenomenological models of tick-by-tick market data.

Continuous-time random walks are rather general and they include compound Poisson processes as particular instances. Well-scaled limit theorems have been presented for a rather general class of CTRWs.

It is the hope of this author that this paper will stimulate further research on high-frequency econometrics based on the concepts outlined above. There are several possible developments.

First of all, one can abandon the hypothesis of i.i.d. log-returns and waiting times and consider various forms of dependence. In this case, it is no longer possible to exploit the nice properties of Laplace and Fourier transforms of convolutions, but, still, Monte Carlo simulations can provide hints on the behaviour of these processes in the diffusive limit.

A second possible extension is to include volumes as a third stochastic variable. This extension is straightforward, starting from a three-valued joint probability density.

A third desirable extension is to consider a multivariate rather than univariate model that includes correlations between time series.

The present author is currently involved in these extensions and is eager to know progress in any direction by other independent research groups. He can be contacted at `scalas@unipmn.it`

4 Acknowledgements

This work was supported by the Italian M.I.U.R. F.I.S.T. Project "High frequency dynamics in financial markets". The author wishes to acknowledge stimulating discussion with P. Buchen, S. Focardi, R. Gorenflo, T. Kaizoji, H. Luckock, F. Mainardi, M. M. Meerschaert and M. Raberto.

References

1. C. Goodhart and M. O'Hara, *High-frequency data in financial markets: Issues and applications*, Journal of Empirical Finance **4**, 73–114 (1997).
2. M. O'Hara *Making market microstructure matter*, Financial Management **28**, 83–90 (1999).
3. A. Madhavan *Market microstructure: A survey*, Journal of Financial Markets **3**, 205–258 (2000).
4. M.M. Dacorogna, R. Gençay, U.A. Müller, R.B. Olsen, O.V. Pictet, *An Introduction to High Frequency Finance*, (Academic Press, 2001).
5. M. Raberto, S. Cincotti, S.M. Focardi, M. Marchesi, *Agent-based simulation of a financial market*, Physica A **299**, 320–328 (2001).
6. H. Luckock, *A steady-state model of the continuous double auction*, Quantitative Finance **3**, 385–404 (2003).
7. R. Engle and J. Russel *Forecasting the frequency of changes in quoted foreign exchange prices with the autoregressive conditional duration model*, Journal of Empirical Finance **4**, 187–212 (1997).

8. R. Engle and J. Russel, *Autoregressive conditional duration: A new model for irregularly spaced transaction data*, Econometrica **66**, 1127–1162 (1998).
9. L. Bauwens and P. Giot *The logarithmic ACD model: An application to the bid-ask quote process of three NYSE stocks*, Annales d'Economie et de Statistique **60**, 117–149 (2000).
10. A. Lo, C. MacKinley and J. Zhang, *Econometric model of limit-order executions*, Journal of Financial Economics **65**, 31–71 (2002).
11. G.O. Zumbach, *Considering time as the random variable: the first hitting time*, Neural Network World **8**, 243–253 (1998).
12. A. Lo and C. MacKinley, *An econometric analysis of nonsynchronous trading*, Journal of Econometrics **45**, 181–212 (1990).
13. S.J. Press, *A compound events model for security prices*, Journal of Business **40**, 317–335 (1967).
14. P.K. Clark, *A subordinated stochastic process model with finite variance for speculative prices*, Econometrica **41**, 135-156 (1973).
15. T.H. Rydberg and N. Shephard, *Dynamics of trade-by-trade price movements: Decomposition and models*, Nuffield College, Oxford, working paper series 1998-W19 (1998).
16. T.H. Rydberg and N. Shephard, *Modelling trade-by-trade price movements of multiple assets using multivariate compound Poisson processes*, Nuffield College, Oxford, working paper series 1999-W23 (1999).
17. T.H. Rydberg and N. Shephard, *A modelling framework for the prices and times of trades made at the New York stock exchange*, in W.J. Fitzgerald, R. Smith, A.T. Walden, P. Young (Editors): *Nonlinear and nonstationary signal processing*, (Cambridge University Press, 2000).
18. E.W. Montroll and G.H. Weiss, *Random walks on lattices, II*, J. Math. Phys. **6**, 167–181 (1965).
19. R. Hilfer, *Stochastische Modelle für die betriebliche Planung*, (GBI-Verlag, Munich, 1984).
20. E. Scalas, R. Gorenflo and F. Mainardi, *Fractional calculus and continuous-time finance*, Physica A **284**, 376–384 (2000).
21. F. Mainardi, M. Raberto, R. Gorenflo and E. Scalas, *Fractional calculus and continuous-time finance II: the waiting-time distribution*, Physica A **287**, 468–481, (2000).
22. R. Gorenflo, F. Mainardi, E. Scalas and M. Raberto *Fractional calculus and continuous-time finance III: the diffusion limit*, in M. Kohlmann and S. Tang (Editors): *Trends in Mathematics - Mathematical Finance*, pp. 171–180 (Birkhäuser, Basel, 2001).
23. J. Masoliver, M. Montero, and G.H. Weiss, *Continuous-time random-walk model for financial distributions*, Phys. Rev. E **67**, 021112/1–9 (2003).
24. J. Masoliver, M. Montero, J. Perello. and G.H. Weiss, *The CTRW in finance: Direct and inverse problem* http://xxx.lanl.gov/abs/cond-mat/0308017
25. R. Kutner and F. Switala, *Stochastic simulation of time series within Weierstrass-Mandelbrot walks*, Quantitative Finance **3**, 201–211 (2003).
26. F. Lundberg, *Approximerad Framställning av Sannolikehetsfunktionen. AAterförsäkering av Kollektivrisker*, (Almqvist & Wiksell, Uppsala, 1903).
27. H. Cramér, *On the Mathematical Theory of Risk*, (Skandia Jubilee Volume, Stockholm 1930).

28. M. Raberto, E. Scalas and F. Mainardi, *Waiting times and returns in high-frequency financial data: an empirical study*, Physica A **314**, 749–755 (2002).
29. E. Scalas, R. Gorenflo, H. Luckock, F. Mainardi, M. Mantelli and M. Raberto, *Anomalous waiting times in high-frequency financial data*, preprint.
30. L. Sabatelli, S. Keating, J. Dudley and P. Richmond, *Waiting time distribution in financial markets*, Eur. Phys. J. B **27**, 273–275 (2002).
31. K. Kim and S.-M. Yoon, *Dynamical behavior of continuous tick data in futures exchange market*, Fractals **11**, 131–136 (2003).
32. T. Kaizoji and M. Kaizoji, *Power law for the calm-time interval of price changes*, Physica A, **336**, 563–570 (2004).
33. P.C. Ivanov, A. Yuen, B. Podobnik, Y. Lee, *Common scaling patterns in inter-trade times of US stocks*, Phys. Rev. E, **69**, 056107 (2004).
34. L.J.B. Bachelier *Theorie de la Speculation*, (Gauthier-Villars, Paris, 1900, Reprinted in 1995, Editions Jaques Gabay, Paris, 1995).
35. P.H. Cootner (ed.) *The Random Character of Stock Market Prices*, (MIT Press, Cambridge MA, 1964).
36. R.C. Merton, *Continuous Time Finance*, (Blackwell, Cambridge, MA, 1990).
37. M. Parkinson, *Option pricing: The american put*, Journal of Business **50** 21–36 (1977).
38. E.F. Fama, *Efficient capital markets: A review of theory and empirical work*, The Journal of Finance **25**, 383–417 (1990) .
39. E.F. Fama, *Efficient capital markets: II*, The Journal of Finance **46**, 1575–1617 (1991).
40. D. Cox and V. Isham *Point Processes*, (Chapman and Hall, London, 1979).
41. J. McCauley, *Dinamics of Markets*, (Cambridge University Press, Cambridge UK, 2004).
42. E. Scalas, R. Gorenflo and F. Mainardi, *Uncoupled continuous-time random walks: Solution and limiting behavior of the master equation*, Physical Review E **69**, 011107(1-8) (2004).
43. I. M Gelfand and G. E. Shilov, *Generalized Functions*, vol. 1. Translated from the Russian. (Academic Press, New York and London 1964. Russian edition: 1958).
44. G.H. Weiss, *Aspects and Applications of Random Walks*, (North-Holland, Amsterdam, 1994).
45. F. Mainardi, R. Gorenflo and E. Scalas, *A Renewal process of Mittag-Leffler type*, in M. Novak (Editor), *Thinking in Patterns: Fractals and Related Phenomena in Nature*, (World Scientific, Singapore, 2004, pp. 35-46).
46. F. Mainardi, R. Gorenflo and E. Scalas, *A fractional generalization of the Poisson processes*, Vietnam Journal of Mathematics, in press (2004).
47. H. Geman, D. B. Madan, and M. Yor, *Time changes for Lévy processes*, Mathematical Finance **11**, 79–96 (2001).
48. F. Mainardi, *Fractional relaxation-oscillation and fractional diffusion-wave phenomena*, Chaos, Solitons & Fractals **7**, 1461–1477 (1996).
49. R. Gorenflo and F. Mainardi, *Fractional calculus: integral and differential equations of fractional order*, in A. Carpinteri and F. Mainardi (Editors): *Fractals and Fractional Calculus in Continuum Mechanics*, pp. 223–276 (Springer, Wien, 1997).
50. R. Hilfer and L. Anton, *Fractional master equations and fractal time random walks*, Phys. Rev. E **51**, R848–R851 (1995).

51. R. Metzler and J. Klafter, *The random walk's guide to anomalous diffusion: a fractional dynamics approach*, Phys. Rep. **339**, 1-77 (2000).
52. F. Mainardi, Yu. Luchko, and G. Pagnini *The fundamental solution of the space-time fractional diffusion equation*, Fractional Calculus and Applied Analysis, **4**, 153–192 (2001). Downloadable from http://www.fracalmo.org.
53. G.M. Zaslavsky *Chaos, fractional kinetics and anomalous transport*, Phys. Rep. **371**, 461–580 (2002).
54. E. Scalas, R. Gorenflo, F. Mainardi, and M. Raberto, *Revisiting the derivation of the fractional diffusion equation*, Fractals **11**, 281–289 Suppl. S (2003).
55. R. Metzler, Y. Klafter, *The restaurant at the end of the random walk: recent developments in the description of anomalous transport by fractional dynamics*, Journal of Physics A: Math. Gen. **37**, R161–R208 (2004).
56. W. Feller, *An Introduction to Probability Theory and its Applications*, Vol. 2 (Wiley, New York, 1971).
57. M.M. Meerschaert, D.A. Benson, H.P. Scheffler, and P. Becker-Kern, *Governing equations and solutions of anomalous random walk limits*, Phys. Rev. E **66**, 060102-1/4 (2002).
58. R. Gorenflo and F. Mainardi, *Non-Markovian random walk models, scaling and diffusion limits*, 2-nd Conference on Lévy Processes: Theory and Applications, 21-25 January 2002, MaPhySto Centre, Department of Mathematics, University of Aarhus, Denmark. Downloadable from http://www.fracalmo.org.
59. R. Gorenflo and E. Abdel-Rehim, *From power laws to fractional diffusion*, Vietnam Journal of Mathematics, in press (2004).

Why Macroeconomic Price Indices are Sluggish in Large Economies ?

Masanao Aoki[1] and Hiroshi Yoshikawa[2]

[1] Department of Economics University of California, Los Angeles, Fax Number
310-825-9528 aoki@econ.ucla.edu
[2] Faculty of Economics, University of Tokyo, Fax 81-3-5841-5629
yoshikawa@e.u-tokyo.ac.jp

Summary. Two new reasons are discussed for sluggish behavior of macroeconomic variables such as price indices.

One is slow spread of the news of microeconomic idiosyncratic shocks in the economy, when the economy is organized into tree structures of heterogeneous subgroups or clusters of agents or goods. Clusters are not symmetrically treated, but the concept of ultrametric distances measure disparities or similarities of clusters.

Another is the effects of uncertainties that affect decision processes, such as about the cost surfaces, or about the shapes of cost landscapes which may have many local minima which are not known precisely. Effectiveness of many search algorithm is reduced in the face of this kind of uncertainty. Flat cost landscapes, called entropic barriers, are discussed as an example.[3]

1 Introduction

The standard approach such as real business cycle theory is based on the premise that the microeconomic behavior of the optimizing agents mimics dynamics of the macroeconomy. This premise is wrong, because we see that macro- and micro-behavior are clearly different in many aspects, such as in their speeds of responses.

Two of the causes for sluggish responses are familiar: complex organizations of macroeconomy, and effects of uncertainty. On complexity of organization, the existing economic literature treats the phenomena of sluggish adjustments or responses of economic variables such as prices, wages, or unemployment rates by treating adjusting variables or agents all on equal footing, that is, without introducing some notions of similarity or closeness between various heterogeneous groups of variables or agents.

It is true that some rudimentary notion of social distances between different clusters of agents is found in economic literature, such as agents being

---

[3] For some other aspects of uncertainty, see Aoki, Yoshikawa, and Shimizu (2002).

placed at sites of lattices. However, no explicit notions of similarity, correlations or distances between various groups of variables or agents are examined. No models with more formal notion of distances among different groups of agents apparently exist.

The manner of how news of idiosyncratic disturbances of some types or price changes spreads through the macroeconomy need be analyzed in more systematic manner by introducing some notion of distance between clusters of agents.

In our view, attempts at dealing with groups of agents in the economic literature do not go far enough. We introduce the notion of hierarchically structured clusters of goods or producers as an essential ingredient in models designed to explain sluggish adjustment processes.

This paper analyzes a particular aspect of the macroeconomy, namely adjustment speeds of macroeconomic price indicies. We use the concept of ultrametrics as measure of distances between clusters.

Since the publication of Keynes' General Theory (1936), inflexibility or ridigity of prices has been always a forcal point of macroeconomics. Many economists take inflexibility of prices as a sign of agents' irrationality. They argue that well organized market forces should make prices flexible. In this paper we explain that prices are necessarily slow to change in large economies.

To improve on the existing literature, we need more appropriate notion than correlation to measure relations among agents or variables, since the notion of correlation is not transitive as has been known, for example, in the numerical taxonomy literature. Feigelman and Ioffe (1991) have an example of three patterns: A=(1, 1, 1, 1), B=(1,1,-1,1) and C=(1,1,1,-1). Calculating correlations by $\rho = (1/4) \sum_i x_i y_i$ where xs and ys are the components of the patterns above, we see that $\rho_{A,B} = \rho_{A,C} = 1/2$ but $\rho_{B,C} = 0$.

To avoid this kind of intransitivity of correaltions, which makes correlation unsuitable as a measure of similarity of patterns, we use the notion of ultrametrics as a measure of distance between clusters of agents. The concept of ultrametrics has been in the literature of mathematics and physics, especially in spin glass models. For these, see Schikhof (1984), or Mézard, Parisi and Virasoro (1986), among others. See Aoki (1996, p. 34) for some elementary explanation of the ultrametrics, and some simple economic applications. Aoki (2002), and Aoki and Yoshikawa (2003) have a more complex example of labor market dynamics and Okun's law,where unemployed workers from different sectors of economy or different human capitals or job experiences form separate clusters and ultrametrics are used to measure distances between clusters. This distance is then used to generate probabilities of unemployed being recalled by a given industry when job openings are created. Taylor's well-known analysis of adjustment of wages is different from ours. He treats groups of workers with different wage contracts as *the* source of slow wage/price dynamics. His model and virtually all multi-sector models treat sectors symmetrically with equal distances between any two sectors. These groups are not hierarchically arranged. There is no notion of adjustment speeds as functions of some similarity measures among clusters, Taylor (1980).

We place clusters of agents as leaves of a tree. Distances between clusters are measured by counting the number of levels one must travel towards the root of the tree to find a common node shared by the two leaves. This is called ultrametric distances.[4] News, such as that of the arrival of some idiosyncratic shocks at some sites, spread throughout the tree as stochastic processes governed by the backward Chapman-Kolmogorov equation, called the master equaation in this paper, in which transition rates are functions of ultrametric distances.

2 Tree Models

We use upside down trees to represent hierarchical structures. At the bottom of a tree we have leaves, also called sites, where each leaf represents a cluster of agents or a (price of) goods, as the case may be. Agents in the same cluster are alike in some sense. They may be producers of some close substitute goods, or they may have similar reaction or decision making delays given a disturbance of some kind in signals they use, and so on. A number of the leaves, denoted by m_1, share a common node of a tree. These leaves are connected to nodes located on level 2 of the tree. There are m_2 of the nodes which branch out from a node at level three, and so on.[5] In general we have K levels in a tree. The top of the tree is the root consisting of a single cluster with $N = m_1 m_2 \cdots m_K$ number of clusters or leaves.

Without loss of generality we assume that an exogenous idiosyncratic disturbance occurs at site 1 at time zero. Let this disturbance be felt at site i at time t with probability $P_i(t)$. The initial condition is $P_1(0) = 1$, and $P_i(0)$, $i \neq 1$. These Ps are governed by the master equation where transition rates are functions of ultrametric distances. See Aoki(1996, 2002) for several examples.

We also use another measure to gauge the speed with which news or disturbances travel through the tree. We define the expected distance reached by the disturbance originated at site 1 by time t, i.e.,

$$< d(t) >= \sum_i d(i,1) P_i(t),$$

where $d(i,1)$ is the ultrametric distance between node i and site 1, which is the source of news or disturbance. This averaged distance indicates how far, on the average, the news of disturbance has spread through the model.

---

[4] See for example, Aoki (1996, p. 31).

[5] Trees need not have symmetric shape or profiles, that is, the number of branches from a node can be different from node to node.

3 Two Types of Lags in Tree Dynamics

3.1 Multiplier Lags

Responses of a macroeconomic price index to shocks to one of its component prices consists of two components. One is the well-known dynamic delays in multipliers or impluse responses. which are familiar in economics and econometrics. The other is called information lag in this paper. It refers to delays in the news or effects of exogenous shocks which originated in one sector of an economy spreading to other sectors stochastically. The multiplier lag is simply illustrated here by dynamic responses of a second order ordinary differential equation to a step input changes. Propagations of shocks are treated in this paper stochastically. This requires solving the master equations for states that are the leaves of trees.

To illustrate this, it is convenient to use Laplace transforms to relate output responses to input changes as

$$H(s)Y(s) = U(s),$$

where s is the Laplace transform variable, $Y(s)$ is the transform of the model output that is response to change in input, and $U(s)$ the Laplace transform of input.

We may write this more directly as

$$Y(s) = \frac{U(s)}{H(s)}.$$

The time domain expression is

$$\frac{d^2 y}{dt} + (a+b)\frac{dy}{dt} + aby = u.$$

Here $1/H(s)$ is called the transfer function. This expression shows how the signal at the input side of the model is transferred to the output side of the model. If the input signal to the system at time zero at full constant force, then, without loss of generality, we can think of the input signal $u(t) = 1$ for t positive. Its Laplace transform is $U(s) = 1/s$. On the other hand, if the signal gradually appear to this system, we may have something like $u(t) = 1 - e^{-\mu t}$ with $\mu > 0$, for $t > 0$ say. This input is initially zero and graudally reach its full force after about the time elapse of $4/\mu$. Its Laplace transform is $U(s) = \mu/[s(s + \mu)]$.

As a simple illustration of the difference of these two types of input signals on $y(t)$, we assume that dynamics are described by a second order ordinary constant coefficient differential equation with zero initial conditions; $y(0) = 0$, and $dy(0)/dt = 0$. To be very concrete suppose that $H(s) = (s + a)(s + b)$ with some positive a and b. This is the transfer function of a dynamic system described by a second order differential equation with two stable eigenvalues $-a$, and $-b$.

With the step input, the dynamic response is obtained by taking the inverse Laplace transform of

$$Y(s) = \frac{1}{s(s+a)(s+b)} = \frac{C}{s} + \frac{A}{s+a} + \frac{B}{s+b},$$

where A, B, and C are the constants, $C = 1/ab$, $A = -1/[(b-a)a]$, and $B = 1/[(b-a)b]$. The time response of the pair of input and output with this transfer function is given by

$$y(t) = C + Ae^{-at} + Be^{-bt}.$$

This $y(t)$ expression shows the multiplier effects of this block or unit of dynamics with the indicated transfer function. If $a < b$, then after the time span of about $4/a$ units of time, the output nearly settles to a constant, $y(t) \approx 1/ab$.[6] It takes about this much time for the effect of a sudden application of a step signal at the input to settle down at the output of the model.

With the other input with a gradually rising magnitude such as $u(t) = 1 - e^{-\mu t}$, with μ a positive constant much smaller than a and b, then $y(t)$ is approximately equal to

$$y(t) = \frac{\mu}{b-a}[-\frac{e^{-at}}{a-\mu} + \frac{e^{-bt}}{b-\mu}] + \frac{\mu}{(a-\mu)(b-\mu)]e^{-\mu t}}.$$

The first two exponential terms are due to the dynamic multiplier effects, and the third term is due to information transmission delay when $u(t)$ gradually appear at the input terminal of this block or unit with the second-order dynamics.

This expression is approximately equal to the last term above when μ is much smaller than a or b. The signal $y(t) \approx (1/ab)(1 - e^{-\lambda t})$, which takes a long time to reach its steady state value. In this case it is the behavior of input, not the dynamics, that causes the sluggish output.

3.2 Stochastic Spread of News in Trees: An example

We next turn to the second type of lags that exist in trees with several levels of nodes.

To illustrate our idea simply, we consider two simple economies with four sectors which are organized in two different ways. One is organized as a one level tree, and the other as a two level tree. Two-level trees are generally more sluggish in response than one-level trees. More generally, the larger the number of levels, the slower the dynamics.

---

[6] If a is larger than b, then the dynamic lag is about $4/b$, that is, with two basic lag structure, it is $4/min(a,b)$ is the multiplier lag.

Without loss of generality we assume that exogenous disturnaves are felt at site 1 at time zero. This disturbance is felt at site i at time t with probability $P_i(t)$. The initial condition is $P_1(0) = 1$, and $P_i(0) = 0$, $i \neq 1$. We pay more attention to the transient behavior than the steady state values of these probabilities becasue the delays are determined by the transient time constants.

The probability at site i is changing over time as the difference of the influx and outflux of probabilities. Denote the transition rate between site i and j by $w(i, j)$. The master equation which describes the dynamics of the probabilities is

$$\frac{dP_i(t)}{dt} = I_i(t) - O_i(t),$$

where the influx to site i is

$$I_i(t) = \sum_{j \neq i} P_j(t) w(j, i),$$

and

$$O_i(t) = P_i(t) \sum_{j \neq i} w(i, j).$$

For the one level tree with four sites

$$I_1(t) = P_2(t) w(2, 1) + P_3(t) w(3, 1) + P_4(t) w(4, 1),$$

and

$$O_1(t) = P_1(t)[w(1, 2) + w(1, 3) + w(1, 4)],$$

and with similar expressions for the other Is and Os. There are similar expressions for the outflows and inflows at other sites as well. We also assumne that $w(i, j) = w(j, i)$ for all i and j.

Next, we posit that the transition rates $w(i, j)$ depends only on the ultrametric distance. Exogenous disturbances is felt first by agents or goods in the same cluster, and then the news or effects will gradually and stochastically propagate to other leaves, that is, to other clusters of the trees. Therefore we speak of the expected changes in price indicies as the results of such shocks to one cluster.

Thus for the one-level tree

$$w(i, j) = q < 1,$$

$i \neq j$, where $q = \exp(-\gamma d(i, j)) = \exp(-\gamma)$, for all i and j between 1 and 4, because the ultrametric distance between any pair of sites is the same, and where γ is some constant. Later we identify it with the inverse of *economic temperature*.

For the two-level tree we have[7]

---

[7] This model is similar to the one in Aoki (1996, p. 38). Details of analysis differ somewhat.

$$w(1,2) = w(3,4) = q,$$

and

$$w(1,3) = w(1,4) = q^2$$

because $d(1,3) = 2$, hence $w(1,3) = \exp(-2\gamma) = q^2$.

The master equation for the probability vector $\mathbf{P}(t)$ of the one level tree consists of probabilities at the four leaves

$$\frac{d\mathbf{P}(t)}{dt} = W\mathbf{P}(t),$$

with

$$W = \begin{bmatrix} W_1 & W_2 \\ W_2 & W_1 \end{bmatrix},$$

where

$$W_1 = \begin{pmatrix} -3q & q \\ q & -3q \end{pmatrix},$$

$$W_2 = qe_2 e_2',$$

where $e_2 = [1\ 1]'$.

This matrix W has eigenvalue 0 with eigenvector $(1\ 1\ 1\ 1)'$, and triple repeated eigenvalue $-4q$ with three independent eigenvectors $(1\ 1\ -1\ -1)'$, $(1\ -1\ 0\ 0)'$, and $(0\ 0\ 1\ -1)'$.

The probabilities evolve with time according to

$$P_1(t) = 1/4 + (3/4)e^{-4qt},$$

and

$$P_2(t) = P_3(t) = P_4(t) = (1/4) - (3/4)e^{-4qt}.$$

Approximately after time span of $1/q$, the probabilities are all about $1/4$.[8] It takes about this time span for the initial shock to propagate to all the sectors. Hence this is the time lag for the shock initialted at sector 1 to spread probabilistically to all the other sectors, i. e., for macroeconomic price index to fully reflect the price shock to one of its sectors.[9]

In the other case, the matrix W is given by

$$W = \begin{bmatrix} W_1 & W_2 \\ W_2 & W_1 \end{bmatrix},$$

where

---

[8] Note that $e^{-1} = 0.018$.

[9] All probabilities eventually stop changing and reach some constants, such as $1/4$ here. This means that the news of disturbance at site 1 has reached all four sites equally by then.

$$W_1 = \begin{pmatrix} -(q+2q^2) & q \\ q & -(q+2q^2) \end{pmatrix},$$

$$W_2 = q^2 e_2 e_2',$$

where $e_2 = [1\ 1]'$.

This matrix W has eienvalues 0, with eigenvector $(1\ 1\ 1\ 1)$, eigenvalue $\lambda_1 = -4q^2$, with eigenvector $(1\ 1\ -1\ -1)'$, and double repeated eigenvalue $\lambda_2 = -2(q+q^2)$, with eigenvectors $(1\ -1\ 0\ 0)'$, and $(0\ 0\ 1\ -1)'$. Note that the magnitude of λ_1 is less than that of λ_2 because q is less than one. The associated with eigenvalue λ_1 is faster than that associated with eigenvalue λ_2. It represents the escape rate of probability from site 1 to site 2.

The probabilities evolve with time as

$$P_1(t) = (1/4 + (1/4)e^{-\lambda_1 t} + (1/2)e^{-\lambda_2 t},$$

$$P_2(t) = (1/4 + (1/4)e^{-\lambda_1 t} - (1/2)e^{-\lambda_2 t},$$

$$P_3(t) = P_4(t) = 1/4 - (1/4)e^{-\lambda_1 t}.$$

After time span of $2/q(1+q)$, the term $e^{-\lambda_2 t}$ is approximately zero. After time span of $1/q^2$ all probabilies are approximately equal to $1/4$. Note however, that the time span $1/q^2$ is much longer than that of $1/q$, that is, the two-level tree is much more sluggish than the one-level tree.

To compare dynamic behavior of this model with the one-level tree, we can aggregate the tree by defining a two-dimensional state vector with components $S_1(t) = P_1(t) + P_2(t)$, and $S_2(t) = P_3(t) + P_4(t)$ by defining

$$\mathbf{Q}(t) = S\mathbf{P}(t),$$

where the aggregation matrix S is given by

$$S = \begin{bmatrix} 1 & 1 & 0 & 0 \\ 0 & 0 & 1 & 1 \end{bmatrix}.$$

The dynamic matrix V for this aggregated vector is given by $V = SWS'(SS')^{-1}$ which has eigenvalues 0 and $-4q^2$.

The vector $Q(t)$ has two components $0.5 + 0.5e^{-\mu t}$, and $0.5 - 0.5e^{-\mu t}$ with $\mu = 4q^2$.

The dynamics of the one-level tree is much simpler. It has eigenvalues 0 and -4q, the latter with multiplicity 3. We can similarly aggregate the first two sites and the second two sites separately to produce a two node tree. The eigenvalue are still 0 and -4q. In other words, after the lapse of time of the order $1/q$, the one-level tree has approximately reached its equilibrium state, while dynamics of the two-level tree has not.

This fact remains true when one-level trees of K sites is compared with k level trees with $K = 2^k$. Suppose that we group l of K sites into one cluster, and the remaining $K - l$ sites into another. The eigenvalues are 0 and $-Kq$, repeated $K - 1$ times, while those of k level tree are 0 and $-(2q)^k$.

We can show that the larger the number of hierarchies the slower the process of disturbance propagation, and response of macro-price index to shocks to one of the sectors. Ogielski and Stein (1985),[10] among several others, have shown that in the limit of the number of hierarchy going to infinity, the response becomes power-law, not exponential decay. See also Paladin et al. (1985)

3.3 Inflexible Macroeconomic Prices: An Example

We present an example of slow adjustments of some macroeconomic price (index) composed of prices of goods of several sectors of economy. To be simple suppose that a price index P_I is the weighted average of two sectoral output prices, Q_A and Q_B. We outline how Sector 1 price Q_A is affected by an exogenous shock to site 1 price, since effects on $Q_B(t)$ are similarly anlalyzed.

For concreteness suppose that node A is composed of a two level tree with two more nodes a and b with two branches each. There are thus four more basic prices at sites 1 through 4, such as factor prices, prices of intermediate goods and so on. The two-level hierarchical tree traces out the relations among these prices. As shown in the previous section, the tree generates spill-over probabilities of an exogenous shock to one of the basic prices.

The Laplace transform of $\mathbf{P}(t)$, dropping subscript b, is

$$\hat{\mathbf{P}}(s) = \frac{1}{4s}u_0 + \frac{1}{2(s + \lambda_2)}u_2 + \frac{1}{4(s + \lambda_1)}u_1,$$

where s is the Laplace transform variable, us are the column vectors shown above.

Consequently we can write down the explicit expression for the expected values of changes in Q_A, denoted by $E[\delta \hat{Q}_A(s)]$. Assuming that transmission lags of the transfer function $h_a(s), h_b(s), h_i(s)$, $i = 1, \ldots 4$ are not as large as $1/q^2$, we can extract the slowest decaying term out of this as

$$E(\delta Q_A(t)) \approx \frac{1}{4}h_a(-\lambda_1)\{h_1(-\lambda_1) + h_2(-\lambda_1)\} - h_b(-\lambda_1)\{h_3(-\lambda_1 +$$

$$h_4(-\lambda_1)\}e^{-\lambda_1 t} + \cdots ,$$

where the slowest term is extracted.

In the case where $h_a(s) = 1/(s + a)$, $h_b(s) = 1/(s + b)$, $h_i(s) = 1/(s + \alpha_i$, $i = 1, \ldots 4$), then a sufficient condition that this term is present is

Proposition When q^2 is negligibly small compared with $a, b, \alpha_i, i = 1, \ldots 4$, price Q_A will exhibit sluggish response to an exogenous price change at site 1 if

---

[10] See Aoki (1996, p. 200) also.

$$\frac{1}{a}(\frac{1}{\alpha_1} + \frac{1}{\alpha_2}) \neq \frac{1}{b}(\frac{1}{\alpha_3} + \frac{1}{\alpha_4}).$$

This condition rules out that effects of exogenous shocks coming from two subtree branches cancel out.

More generally, structures of interconnections between the basic prices and Q_A are conveniently expressed in terms of the Laplace transforms as

$$E[\delta\hat{Q}_A(s)] = \sum_{i=1}^{4} \hat{H}_i(s)E[\delta\hat{q}_i(s)],$$

with $\hat{H}_i(s)$ being the Laplace transform of the transfer function from site i to the price Q_A. The symbol E is the expectation operator, that is the expected values of changes in the basic price q_i with the probabilities of spill-over.

For example, $\hat{H}_i(s)$ could be a simple first order transfer function such as $(s + c_i)$ with some positive constant c_i, or more complex second or third order transfer functions, possibly with complex as well as real roots.

The example of this section illustrates the effects of spill-over delays due to hierarchical tree structure, in addition to the usual delays due to dynamics of transmission which are present in the transfer functions.

The macroeconomic price indices thus have two sources of sluggishness; one is the usual dynamic lags of transfer functions , and the other information spread or spill-over lags, that is, the lags for the signal to arrive as inputs to some transfer function that are connected with the price index expressions.

4 Effects of Uncertainty

4.1 Rugged Landscape Problems

Thus far, we have focussed on the tree structure of the economy that is responsible for producing slow dynamic behavior. This result is generic since it does not depend on any specific assumptions on the model. We now turn to another reason for sluggish macroeconomy that has to do uncertainty.

The standard analysis in economics assumes explicitly or implicitly that agents know the global shape of objective functions, which are smooth and well-behaved, and constraints. In reality, agents have only local knowledge at best, and must try to improve their performance by guessing the right directions to adjust their decision varibles. To do so they face complicated and often hard optimization problems.

Agents are thus often stuck at some local optimal point or in basins associated with local optima, and they may not know of the existence of better local optima or global ones. "Rugged or flat landscapes" are the words often used to indicate that agents do not know which directions they should adjust their decision variables.

Suppose that agents conduct exploratory moves and evaluate the results with the Metropolis algorithm, Metropolis et al. (1953). See Ripley (1987)for exposition of the method, for example. In the Metropolis algorithm, the move is accepted with probability one if it results in lower cost. In order to move from a local minimum to a better on, the move is accepted with probability $exp(-\beta \Delta c)$, even when the cost increases by $\Delta c > 0$, where β is some positive parameter. In summary, a trial move is accepted even if a move is associated with an immediate increas in cost, in the hope that the direction of the move is correct in eventually it leads to leaving the current basin to a new basin with lower cost.

If agents are confident of the directions of move, then they will choose them with high probability, that is, the value of β is small in economies with high levels of activities. However, if they are uncertain of the moves, then the move will be chosen with small probability, that is, β has a large value in case of uncertainty, i.e., of low level of economic activities.

What is important to recognize is the fact that the parameter β may become very large, especially wen macroeconomy is in recession. With a very large values of β any move which does not result in immediate cost reduction will be rejected with very high probability. This means that agents will not revise their current position with almost probability one. Tis will prolong the recessions. Put differently, adjustment paths out of the current states are rarely taken, or become inaccessible with high values of β. Some such examples are found in Aoki (2002, p.48, p.53; 1996 p.133, p.179) which illustrate effects of uncertainty associated with an action or decision.

4.2 Economic Temperature

We often say that "economy is heating up" or "cooling down". We propose a notion of economic temperature to formalize this to convey the idea that economic activities increase as agents increase their activities by intensifying levels of existing economic transactions and establishing or creating new links between agents. In tree models, more sites are introduced and/or more levels are added to hierarchical trees. When economic activities slow down, intensities of existing activities reduce and some of the existing links may even disappear.

One way to capture these effects is to change the transition rates, that is, probability intensities between between some sites. In the above we used $w(i,j) = \exp[-\gamma d(i,j)]$, where γ is some posiitve constant. We now set $\gamma = 1/T_e$ where T_e is the economic temperature. For example, as T_e becomes lower, the slower the transition rate, and $< d(t) >$ will be smaller.

More generally, as degree of uncertainty increases, the economic temperature becomes cooler because of reduced economic activities.

4.3 Entropic Barriers

This section presents a model that focusses on possibilities of losing adjustment path as β becomes large, and not on the aspect of cost minimization. Agents' adjustment problem is thus cast as a random walk problems in one dimension, that is, as a birth-death process. In this setting we recognize that $\Delta c = 0$. We have a flat cost landscape. See Godréche and Luck (2002) for a similar model.

Suppose that agents face 'cost' or 'utility' surfaces with full of local minima. Agents are not sure whether the minima they find themselves are global minimum or they are stuck in some local minima. They do not know which directions, if any, they should move in order to improve their performances. They must modify steepest descent or some gradient procedure to take account of a possiblity that they need to overcome some barriers to leave the current basin of attraction to another one with smaller costs. For this reason we assume that they employ the Metroplis mehtod or some variant of it, Metropolis (1953). This method gives some probability to go in directions which increases the cost in the hope that the path leads to a local peak of the barrier separating the basin from another with smaller local minima. We show that probabilities with which this type of paths is tried become smaller as economic activity levels become lower.

We suppose there are K boxes (cluster, categories, or types) into which agents belong. To be concrete, we may think of the boxes as representing strategies or directions of descents that agents try. In this model economy-wide optimum corresponds to a situation in which all boxes are empty. Presence of an agent in some box indicates that agent has not achieved its optimal operating condition. At a given time we pick an agent at random uniformly and place it in another box at random. If there are k agents in total, one agent is chosen with probability k/N where N is the total number of agents in all the boxes. The chosen agent exits or depart from the box he is in, and will go to one of the remaining box with probability $1/(K-1)$. Call the box he goes to as the arrival box. Let S denotes (Boltzmann) entropy of the configuration. The difference

$$\Delta S = S(n_d - 1, n_a + 1) - S(n_d, n_a),$$

is the change in entropy when one agent departs box d and goes to box a. We assume that

$$\Delta S = -\beta, n_d = 1, n_a \neq 0,$$

$$\Delta S = \beta, n_d \neq 1, n_a = 0,$$

and ΔS is zero in all other cases. In other words, entropy changes only when the number of empty boxes, that is, successful types of firms is reduced by one ($n_a = 0, n_d \neq 1$), which increases S by β, or the number of empty boxes increases by one, ($n_a \neq 0, n_d = 1$), which increases the number of empty boxes by one.

To treat the simplest cases we focus on one of the boxes, called box 1, and let $n_1(t)$ denotes the number of agents in it,

$$p_k(t) = \Pr(n_1(t) = k).$$

This and related models have been analyzed by several physicists. We follow Codrech and Lux 91997). We write the master equation for it as

$$\frac{dp_k}{d\tau} = (k+1)p_{k+1} + \lambda(\tau)p_{k-1} - (\lambda(\tau) + k)p_k, k \geq 2,$$

and similar equations for $k = 2$ and $k = 1$. To simplify equations we change the time from τ to t by defining

$$\frac{dt}{d\tau} = \lambda(\tau).$$

With this change of variable we have

$$\frac{dp_k}{dt} = \frac{k+1}{\lambda(t)} + p_{k-1} - [1 + \frac{k}{\lambda(t)}]p_k, \ k > 2,$$

$$\frac{dp_1}{dt} = \frac{2}{\lambda(t)}p_2 + \mu(t)p_0(t) - 2p_1(t),$$

and

$$dp_0(t)dt = p_1(t) - \mu(t)p_0(t),$$

where

$$1/\lambda(t) = 1 + (e^{-\beta} - 1)p_0(t),$$

and

$$\mu(t) = e^{-\beta} + (1 - e^{-\beta}p_1(t).$$

See Appendix for the derivation of the transition rates.

We easily verify that these $p's$ sum to one, and the mean of k, $\sum_k kp_k(t) = N/K := \rho$. We take ρ to be 1 for simplicity.

This set of equations can be used to calculate the generating function

$$F(z,t) = \sum_k p_k(t)z^k,$$

with the initial condition assumed to be $F(z,0) = z$, that is $p_1(0) = 1$ and all other ps are zero.

This generating function satisfies the partial differential equation

$$\frac{\partial F}{\partial t} = (z-1)F(z,t) - \frac{z-1}{\lambda(t)}\frac{\partial F}{\partial z} - (z-1)Y(t),$$

where $Y(t) = (1 - e^{-\beta}p_0(t) = 1 - 1/\lambda(t)$. See Aoki (2002, p. 70) for deriving the partial differential equation for the generating function. See also Aoki

(2002, App. A.1) for solving the partial differential equation by the method of characteristic curves. The sollution is obtained by solving

$$\frac{dt}{1} = \frac{dz}{(z-1)/\lambda(t)} = \frac{dF}{(z-1)(F-Y)}.$$

When $\beta = 0$ which corresponds to high level of economic activities and a well-behaved cost curve, the equations is especially simple, because $\lambda(t) = \mu(t) = 1$. We obtain

$$F(z,t) = \{1 + (z-1)e^{-t}\} \exp\{(z-1)(1-e^{-t}\},$$

and from it

$$p_0(t) = (e^{-t} - 1)\exp(e^{-t} - 1) = -\frac{1}{e} + \frac{1}{2e}e^{-2t} + \cdots.$$

This shows that $p_0(t)$ approaches the equilibrium value with time constant $1/2$.

With $\beta large$ Godrech and Lux show that the time constant becomes e^{β}/β^2, a much larger number, indicating a sluggish approach to the equilibrium.

5 Concluding Discussion

Elementary examples and discussion of ultrametric trees and spaces are given in Aoki (1996). See Ogielski and Stein (1985) or Paladin, Mezard, and deDominicis (1985) for additional discussion.

Uncertainty and some macroeconomic policies may affect the adjustment speed parameters. One simple way to incorporate these into q_i is to set $q_i = \exp(-i\omega q_0/D), i = 1, \ldots R$, where β is a parameter to reflect the degree of uncertainty. Larger degree of uncertainty, slower the speed of adjustment. This implies that ω gets larger as economy faces more uncertainty. Parameter D stands for some macroeconomic variables. More positive values of D causes q_i to become larger. The formulation presented in this paper can accommomodate these changes by incorporating these Gibbs factors into the adjustment speed parameters. See Paladin et al, or Aoki (1996, p. 156).

References

1. Aoki M (1996) New Approaches to Macroeconomic Modeling: Evolutionary stochastic dynamics, multiple equilibria, and externalities as field effects, Cambridge Univ. Press, New York
2. Aoki M (2002) Modeling Aggregate Behavior and Fluctuations in Economics: Stochastic View of Interacting Agents Cambridge Univ. Press, New York

3. Aoki M, Yoshikawa H (2003) A new model of labor dynamics: ultrametrics, Okun's law, and transient dynamics, forthcoming Wehia 2003 proceedings, Springer-Verlag
4. Aoki M, Yoshikawa H, Shimizu T(2002) Long Stagnation and Monetary Policy in Japan:A Theoretical Explanation, presented at a conference in honor of James Tobin, a conference volume forthcoming from Springer-Verlag
5. Feigelman MV, Ioffe CB (1991) Hierarchical Organization of Memory in E. Domany et al (eds) Models of Neural Network, Springer-Verlag, Berlin
6. Godréche C, Luck JM (2002) Nonequilibrium Dynamics of Urn Models, J. Phy. Cond. Matt 14:1601–1615
7. Metropolis N, Rosenbluth AW, Rosenbluth N, Teller AH, Teller E (1953) Equation of state calculation by fast computing machine, it Journal of Chem. Phys 21:1087-92
8. Ogielski AT, Stein DL (1985) Dynamics on ultrametric spaces, Phy. Rev. 55:1634–1637.
9. Paladin O, Mézard J, deDominicis C (1985) Diffusion in an ultrametric space : a simple case." J. Phys. Lett. **46** L-985–L-989.
10. Ripley BD (1987) Stochastic Simulation Wiley, New York
11. Schikhof WH (1984) Ultrametric Calculus: An Introduction to p-adic Analysis Cambridge Univ. Press, London.
12. Taylor, JB (1980) Aggregate Dynamics and Staggered Contracts, Jour. Pol. Econ. 88:1–23

Appendix

For dynamics of trees see also Paladin, Mézard and deDominicis (1985), and Schreckenberg in Aoki (1996, Sec.7.2.3).

Power-Law Adjustment Behavior

We now discuss general R level trees with large R. The time constants are getting larger as you go up the levels towards the root. At level i it is given by

$$1/\tau_i = \sum r = i^R(q_r - q_{r+1})N_r \approx q_i N_i.$$

Then for time t between the two sucessive time constants, τ_{l_0-1} and τ_{l_0}, and closer to the latter, then t is much larger than the former and we have an approximate expression for x_1 in this time interval

$$x_1(t) \sum_i \exp(-t/\tau_i)/N_i + 1/N \approx 1/N_{l_0},$$

where $1/\tau_{l_0} \approx 1/t \approx q_{l_0} N_{l_0}$.

To present simple cases, assume that $q_i = q^{-i}$, and $m_i = m^i$, $i = 1, \dots R$. Then, it is well known that we observe a power-law behavior for the prices at level 1

This heuristic argument can be made more precise by evaluating the expression for the sum by approximating it by an integral. See Paladin, Mezard, and de Cominicis (1985) or Ogielski and Stein (1985) for detail. In Appendix we outline the argument. One can use steepest descent argument or change of variables to approximate the sum by an incomplete Gamma function. The conclusion that power-law emerges does not change.

$$x_1(t) \approx t^{-\theta},$$

with $\theta = lnm/ln(q/m)$.[11]

Tree Dynamics

A K-site one level tree has dynamics

$$\frac{dP(t)}{dt} = WP(t),$$

where $W = -KI_K + qe_Ke'_K$. This matrix has eigenvalue 0, and K_1 repeated eigenvalues of $-Kq$, where q is the basic transition rate between any two sites.

Let $K = 2^k$ for some positive integer k. For example with $k = 2$ the one-level 4 branch dynamics have one 0 and three repeated eigenvalue -4q.

To consider a cluster composed of three sites, define a matrix S with the first row (1110), and the second row (0011). Define

$$Q = SWS'(ss')^{-1},$$

as the aggregation matrix. This matrix aggregates four sites into 2 clusters of 2 sites each. The dynamics for the two-dimensiona vector $r = Sp$ is given by

$$\frac{dr}{dt} = Rr,$$

where R has eigenvalues 0 and $-4q$.

We state two propositions:

Proposition 1 With one level tree, the eigenvalues are invariant with the manner the sites are aggregated into 2 super-sites of size k and $K - k$.

Proposition 2 Consider a tree with k levels where $K = 2^k$. This tree aggregted into one level of two clusters of sizes 2^{k-1} each, has eigenvalues 0 and $-(2q)^k$.

Example of Dynamics of a Two-Level Tree

This tree has three nodes at level 2, that is, three branches come out of the root, $m_2 = 2$. Each node at level 2 separates into three branches, that is

---

[11] Note that the relation $x(t) = Ct^\theta$, and $\mu x(\lambda t) = x(t)$ implies that $\theta = -ln\mu/\lambda$.

$m_1 = 3$. At the end of each branch, there is a leaf. Leaves at the bottom of the tree, that is at level 1, are the states of a continuous-time Markov chain. For simpler notation, name the states as a, b, c, d, e, f, g, i. Transition rate between states a and b is denoted by $q_{a,b}$. The expression $q_{a,f}$ means the transition rate between state a and f. These two states belong to a different clusters. Transition rates are assumed to be symmetric, i.e., $q_{a,i} = q_{i,a}$, and depends only on the ultrametric distance between the states.

The ultrametric distance between two states is the levels of trees that these states need to go up the tree toward the root to reach the common node, that is node from which both states branch out. If we regard the tree as a genealogical chart with a node representing an ancester, ultrametric distance 1 means that the states share the same parent, distance 2 grandparent.

Denote the three nodes at level 2 by α, β, and γ. These are superstates or superclusters. leaves d, e, f branch out from node β.

Therefore, states among the same cluster at level 1 have the same transition rate, q_1. Transition rates between nodes at level 2 are the same and denoted by q_2. For example, $q_{a,c} = q_1$, but $q_{a,i} = q_2$.

The master equation for state a is

$$\frac{dp_a}{dt} = I_a - O_a,$$

where the infow of probability flux is

$$I_a = p_b q_{b,a} + p_c q_{c,a} + p_d q_{d,a} + \cdots p_i q_{i,a},$$

and

$$O_a = p_a[(q_{a,b} + q_{a,c}) + (q_{a,d} + \cdots q_{a,i})].$$

From our discussion of the ultrametrics above we can write these more simply as

$$I_a = q_1(p_b + p_c) + q_2(p_d + \cdots p_i),$$

and

$$O_a = p_a(m_1 - 1)q_1 + (2-1)m_1 q_2.$$

Noting that $p_b + p_c = (p_a + p_b + p_c) - p_a = p_\alpha - p_a$, and that $p_d + \cdots p_i = p_\beta + p_\gamma = 1 - p_\alpha$, we obtain

$$\frac{dp_a}{dt} = q_2 + (q_1 - q_2)p_\alpha - p_a[m_1(q_1 - q_2) + Nq_2,$$

where $N = m_1 m_2$.

The master equation for p_b and p_c have the same structure. Recall that p_α is the sum of these three probabilities. Adding the three equations together, we obtain

$$\frac{dp_\alpha}{dt} = -N(p_\alpha - \frac{1}{m_2}).$$

The solutions of these equations are

$$p_\alpha(t) = \exp(-t/\tau_2)[p_\alpha(0) - \frac{1}{m_2} + \frac{1}{m_2},$$

where

$$1/\tau_2 = m_1 m_2 q_2,$$

and

$$p_a(t) = \exp(-t/\tau_1)[p_a(0) - \frac{1}{m_1}p_\alpha] + \exp(-t/\tau_2)[\frac{1}{m_1}p_\alpha(0) - \frac{1}{m_1 m_2}] + \frac{1}{m_1 m_2},$$

with

$$1/\tau_1 = m_1 m_2 q_2 + m_1(q_1 - q_2).$$

In a R-level tree with transition rates between two states depending only on the ultrametric distance between them, that is, the difference in levels of the lowest nodes common to the two states, the structure of these solutions remain. Denoting what corresponds to level 2 by p_{j_R}, where j_R runs from 1 to m_R, which is the number of branches from the root, its differential equation is

$$\frac{dp_{j_R}}{dt} = -Nq_R(p_{j_R} - \frac{1}{m_R}).$$

The expression for the leaves at level 1 is

$$p(j_1, \ldots, j_R) = \sum_{i=1}^{R} \exp(-t/\tau_i)[\frac{p(j_i, \ldots, j_R)(0)}{N_i} - \frac{p(j_{i+1}, \ldots, j_R)(0)}{N_{i+1}}] + \frac{1}{N},$$

where

$$1/\tau_i = \sum_{r=i}^{R}(q_r - q_{r+1})N_r,$$

where $N_r = m_1 \cdots m_r$, and $q_{R+1} = 0$.

This sum equals

$$1/\tau_i = \sum_{r=i}^{R}(m/q)^{i-1} = (\lambda^{-i+1} - \lambda^{-R})/(1 - \lambda),$$

where $\lambda = q/m$. With *lambda* less than 1 we obtain the usual exponential dynamic behavior. With the value of λ larger than 1, by letting $\hat{t} = t/\tau_1$ for convenience, we obtain

$$x_1(t) = \sum m^{-i+1}e^{-\hat{t}}\lambda^{-i+1} + 1/N.$$

The summand is of the form $\exp[(1/m)^i - c\exp(-iln\lambda]$. Define

$$h(y) = \exp[-\frac{lnm}{ln\lambda}y - e^y].$$

Then the sum can be approximated by $\int h(lny)dy$ which can be put into the form of an incomplete Gamma function.

Ultrametric Dynamics

This section describes dynamics on a two-level ultrametric tree of Fig. 1 as a simple example. A general R level tree dynamics is sketched in appendix.

In the example, the leaves at level 1 are grouped into three clusters, each cluster composed of three agents (prices). Denote by $d(i,j)$ the distance between leave i and j. In the two-level tree, this distance takes values of 1 or 2, depending on the clusters to which i and j belong. If they belong to the same cluster, the distance is one, else two.

Associate variable x_i with leaf i, $i = 1, \ldots m_i, i = 1, 2$. This variable could be price charged by producer or store i, wage rate or some other economic varibles. On level two each node is labeled by the averages of xs of leaves in the cluster associated the node. For simpler notation we use $p_\alpha = (x_1 + \cdots + x_{m_1})/m_1$ to label the first node on level 2, and so on.[12]

We assume that xs adjsuts to reduce differences among them. For example, x_1 adjusts with speed q_1 to reduce the gap between it and other xs in the same cluster, and with speed 2 to reduce differences of the averages of xs in other clusters, since they are at distance 2 from cluster 1. In other words, $x_1(t)$ adjusts according to the dynamics of

$$\frac{dx_1}{dt} = q_1(x_2 - x_1) + q_1(x_3 - x_1) + q_2(x_4 - x_1) + \cdots + q_2(x_9 - x_1)$$

$$= q_1(x_2 + x_3) + q_2(x_3 + \cdots + x_9) - x_1(2q_1 + 6q_2).$$

Define the average (price) of cluster 1 by

$$m_1 p_\alpha = x_1 + x_2 + x_3.$$

Define p_β as the arithmetic average of x_4, x_5, x_6 and p_γ the average of x_7, x_8, x_9.

Noting that $x_2 + x_3 = m_1 p_\alpha - x_1$, we rewrite the adjustment equation for x_1 as

$$\frac{dx_1}{dt} = q_1(m_1 p_\alpha - x_1) + m_1 q_2(m_2 \bar{P} - p_\alpha) - x_1(m_1(q_1 - q_2) + Nq_2,$$

where we denote the total number of agents or good prices by $N = +m_1 m_2$, and define the over-all average

$$\bar{P} = (p_\alpha + p_\beta + p_\gamma)/m_2 = (x_1 + \ldots x_9)/N,$$

which is a constant by definition.

This equation may be rewritten as

---

[12] More generally in a R-level tree, leaves are labelled by specifying the nodes at all levels as $(j_1, j_2, \ldots, j_R)$, $j_i = 1, \ldots, m_i$. In this notation $x_1 = x(j_1 = 1, j_2 = 1)$. Suming $x(j_1, j_2)$ over $j_1 = 1, \ldots, m_1$ and dividing the sum by m_1 we obtain the average over a level 1 cluster, $(x(j_2 = 1)$, say.

$$\frac{dx_1}{dt} = Nq_2\bar{P} + m_1(q_1 - q_2)p_\alpha - x_1[m_1(q_1 - q_2) + Nq_2]. \tag{1}$$

The time constant of this differential equation is

$$1/\tau_1 = m_1(q_1 - q_2) + m_1 m_2 q_2.$$

Since the differential equations for x_2 and x_3 are of the same structure as that for x_1, we obtain the differential equations for p_α by addding them up and dividing the sum by m_1

$$\frac{dp_\alpha}{dt} = Nq_2\bar{P} - Nq_2 p_\alpha. \tag{2}$$

The time constant of this average dynamics is

$$1/\tau_2 = m_1 m_2 q_2.$$

We assume that the tree is such that $q_1 \geq q_2$. For example information about prices posted by agents in the same cluster, that is those on level 1 may become available more quickly than information about price averages on level 2. Then time constant of adjustemt dynamics on level 1 is shorter than that of level 2. If m_1 is large and q_1 and q_2 differs by an order of magnitude, then the difference in the two time constants could be very large. This type of topics is discussed next. In other words, note that $1/\tau_1 - 1/\tau_2 = m_1(q_1 - q_2)$ is positive and large when m_1 is large and q_1 is much larger than q_2.

Equations (1) and (2) are not the only possible dynamics. We could assume that a producer reacts to the average price of his cluster, rather than to individual prices, due to cost of information gathering or some such factors. Then we have

$$\frac{dx_1}{dt} = q_1(p_\alpha - x_1) + q_2[p_\beta - x_1] + (p_\gamma - x_1)$$

$$= m_2 q_2\bar{P} + (q_1 - q_2)p_\alpha - [q_1 + (m_2 - 1)q_2]x_1.$$

The equation for the average price is

$$\frac{dp_\alpha}{dt} = m_2 q_2\bar{P} - m_2 q_2 p_\alpha. \tag{3}$$

In this model, $1/\tau_1 = q_1 + (m_2 - 1)q_2$, and $1/\tau_2 = m_2 q_2$. The difference in time constants is $q_1 - q_2$. It is still positive, but smaller than that of the previous model.

Growth Volatility Indices*

Davide Fiaschi[1] and Andrea Mario Lavezzi[2]

[1] University of Pisa, Dipartimento di Scienze Economiche dfiaschi@ec.unipi.it
[2] University of Pisa, Dipartimento di Scienze Economiche lavezzi@ec.unipi.it

Summary. We study the determinants of growth rate volatility in a multisector economy where sectors are heterogeneous in their individual volatility. We propose a model where aggregate volatility is explained by structural change and the size of the economy. We present a first attempt to test these predictions measuring growth volatility by indices based on Markov transition matrices. Growth volatility appears to (i) decrease with total GDP and (ii) increase with the share of the agricultural sector on GDP, although some nonlinearities appear. Trade openness, which we relate to the size of the economy, also plays a role. In accordance with our model, the explanatory power of per capita GDP, a relevant variable in other empirical works, vanishes when we control for these variables.

Key words: growth volatility, Markov transition matrix, structural change, nonparametric methods.
JEL classification numbers: O11, O40, C14, C21.

1 Introduction

The relationship between growth rate volatility (*GRV* henceforth), income levels and other explanatory variables has recently started to receive attention. Contributions can be divided into two main groups. The first highlights that development is accompanied by a sharp reduction in *GRV* (see [1] and [14]), while the second refers to a negative (power-law) relationship between the *size* of an economy and *GRV* (see [6]). In addition, [2] identify another possible causal explanation of volatility in the lack of strong "institutions" (which may, e.g., enforce property rights, reduce corruption and/or political instability), while [7] focus on the development of the financial sector as a cause for the reduction in volatility. Therefore, identifying the main determinants

* We are very grateful to Carlo Bianchi, Eugene Cleur and seminar participants at Siena and Guanajuato for helpful comments. The usual disclaimers apply.

of growth volatility of a country can be relevant for a better understanding of the development process, especially in low-income countries, as well as for the design of economic policies aiming at stabilizing the growth path in underdeveloped economies, as high volatility may be associated to low long-run growth (see [18]).

In this paper we propose a first attempt to shed light on the determinants of GRV by means of *volatility indices*. To measure growth volatility we present a new statistical methodology based on Markov transition matrices, in contrast to the use of measures of volatility based on the standard deviation of growth rates.[3] In particular we propose some *growth volatility indices* based the literature on mobility indices (see, e.g. [3]). Specifically, we reinterpret a set of indices generally utilized to measure intergenerational mobility as measures of volatility, and propose two new indices.

Since development is generally proxied by per capita GDP, a first possible empirical investigation regards the relationship between GRV and per capita GDP. In this context, we also analyse structural change, a typical phenomenon associated to development. In fact, a plausible explanation of the reduction in GRV as development proceeds, resides in the decreasing weight of sectors with more volatile output, like agriculture and primary sectors, with respect to sectors with less volatile output, like manufacturing and services.[4] Differently, the increase in the number of sectors (or productive units) associated to a growing size of the economy is the most common explanation of the relationship between the size of the economy and GRV. In this case, a reduction in aggregate GRV may derive from averaging an increasing number of sectoral growth rates, since idiosyncratic shocks to each sector would tend to cancel out by the law of large numbers. For example, [6] find that the relation between growth volatility and the size of the economy takes the form of a power law.

In this paper we introduce a simple analytical framework and then test for the existence of these relationships in a large sample of countries from [12]'s dataset. In particular we focus on the effect of three variables on GRV: (i) the level of per capita GDP (GDP henceforth) as proxy for the level of development, (ii) the share of agriculture on GDP (AS henceforth) as proxy for structural change and (iii) total GDP ($TGDP$ henceforth) as proxy for the size of the economy. We also consider a measure of trade openness (TR henceforth), to proxy the effective dimension of an economy which may not be entirely captured by $TGDP$ alone. Therefore, we limit our analysis to real factors, without considering monetary factors, e.g. inflation, which can have explanatory power. In a companion paper (see [9]) we consider a monetary factor, that is the size of the credit market, and the composition of trade.

---

[3] In a companion paper we measure growth volatility by standard deviations. See [8].

[4] So far the literature on structural change has not paid attention to this issue (see e.g. [13]).

Individually, we find evidence of an inverse relationship between GRV and both GDP and $TGDP$, and of a positive relationship between GRV and AS as we expected. However, we show that the relationship between $TGDP$ and GRV appears more complex than the one found by [6], and that some nonlinearities appear in the case of AS. When we consider the interaction among the variables, the effect of GDP on GRV vanishes when it is considered jointly with $TGDP$ and AS. The effect of TR, when considered with respect to that of $TGDP$, appears also negative. These findings agree with the predictions of our model, in which GRV is explained by structural change and, especially, by the size of the economy.

The paper is organized as follows. Section 2 proposes a simple model to explain the growth volatility of a multisector economy. Section 3 contains a graphical analysis of GRV based on nonparametric methods; Section 4 discusses the issue of the measurement of growth volatility and introduces the GRV indices; Section 5 presents and discusses the results; Section 6 concludes.

2 A Basic Analytical Framework

In this section we present a simple model to highlight the key factors which can account for GRV in a country. In particular our focus is on the composition of output and on the size of the economy.

Consider an economy with N_t sectors, where t indexes time. Sector i's output grows according to the following rule:

$$y_t^i = y_{t-1}^i \left(1 + g_t^i \varepsilon_t^i\right),$$

where y_t^i is output of sector i in period t, g_t^i is the exogenous growth rate of sector i, and ε_i^t is a random shock.

We assume that shocks are normally distributed with mean 1 and variance $\left(\sigma^i\right)^2$, that is:

$$\varepsilon_t^i \sim N\left(1, \left(\sigma^i\right)^2\right).$$

Let Γ_t be the $N_t \times N_t$ covariance matrix, where γ_t^{ij} is an element. Note that assuming a nonzero covariance among shocks is a simple way to model sectoral interdependence.[5] We assume that the autocorrelation of the shocks is zero, that is $cov\left(\varepsilon_t^i, \varepsilon_{t-1}^i\right) = 0$, $\forall i = 1, ..., N_t$ and $\forall t$. Finally, we assume that $\sigma^{i-1} \geq \sigma^i$, $i = 2, ..., N_t$, that is we order sectors on the basis of GRV.[6]

---

[5] [11] shows that a multisector model with intermediate goods and idiosyncratic shocks to individual sectors can generate an aggregate dynamics for certain structures of sectoral outputs' correlation.

[6] [10] endogenize the different volatility of sectors on the basis of rent-seeking theory. Predation may be actually relevant in countries with large primary sectors (i.e. rich in natural resources). Here, we argue that primary sectors are intrin-

Note that shocks are assumed to be normally distributed for analytical convenience. In fact, this allows us to measure aggregate GRV by the standard deviation of the aggregate growth rate. If we relax this assumption, the measurement of GRV of a country can become complex.

Let Y_t be aggregate output in period t, that is:

$$Y_t = \Sigma_{i=1}^{N_t} y_t^i.$$

Therefore the aggregate growth rate is given by:

$$\mu_t = \frac{Y_t - Y_{t-1}}{Y_{t-1}} = \frac{\Sigma_{i=1}^{N_t} y_{t-1}^i \left(1 + g_t^i \varepsilon_t^i\right)}{\Sigma_{i=1}^{N_t} y_{t-1}^i} - 1 = \Sigma_{i=1}^{N_t} \alpha_{t-1}^i g_t^i \varepsilon_t^i, \tag{1}$$

where $\alpha_{t-1}^i = \frac{y_{t-1}^i}{\Sigma_{i=1}^{N_t} y_{t-1}^i}$ is the share of output of sector i with respect to total output, so that $\Sigma_{i=1}^{N_t} \alpha_{t-1}^i = 1$, $\forall t$.

From definition (1) we have that the expected value and variance of μ_t are given by:

$$\bar{\mu}_t = E_t\left[\mu_t\right] = \Sigma_{i=1}^{N_t} \alpha_{t-1}^i g_t^i \tag{2}$$

$$\bar{\sigma}_t^2 = E_t\left[\left(\Sigma_{i=1}^{N_t} \alpha_{t-1}^i g_t^i \eta_t^i\right)^2\right], \tag{3}$$

where $\eta_t^i = \varepsilon_t^i - 1$. Trivially, η has the same properties of ε, but its mean is equal to 0 (that is $\eta_t^i \sim N\left(0, \left(\sigma^i\right)^2\right)$). It follows that μ_t is normally distributed, that is $\mu_t \sim N\left(\bar{\mu}_t, \bar{\sigma}_t^2\right)$. From (3) we obtain the following expression for $\bar{\sigma}_t^2$:

$$\bar{\sigma}_t^2 = \Sigma_{i=1}^{N_t} \left(\alpha_{t-1}^i g_t^i \sigma^i\right)^2 + \Sigma_{i=1}^{N_t} \Sigma_{j=1, j\neq i}^{N_t} \alpha_{t-1}^i \alpha_{t-1}^j g_t^i g_t^j \gamma_t^{ij}, \tag{4}$$

where $\gamma_t^{ij} - 1$ is the covariance between η_t^i and η_t^j.

The functional form of Equation (4) does not allow for a simple identification of the effects of the elements on the right-hand side on $\bar{\sigma}_t^2$, except for g_t^i. An increase of g_t^i, *ceteris paribus*, increases both $\bar{\mu}_t$ and $\bar{\sigma}_t^2$. However, the effects of the other variables involved, in particular the number of sectors N_t and the structure of the economy $\left(\alpha_{t-1}^1, ..., \alpha_{t-1}^{N_t}\right)$, may not be so easily identifiable.

---

sically more subject to random shocks, e.g. changes in terms of trade, climatic changes and the like. "Good" institutions (or other variables included in the concept of "social capital") indicated as factors reducing volatility generally emerge where predation is less important. In this respect, institutions are the endogenous result of the structure of the economy. In addition, papers such as [19] and [11], which study the emergence of aggregate fluctuations from sectoral shocks, do not consider differences in volatility across sectors and thus are not concerned with structural change.

To proceed, suppose that Y_t comes from the agricultural sector A (sector 1), and from the rest of economy R (sectors $2, ..., N_t$), which includes secondary and tertiary sectors (we will use this distinction in the empirical analysis). Denoting by α_A and α_R the shares from the two macrosectors, equation (4) becomes:

$$\bar{\sigma}_t^2 = \left(\alpha_{t-1}^A g_t^A \sigma^A\right)^2 + \left(\alpha_{t-1}^R g_t^R \sigma_t^R\right)^2 + \alpha_{t-1}^A \alpha_{t-1}^R g_t^A g_t^R \gamma_t^{AR}. \tag{5}$$

It is plausible to assume that $\gamma_t^{AR} = 0$ because shocks to A and R are likely to be of different nature and uncorrelated.[7] Therefore we have:

$$\bar{\sigma}_t^2 = \left(\alpha_{t-1}^A g_t^A \sigma^A\right)^2 + \left[\alpha_{t-1}^R g_t^R \sigma_t^R\right]^2. \tag{6}$$

Generally, a change in α_{t-1}^A and $\alpha_{t-1}^R = 1 - \alpha_{t-1}^A$, and/or a change in the number of sectors N_t have an ambiguous effect on aggregate variance. Let us analyse first the role of N_t.

Number of Sectors and Growth Volatility Some authors argue that aggregate *GRV* is affected by the size of the economy, when measured in terms of number of sectors or units of production (see e.g. [19]). In our model, the possible negative correlation between *GRV* and N can derive from an inverse correlation between σ^R and N. We can identify simple conditions under which $d\sigma^R/dN < 0$. Assume that $g_t^i = g^R$, $\alpha_t^i = \alpha_0^i = \frac{1}{N_0-1}$, for $i, j = 2, ..., N_t$. Thus, from Equation (4) written with respect to R, we have:

$$\left(\bar{\sigma}_t^R\right)^2 = \left(\frac{g^R}{N_t-1}\right)^2 \left[\Sigma_{i=2}^{N_t} \left(\sigma^i\right)^2\right] + \left(\alpha_0^i\right)^2 \left(g^R\right)^2 \Sigma_{i=2}^{N_t} \Sigma_{j=2, j\neq i}^{N_t} \gamma_t^{ij}, \tag{7}$$

If $\gamma_t^{ij} = 0$ for $i, j = 2, ..., N_t$, then $\left(\bar{\sigma}_t^R\right)^2$ is decreasing in N_t and increasing in g^R, given that $\sigma^{i-1} \geq \sigma^i$.

Hence, the higher is the number of sectors in R, the lower is the variance of its growth rate, if the covariance between sectors is negligible (this is an application of the law of large numbers). If the size of an economy is positively related to the number of sectors N, then the size of an economy and its growth volatility are inversely related.

Moreover, higher g^R leads to higher *GRV* but, if the output of some sectors has a strong positive correlation with the output of others, then *GRV* can nonetheless increase if the latter effect is stronger than the effect of the increase in N. An example can be the emergence of a financial sector, whose output is correlated to many sectors through the capital market. That is, the appearance of a global capital market may increase the interdependence among sectors and affect *GRV*.[8]

To conclude, if $\sigma^R = \sigma^R(N)$, where $d\sigma^R/dN < 0$, then from Eq. (6) we obtain:

---

[7] For a discussion of the relationship between $\bar{\sigma}^2$ and Γ see [11].
[8] In [9] we study the effects of the financial sector on *GRV*.

$$\frac{\partial \bar{\sigma}_t^2}{\partial N_t} = \left[\alpha_{t-1}^R g_t^R\right]^2 \frac{d\left(\sigma^R\right)^2}{dN_t} < 0. \tag{8}$$

We show below that this relationship finds an empirical support when we proxy for N by the dimension of the economy.[9]

Composition of Output and Growth Volatility In a typical process of development and structural change, primary sectors grow less than industrial and service sectors. This implies that the share of sectors with higher variance declines over time. The overall result would be a decrease in aggregate GRV, as the latter is a weighted sum of sectors' variances, and weights are proportional to sectors' shares.

From Eqs. (2) and (6) we have:

$$\bar{\sigma}_t^2 = \left(\alpha_{t-1}^A g_t^A \sigma^A\right)^2 + \left[\left(\bar{\mu}_t - \alpha_{t-1}^A g_t^A\right) \sigma^R\right]^2.$$

Calculations lead to:

$$\frac{\partial \bar{\sigma}_t^2}{\partial \alpha_{t-1}^A} > 0 \Leftrightarrow \alpha_{t-1}^A > \frac{\bar{\mu}_t}{g_t^A} \left[1 + \frac{\left(\sigma^A\right)^2}{\left(\sigma^R\right)^2}\right]^{-1} = \bar{\alpha}_t. \tag{9}$$

This means that for $\alpha_{t-1}^A < \bar{\alpha}_t$ ($\alpha_{t-1}^A > \bar{\alpha}_t$) GRV is decreasing (increasing) in the share of the agricultural sector α^A. That is, the relationship between α_{t-1}^A and $\bar{\sigma}_t^2$ is U-shaped.[10] The dependence of the threshold on time introduces some complications, as the threshold itself changes with α_{t-1}^A.

Consider two cases: i) $g_t^A = g_t^R = g$ and ii) $g_t^A = g^A < g_t^R = g^R$. Case i) refers to the absence of structural change, as the weights of the sectors remain unchanged. In this case $\bar{\mu}_t = g_t^A = g$ and the value of the threshold is the constant term in square brackets. In case ii), we have $\alpha_{t-1}^A \to 0$ and $\bar{\mu}_t \to g^A$, and

$$\bar{\alpha}_t \to \frac{g^R}{g^A} \left[1 + \frac{\left(\sigma^A\right)^2}{\left(\sigma^R\right)^2}\right]^{-1}. \tag{10}$$

In case ii) the threshold value is higher than in case i). In general, the threshold value is lower (and therefore the range of values where GRV is decreasing in α^A is wider), the higher is $\left(\sigma^A\right)^2$ with respect to $\left(\sigma^R\right)^2$. Moreover, if $\sigma^R = \sigma^R(N)$ and $d\sigma^R/dN < 0$, then the threshold value $\bar{\alpha}$ decreases in N_t.

To summarize our results consider the following equation, derived from Eq. (6):

---

[9] Note that to isolate the effect of N_t, we ruled out the presence of structural change but, clearly, the two factors interact.

[10] Note that the U-shaped relation between α_{t-1}^A and $\bar{\sigma}_t^2$ resembles the relation between the variance of a portfolio and the share of the more volatile asset. In the problem of portfolio choice, the variance of portfolio decreases with the share of the more volatile asset until a positive threshold value is reached, then increases.

$$\bar{\sigma}_t^2 = \left(\bar{\mu}_t \sigma^R\right)^2 + \left(\alpha_{t-1}^A g_t^A\right)^2 \left[\left(\sigma^A\right)^2 + \left(\sigma^R\right)^2 - \frac{2\sigma^R}{\alpha_{t-1}^A g_t^A}\right]. \tag{11}$$

In Eq. (11) aggregate variance depends on two terms: the first term captures the effect of the variance of the "rest of the economy", which we argue depends negatively on the number of sectors N (see Eq. (8)); the second term represents the effect of the share of agriculture α^A, whose sign depends in a non-trivial way on the relative values of sectoral growth rates and on the interaction with N, via σ^R (see condition (9)). Note finally that GDP does not play any role in the model.

In our empirical analysis we focus on Eq. (11) and try to shed some light on the complex relations it highlights. In particular we will consider our variables in isolation and study their interactions by pairwise relations.[11]

3 Graphical Analysis

We use data on GDP and $TGDP$ from [12]'s database and data on agriculture and trade from the World Bank's *World Development Indicators 2002*. Our sample includes 119 countries for the period 1960–1998.[12] As noted, we proxy for the structure of the economy by the share of the agricultural sector in aggregate value added, AS, and measure the *effective* dimension of the economy, related to the number of sectors N in the model, both by the *total* GDP (*TGDP*) *and* trade openness (*TR*), which is the ratio of the sum of imports and exports on GDP. The latter, jointly with *TGDP*, provides a more exact measure of the extent of the overall market for an economy.

We consider both the cross-country and the time-series dimension of growth volatility. In particular, to evaluate the relation between GRV and level of development we separate all observations on GDP and $TGDP$ into 151 classes with a similar number of observations (approximately 30), while to evaluate the relation between GRV and structural change we separate all observations on AS into 109 classes. Finally, for the relation between GRV and TR we have 125 classes of observations.[13]

For every observation on GDP in year t we calculate the growth rate from t to $t+1$.[14] Figure 1 reports the standard deviation of growth rates, GRV, relative to the observations in a class against, respectively, the log of the

---

[11] This choice depends on our methods of empirical investigation. In [8] we estimate directly Eq. (11) by nonparametric methods.

[12] Data on GDP and $TGDP$ are in 1990 international dollars. Not all observations on agriculture and trade openness were available for each country for all years. See Appendix A for the country list.

[13] In [9] we discuss the methodology for the empirical investigation in more detail.

[14] For data on AS, $TGDP$ and TR we consider the corresponding observation on GDP and calculate the associated growth rate.

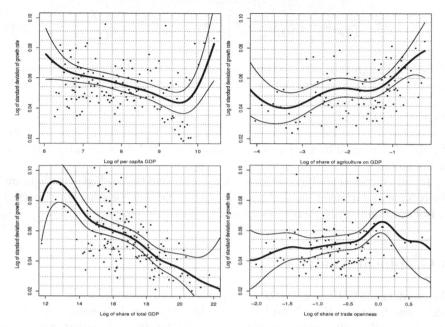

Fig. 1. *GRV* estimated by *STD* vs, respectively, log of *GDP*, *AS*, *TGDP* and *TR*.

average *GDP*, *AS*, *TGDP* and *TR* in that class, and run a nonparametric estimation of these relationships.[15]

Figure 1 is the counterpart of Figure 1 in [1], where only cross-country variation in growth volatility is considered. They estimate an OLS regression and find a decreasing relationship between growth volatility and development, proxied by the initial level of *GDP*.

In our case, we see at first glance that *GRV* tends to fall with *GDP*. The high volatility at the lowest and, especially, highest *GDP* levels is associated with a much wider variability band, meaning that there the estimate is not precise. In Figure 1 growth volatility appears to be increasing with *AS*. This relation is not monotonic, but the variability band is tighter where the upward sloping portion is steeper, indicating that the estimation is more precise where the curve is sharply increasing (we return on this below). In Figure 1 *GRV*

---

[15] For all the nonparametric estimates we used [17], in particular the statistical package *sm*, see [5]. We used the standard settings suggested by the authors (i.e. optimal normal bandwidth). To test the robustness of this estimate, we ran an alternative nonparametric regression using the plug-in method to calculate the kernel bandwidth, and obtained a similar picture. We refer to [4] for more details. We report the variability bands representing two standard errors above and below the estimate. They give a measure of the statistical significance of the estimate (see [4, pp. 29–30] for details on variability bands vs confidence bands). Data sets and codes used in the empirical analysis are available on the authors' websites (http://www-dse.ec.unipi.it/fiaschi and http://www-dse.ec.unipi.it/lavezzi).

clearly decreases with $TGDP$, as the extreme portions of the estimate have a wide variability band.

Finally, the relationship between GRV and TR in Figure 1 appears inversely U-shaped. In particular, the estimate of both the decreasing parts has a wide variability band. As noted, the impact of TR on GRV does not interest us *per se*, but in conjunction with $TGDP$ when we proxy for the effective size of an economy. In our view, the effective size of the economy increases if it is highly integrated with other economies.[16]

Note that in Figure 1 we have studied the effects on GRV of the variables taken individually. However, from our model, these variables are expected to have a joint effect on GRV, that is their effect should be evaluated given the presence of other variables and of possible interactions among them.[17]

4 Growth Volatility Indices

In this section we propose a set of synthetic indices to measure GRV.[18] In particular, the measurement of GRV first requires the estimation of a Markov transition matrix, whose states $S = \{1, 2, ..., n\}$ represent growth rate classes. A transition matrix summarizes the information on the dynamics of growth rates (for more details see [15]), and is the basis to calculate GRV indices.

Heuristically, the indices quantify volatility by the *intensity of switches across growth rate classes*. The advantage of the approach based on transition matrices is that we can keep track of the dynamics of individual countries in the sample. This is an important aspect which is obscured by measures of volatility based on standard deviation of growth rates, often found in the literature (and used in [8]).

For example, consider two countries for which we have observations for, say, two AS classes and three growth rate classes. In the first AS class both countries have highly unstable time series, crossing all levels of growth rate classes. In the second (lower) AS class, both countries have more stable time series which span only two growth rate classes. However, the first country has observations in the first and second growth rate class, while the second has observations in the second and the third growth rate class (in other words they can have different average growth rates). If we wanted to measure growth

---

[16] In general, as noted by [7, p. 10], the effect of trade openness is ambiguous: trade may reduce volatility if it represents an increase in the size of the economy, but it may also increase it by making a country more vulnerable to external shocks. We remark that a relevant issue is the *composition* of trade. For instance, a high level of trade openness associated to a concentration of export in few goods (e.g. primary or oil) may result in a positive relation between trade openness and growth volatility. We study the effect of trade composition in [9].

[17] In the nonparametric estimation in [8] we explicitly consider the interactions by introducing interaction terms in the regressions.

[18] The statistical properties of the indices are derived in [9].

volatility in the two AS classes, we would not be able to detect the decrease in volatility if we used the standard deviation, while we would correctly detect the reduction in volatility if we used a transition matrix. To evaluate the relationship between GRV and, for instance, AS we calculate the values of the indices for different classes of AS.

To define indices of GRV we draw on studies of inter- and intragenerational mobility of individuals (see, among others, [3, pp. 24–30] and [20]), and propose two new indices. Basically, these indices are functions of the elements of a transition matrix. In a transition matrix high values on the principal diagonal indicate low mobility, while the values of off-diagonal elements refer to changes of state and, therefore, high values of the latter are associated to high mobility.

A simple mobility index is the following, proposed by [20]:

$$I^S(\mathbf{P}) = \frac{n - trace(\mathbf{P})}{n - 1}, \tag{12}$$

where $\mathbf{P}$ is a transition matrix of dimension n. The range of the index is $[0, n/(n-1)]$ and a high value means high mobility. However, I^S is not well-suited to measure growth volatility because it is not affected by the value of off-diagonal elements, a key point for the present analysis, but we refer to it as a term of comparison with the other indices discussed below.

[3, p. 28], proposes the following index which takes explicitly into account the *distance* covered by a transition from i to j, $(i, j \in S)$, when the states correspond to increasing or decreasing values of a variable:

$$I^B(\mathbf{P}) = \frac{1}{n-1} \sum_{i=1}^{n} \sum_{j=1}^{n} \pi_i p_{ij} |i - j|. \tag{13}$$

In I^B, p_{ij} is an element of the transition matrix $\mathbf{P}$, while π_i is an element of the associated ergodic distribution.[19] The range of I^B is $[0, 1]$: a higher value means higher mobility.

In this case only the absolute value of the difference between i and j is taken into account. It is worth verifying the effect of increasing the weight attached to "longer" jumps, in order to better appreciate the magnitude of the fluctuations. Therefore we introduce the following index:

$$I^{BM}(\mathbf{P}) = \frac{1}{(n-1)^2} \sum_{i=1}^{n} \sum_{j=1}^{n} \pi_i p_{ij} (i - j)^2, \tag{14}$$

in which the distance of the transition enters in a quadratic form. As before $I^{BM} \in [0, 1]$ and a higher value means higher mobility/volatility.

---

[19] The ergodic distribution represents the long-run distribution of the Markov process. For more details see [3].

Indices I^B and I^{BM} weight the transitions from growth rate class i by the corresponding mass in the long-run equilibrium (i.e. in the ergodic distribution). In other words, considering the elements of the ergodic distribution as weights amounts to measuring GRV in the long-run equilibrium. However, also the volatility along the transition path can reveal very interesting information. The following indices fill this gap:

$$I^{FL}(\mathbf{P}) = \frac{1}{A} \sum_{i=1}^{n} \sum_{j=1}^{n} p_{ij} \left| i - j \right|; \tag{15}$$

$$I^{FLM}(\mathbf{P}) = \frac{1}{A^2} \sum_{i=1}^{n} \sum_{j=1}^{n} p_{ij} (i - j)^2. \tag{16}$$

I^{FL} and I^{FLM} respectively correspond to I^B and I^{BM}, except for the absence of the elements of the ergodic distribution. The constant A normalizes both indices to the range $[0, 1]$.[20] A higher value still means higher mobility/volatility.

5 Empirical Results

In the following we study the relation between GRV, GDP, AS and $TGDP$ by calculating the values of the indices described in Section 4. As in Section 3, we first evaluate the individual effects of our variables, and then study their interactions, with particular attention to the explanatory power of GDP.

In particular, in a first stage: (i) we separate the observations on GDP, AS and $TGDP$ in four classes for each variable, from "low" to "high" values; (ii) we calculate the transition matrix with five growth rate classes for each class, (iii) we compute indices (12), (13), (14), (15), (16) for every transition matrix and, finally, (iv) we make inference on these estimates. In a second stage we evaluate the interactions of the variables in this framework.

First we define the five growth rate classes common to all three variables. We set the central class to include the average growth rate of the sample, equal to 2%, and define the other classes symmetrically around this central class.[21] With this criterion we obtain the state space:[22]

$$S = \{[-\infty, -2\%), [-2\%, 1\%), [1\%, 3\%), [3\%, 6\%), [6\%, +\infty)\}. \tag{17}$$

---

[20] In particular:

$$A = \begin{cases} 2\Sigma_{i=\frac{n-1}{2}+1}^{n-1} i + \frac{n-1}{2} & \text{for } n \text{ odd;} \\ 2\Sigma_{i=\frac{n}{2}}^{n-1} i & \text{for } n \text{ even.} \end{cases}$$

[21] In [9] we propose an alternative method of calculating the growth rate classes based on the density of observations.

[22] Results are not affected by slight changes of the classes' limits.

Index\GDP class	I	II	III	IV
I^S	0.8313	0.7831	0.7417	0.7793
	(0.0195)	(0.0192)	(0.0193)	(0.023)
I^B	0.2583	0.2474	0.2272	0.1983
	(0.0108)	(0.0093)	(0.0086)	(0.0077)
I^{BM}	0.1548	0.1382	0.1206	0.0888
	(0.0099)	(0.008)	(0.0074)	(0.0062)
I^{FL}	0.4178	0.3658	0.3136	0.3253
	(0.0149)	(0.0126)	(0.0115)	(0.0162)
I^{FLM}	0.3199	0.259	0.2083	0.2148
	(0.0169)	(0.0138)	(0.0122)	(0.0188)

Table 1. Growth volatility indices. Standard errors in parenthesis. *GDP*

Alternatively, the state space for the calculation of the transition matrices could be based not on absolute values of growth rates, but on deviations from the trend, as in [9].

5.1 Per Capita GDP

We define four *GDP* classes (in logs) which contain the same number of observations (≈ 1100), obtaining the following:

$$I = [0, 6.98), II = [6.98, 7.9), III = [7.9, 8.82), IV = [8.82, +\infty).$$

For every *GDP* class we estimate a transition matrix relative to the state space S.[23] Table 1 contains the values of the indices calculated for each of the four transition matrices.

We observe that in all cases the value of the index is generally decreasing with respect to the *GDP* class and that, in particular, the value of the index in the first *GDP* class is always higher than in the last. This result broadly agrees with Figure 1 in which volatility is measured by the standard deviation of growth rates.

We tested for the joint equality of the estimated value and strongly rejected the null-hypothesis.[24]

---

[23] The four transition matrices and the four ergodic distributions (one for each *GDP* class) are obtainable with the codes available on the authors' websites.

[24] [8] show that these indices are asymptotically normally distributed. The test statistic:

$$F_{r-1, r(n-1)} = \frac{n\Sigma_{i=1}^r \left(I\left(\hat{P}^i\right) - I\overline{(P)}\right)^2 / (r-1)}{\Sigma_{i=1}^r \frac{\hat{\sigma}_{Ii}^2}{n} / r},$$

where r is the number of indices ($r = 4$), n is the number of observation used for the computation of every index $\hat{\sigma}_{Ii}^2/n$, $i = 1, 2$, is the variance of the value of the index calculated from transition matrix $\hat{P}^i$ and $I\overline{(P)}$ is the average value of index, is distributed as a "F" distribution with $r - 1$, $r(n-1)$ degrees of freedom.

Index\GDP class	I vs II	I vs III	I vs IV
I^S	0.04*	0*	0.04*
I^B	0.22	0.01*	0*
I^{BM}	0.10	0*	0*
I^{FL}	0*	0*	0*
I^{FLM}	0*	0*	0*

Table 2. Test of equality between the GRV index of GDP class I versus its value in the other classes. * means rejection of the null hypothesis of equality at 5% confidence level.

Table 2 reports the p-values of tests of a null hypothesis of equality between the value of the index in the first GDP class versus its value in each of the other GDP classes, for all the indices.

Tests confirm that GDP class I generally has a statistically significant higher GRV. At a conventional 5% level, the null hypothesis is not rejected only in the comparison between the value of the index in the first and in the second GDP class for indices I^B and I^{BM} (but it is rejected at 10% for the latter).

To check if there is a monotonic decreasing relationship among the values of the indices at different GDP levels we also tested the following hypotheses of equality (details omitted): (i) GDP class II vs GDP class III; (ii) GDP class III vs IV. In case (i), the hypothesis is strongly rejected for I^{FL} and I^{FLM}, and is rejected at approximately 6% level for the other indices; in case (ii) the hypothesis is rejected only for I^B and I^{BM}. However, in the other cases we do not reject the null hypothesis that the indices in GDP classes III vs IV are equal (note that in Table 1 the value of the index in GDP class IV is actually higher than in GDP class III for I^S, I^{FL} and I^{FLM}). Hence, according to indices I^B and I^{BM}, the decrease is statistically significant when we move from class II onwards, while with I^{FL} and I^{FLM} the decrease is statistically significant from class I, but for the two higher GDP classes the relation may be flat. Also, for I^S we do not find evidence of a monotonic decrease as the value of the index in GDP classes III and IV may be equal.

Overall, the indices indicate the presence of a negative relationship between GRV and GDP, which may become flat in some GDP ranges. In any case, a comparison between the first and the last GDP classes always shows a significantly higher volatility in the former.

5.2 The Dimension of the Economy

In this section we repeat the exercise considering $TGDP$ to proxy for the dimension of the economy.[25] We define four $TGDP$ classes (in logs) with the same number of observations (≈ 1100):

---

[25] In the next section we examine the interaction of $TGDP$ with TR. In [9] we also compute the indices for TR classes.

Index\TGDP class	I	II	III	IV
I^S	0.7721 (0.0185)	0.7998 (0.0182)	0.7799 (0.0186)	0.7339 (0.0193)
I^B	0.2904 (0.0104)	0.2749 (0.0085)	0.2645 (0.0088)	0.2155 (0.0072)
I^{BM}	0.18 (0.01)	0.1524 (0.0076)	0.1462 (0.0081)	0.1001 (0.0055)
I^{FL}	0.3682 (0.0116)	0.3478 (0.0107)	0.3449 (0.0115)	0.2846 (0.0103)
I^{FLM}	0.2725 (0.0124)	0.2326 (0.011)	0.2375 (0.0121)	0.1679 (0.01)

Table 3. Growth volatility indices. Standard errors in parenthesis. $TGDP$

Index\TGDP class	I vs II	I vs III	I vs IV
I^S	0.14	0.38	0.08
I^B	0.12	0.03*	0*
I^{BM}	0.01*	0*	0*
I^{FL}	0.10	0.07	0*
I^{FLM}	0*	0.02*	0*

Table 4. Test of equality between the GRV index of $TGDP$ class I versus its value in the other classes. * means rejection of the null hypothesis of equality at 5% confidence level.

$$I = [0, 15.45), II = [15.45, 16.6), III = [16.6, 18.2), IV = [18.2, 2.90].$$

With this class definition, and with the same state space for growth rate classes, we obtain the volatility indices in Table 3.

In this case we observe a monotonic decrease for indices I^B, I^{BM} and I^{FL} across the $TGDP$ classes, while for I^{FLM} the value is higher in $TGDP$ class III than in II. Finally, no clear relation emerges from I^S.

From Table 4 we see that, at 5% confidence level, the value of the index in class I is statistically different from the value in class II in two cases, from the value in class III in three cases, and from the value in class IV in four cases. However, (i) in neither case we can reject the hypothesis of equality between the values in $TGDP$ classes II and III (details omitted), (ii) we always reject the hypothesis of equality between the values of the indices in classes III and IV.

This is in agreement with Figure 1, in which the relation between GRV and $TGDP$ is slightly flatter at intermediate $TGDP$ levels, and is nonmonotonic (although with a relatively large variability band) at low $TGDP$ levels. Therefore, the relationship is clearly negative when we compare the extreme $TGDP$ classes, but it appears overall more complex than the one identified by [6], who find a linear, decreasing relationship.[26]

---

[26] They use a smaller dataset and a lower number of $TGDP$ classes, and consider as an outlier the only observation for low $TGDP$ which would produce nonlinear behavior.

Index\AS class	I	II	III	IV
I^S	0.7428 (0.0229)	0.8152 (0.0214)	0.7834 (0.0223)	0.8756 (0.0207)
I^B	0.2297 (0.0088)	0.2735 (0.0105)	0.2576 (0.0102)	0.3199 (0.0115)
I^{BM}	0.1105 (0.007)	0.1501 (0.0094)	0.1377 (0.0092)	0.1915 (0.0114)
I^{FL}	0.3037 (0.0132)	0.3611 (0.0133)	0.3364 (0.0136)	0.4119 (0.0138)
I^{FLM}	0.1905 (0.0134)	0.2478 (0.014)	0.2249 (0.0142)	0.2975 (0.0151)

Table 5. Growth volatility indices. Standard errors in parenthesis. AS

Therefore, keeping in mind these remarks, we find a broad confirmation of the existence of a negative relation between GRV and $TGDP$, in particular when comparing the extreme classes.

5.3 Structural Change

In this section we address the relationship between GRV and structural change proxied by AS. We first define four AS classes with the same number of observations (≈ 784). The resulting classes' limits are:

$$I = [0, 0.08), II = [0.08, 0.2), III = [0.2, 0.33), IV = [0.33, 1].$$

Table 5 contains the volatility indices calculated with this class definition. Results seem to be in accordance with the pattern in Figure 1. Moving from high to low levels of AS, that is following a typical development path, volatility decreases from class IV to class III, then increases in class II and decreases again in class I. However, tests of equality between the value of the indices in class III and in class II do not allow to reject the null hypothesis at conventional 5% level.[27] Finally, volatility is significantly higher in class IV than in class I: the hypothesis of equality between the value of the index in class I and in class IV is strongly rejected for all indices (we omit the details of the tests).

At this stage, we take this result as indicating the possible presence of a more complex behavior at intermediate levels of AS, which is in accordance with the non–monotonic pattern of GRV found in Figure 1. Note that indices calculated for classes II and III are not significantly different at 5% level, but only at about 15%.

5.4 On Conditioning

In [8] we present the results of nonparametric estimations suggesting that GDP is not informative when $TGDP$, TR and AS are considered. In other

---

[27] The p-values of the tests for I^S, I^B, I^{BM}, I^{FL}, I^{FLM} are, respectively, 0.15, 0.14, 0.17, 0.10, 0.13.

words, when the latter explanatory variables are present in a (nonparametric) regression, *GDP* does not provide further information on *GRV*.

Here we address this issue in the approach based on the volatility indices computed from a Markov transition matrix. First, note that the analysis in the previous section can be considered as deriving from the estimation of *conditioned* transition matrices. In fact, the basis for the calculation of each index is a transition matrix indicating the probabilities to observe transitions across growth rate classes, starting from a given growth rate class *and* a given e.g. *TGDP* class.[28]

The formal definition of a transition probability from growth rate class S_i to growth rate class S_j, given that the observation is in *TGDP* class I is:

$$p(g_t \in S_j | g_{t-1} \in S_i, TGDP_{t-1} \in I) =$$
$$= p(g_t \in S_j | g_{t-1} \in S_i) \left[\frac{p(TGDP_{t-1} \in I | g_t \in S_j, g_{t-1} \in S_i)}{p(TGDP_{t-1} \in I | g_{t-1} \in S_i)} \right]. \quad (18)$$

In Equation (18) the term on the left–hand side is an element of the conditioned transition matrix from which we derived our indices relative to *TGDP* class I. The first term on the right-hand side is an element of the transition matrix for growth rate classes only; the second term is the probability that, given a transition from growth rate class S_i to S_j, the initial growth rate is associated to a *TGDP* in class I.

If the conditioning variable *TGDP* is not relevant, $p(g_t \in S_j | g_{t-1} \in S_i, TGDP_{t-1} \in I) = p(g_t \in S_j | g_{t-1} \in S_i)$: any transition matrix calculated considering alternative values of *TGDP* would not be statistically different from the unconditioned transition matrix.[29] Therefore, *GRV* indices calculated from the former would not be statistically different from each other, and from those calculated from the unconditioned transition matrix. In the same manner, if we condition on two variables, e.g. *TGDP* and *GDP*, we have:

$$p(g_t \in S_j | g_{t-1} \in S_i, TGDP_{t-1} \in I, GDP_{t-1} \in I) =$$
$$= p(g_t \in S_j | g_{t-1} \in S_i, TGDP_{t-1} \in I) *$$
$$* \left[\frac{p(GDP_{t-1} \in I | g_t \in S_j, g_{t-1} \in S_i, TGDP_{t-1} \in I)}{p(GDP_{t-1} \in I | g_{t-1} \in S_i, TGDP_{t-1} \in I)} \right] \quad (19)$$

and the same reasoning for the relevance of *GDP*, given *TGDP*, applies.

Here we do not provide a complete discussion of this issue, but only some evidence on the relevance of *TR*, *AS* and *GDP* in explaining *GRV* given *TGDP*. Namely, we compare the values of two indices, I^B and I^{FL}, computed for *TGDP* classes only, with the values obtainable when the transition matrix

---

[28] We could have estimated an *unconditioned* transition matrix for growth rate classes only, which does not distinguish among the *TGDP* levels associated to each transition. An example of conditioned Markov chains is in [16].

[29] From the condition $p(g_t \in S_j | g_{t-1} \in S_i, TGDP_{t-1} \in I) = p(g_t \in S_j | g_{t-1} \in S_i)$ derives $p(TGDP_{t-1} \in I | g_t \in S_j, g_{t-1} \in S_i) = p(TGDP_{t-1} \in I | g_{t-1} \in S_i)$.

	Uncond.	$AS(I)$	$AS(II)$	$AS(III)$	$AS(IV)$
$TGDP(I)$	0.2904 (0.0104)	0.1923** (0.02921)	0.3079 (0.02720)	0.3510 ** (0.02908)	0.3419** (0.01852)
$TGDP(II)$	0.2749 (0.0085)	0.3174** (0.02348)	0.2725 (0.02304)	0.2503 (0.01817)	0.2958 (0.01962)
$TGDP(III)$	0.2645 (0.0088)	0.2517 (0.02355)	0.2989** (0.01961)	0.2188* (0.01912)	0.2876 (0.03172)
$TGDP(IV)$	0.2155 (0.0072)	0.2027 (0.01091)	0.2204 (0.01741)	0.2307 (0.01859)	–

Table 6. Values of I^B for $TGDP$ conditioned on AS. ** indicates rejection of null hypothesis of equality to the index in the first column at 5% level; * at 10% level

	Uncond.	$AS(I)$	$AS(II)$	$AS(III)$	$AS(IV)$
$TGDP(I)$	0.3682 (0.0116)	0.2650** (0.03750)	0.3812 (0.03108)	0.4679** (0.03331)	0.4377** (0.02014)
$TGDP(II)$	0.3478 (0.0107)	0.3935* (0.02679)	0.3641 (0.02966)	0.3532 (0.02738)	0.4010** (0.02570)
$TGDP(III)$	0.3449 (0.0115)	0.3216 (0.02692)	0.4099 (0.02415)	0.3037* (0.02582)	0.3932 (0.03876)
$TGDP(IV)$	0.2846 (0.0103)	0.3330* (0.02834)	0.2929 (0.02299)	0.2984 (0.02618)	–

Table 7. Values of I^{FL} for $TGDP$ conditioned on AS. ** indicates rejection of null hypothesis of equality to the index in the first column at 5% level; * at 10% level

is calculated for every $TGDP$ class conditioned to each class of AS, TR and GDP. The choice of $TGDP$ as the "pivotal" variable is dictated by its high explanatory power, as assessed in [8].[30]

Tables 6 and 7 consider the relevance of the information provided by AS when $TGDP$ is the principal variable, respectively, for index I^B and index I^{FL}.

The first column of the tables contains the volatility index obtained for $TGDP$ classes (see Table 3). The other columns contain the values of the index when, for each $TGDP$ class, we condition on each AS class. If AS matters, then GRV should increase with AS, and the conditioned indices should be statistically different from the unconditioned indices in the first column.[31]

What we observe is that: given a $TGDP$ level, an increase in AS is generally associated to an increase in GRV (at least if we compare $AS(I)$ to $AS(IV)$), confirming the insight that a higher agricultural share causes a

---

[30] See in particular Table 1, although present results are not completely comparable with those of [8]. In that case more variables were considered jointly, while here we investigate pairwise relations, in which there is one principle variable, $TGDP$ or AS, and only another one interacting with it.

[31] Each conditioned indices is calculated starting from the observations belonging to a $TGDP$ class and an AS class. We chose to consider only indices calculated from a minimum number of observations, at least 75. An alternative test can be conducted, aiming at testing the *joint* significance of the differences between the conditioned and unconditioned indices. We leave this issue on a side for future research.

	Uncond.	$TR(I)$	$TR(II)$	$TR(III)$	$TR(IV)$
$TGDP(I)$	0.2904 (0.0104)	0.3494* (0.03658)	0.3643** (0.03172)	0.2838 (0.02427)	0.3080 (0.01845)
$TGDP(II)$	0.2749 (0.0085)	0.2875 (0.02191)	0.2668 (0.02064)	0.2810 (0.01773)	0.2900 (0.01771)
$TGDP(III)$	0.2645 (0.0088)	0.2901 (0.02303)	0.2581 (0.01893)	0.2363* (0.01873)	0.2399 (0.01828)
$TGDP(IV)$	0.2155 (0.0072)	0.2404** (0.01331)	0.2000 (0.01275)	0.1936* (0.01453)	0.1825* (0.01903)

Table 8. Values of I^B for $TGDP$ conditioned on TR. ** indicates rejection of null hypothesis of equality to the index in the first column at 5% level; * at 10% level

	Uncond.	$TR(I)$	$TR(II)$	$TR(III)$	$TR(IV)$
$TGDP(I)$	0.3682 (0.0116)	0.4671** (0.03962)	0.4684** (0.03220)	0.3691 (0.02738)	0.3871 (0.02101)
$TGDP(II)$	0.3478 (0.0107)	0.3817 (0.02819)	0.3600 (0.02774)	0.4064** (0.02600)	0.3632 (0.02177)
$TGDP(III)$	0.3449 (0.0115)	0.3678 (0.02748)	0.3551 (0.02418)	0.3375 (0.02755)	0.3120* (0.02231)
$TGDP(IV)$	0.2846 (0.0103)	0.3380** (0.01889)	0.2722 (0.02386)	0.2578 (0.02199)	0.2359** (0.02489)

Table 9. Values of I^{FL} for $TGDP$ conditioned on TR. ** indicates rejection of null hypothesis of equality to the index in the first column at 5% level; * at 10% level

higher GRV. However, the conditioned indices seem to be statistically different from the unconditioned ones especially for $TGDP(I)$. In fact, for both I^B and I^{FL}, in three out of four cases the difference is statistically significant at 5% level. This indicates that structural change may be particularly important if the economy is "small". "Large" economies do not seem to be particularly affected by a reduction in AS.

We have observed that the integration in world markets may reduce GRV, if the effective size of an economy results increased, or it can increase GRV if it allows for the "import" of external shocks (recall the remark from [7]). In general, we need a way to measure the effective size of an open economy, and therefore something that takes jointly into account $TGDP$ and TR.

Here we suggest to consider TR as a conditioning variable, with respect to $TGDP$.[32] This is motivated by the fact that, in general, it is possible to find economies with all possible situations: high $TGDP$ and high TR, high $TGDP$ and low TR, etc. so that the definition of a joint measure can be difficult. Tables 8 and 9 analyse the relation between $TGDP$ and TR.

We find that, comparing the values of the indices for $TR(I)$ and $TR(IV)$, GRV generally decreases. However, the conditioned indices are statistically different from the unconditioned ones especially for $TGDP(I)$ and $TGDP(IV)$

---

[32] In [8] we consider $TGDP$ and TR in interaction terms in a regression.

	Uncond.	$GDP(I)$	$GDP(II)$	$GDP(III)$	$GDP(IV)$
$TGDP(I)$	0.2904 (0.0104)	0.3197* (0.01579)	0.2956 (0.01855)	0.2493** (0.02079)	–
$TGDP(II)$	0.2749 (0.0085)	0.2763 (0.01673)	0.2706 (0.01468)	0.2559 (0.01424)	0.3272 (0.031413)
$TGDP(III)$	0.2645 (0.0088)	0.3214** (0.02679)	0.2661 (0.01948)	0.2414* (0.01520)	0.2378* (0.014905)
$TGDP(IV)$	0.2155 (0.0072)	–	0.2183 (0.02001)	0.2275 (0.01546)	0.1949** (0.008536)

Table 10. Values of I^B for $TGDP$ conditioned on GDP. ** indicates rejection of null hypothesis of equality to the index in the first column at 5% level; * at 10% level

	Uncond.	$GDP(I)$	$GDP(II)$	$GDP(III)$	$GDP(IV)$
$TGDP(I)$	0.3682 (0.0116)	0.4102** (0.01731)	0.3770 (0.02158)	0.3435 (0.02594)	–
$TGDP(II)$	0.3478 (0.0107)	0.3939** (0.02278)	0.3584 (0.01889)	0.3257 (0.01936)	0.4427** (0.04050)
$TGDP(III)$	0.3449 (0.0115)	0.4263 (0.03413)	0.3507 (0.02367)	0.3074* (0.01851)	0.3173 (0.02052)
$TGDP(IV)$	0.2846 (0.0103)	–	0.3067 (0.03041)	0.2891 (0.02013)	0.2900 (0.01776)

Table 11. Values of I^{FL} for $TGDP$ conditioned on GDP. ** indicates rejection of null hypothesis of equality to the index in the first column at 5% level; * at 10% level

[33]. At 5% or 10% level, the difference is statistically significant for two or three TR classes.

Conditioning on GDP

We concentrate now on the role of GDP. Our hypothesis is that the information on GDP is not relevant when we control for the dimension of the economy and for structural change, proxied by the share of the agricultural sector. We first evaluate GDP against $TGDP$ in Tables 10 and 11.

We expect to find a decreasing GRV with GDP, and this tendency can be partially found in the results.[34] On the other hand, there appear to be no $TGDP$ class for which the inclusion of GDP produces statistically different values for the GRV indices, with the exception of $TGDP(III)$ for I^B, in which three out of four values are statistically different from the unconditioned value.

However, the most important test for the relevance of GDP is the analysis of its role in presence of AS, which is directly connected to economic develop-

---

[33] Note that the remark that all possible combinations between the levels of $TGDP$ and TR are possible is confirmed by the fact that there are enough observations to compute all values of the conditioned indices.

[34] The value of 0.4427 in Table 11, which would in any case contradict the hypothesis of decrease of GRV with GDP, is based on only 88 observations and is therefore scarcely relevant.

	Non cond.	$GDP(I)$	$GDP(II)$	$GDP(III)$	$GDP(IV)$
$AS(I)$	0.2297 (0.0088)	–	–	0.2464 (0.01646)	0.2258 (0.01071)
$AS(II)$	0.2735 (0.0105)	–	0.3010 (0.02414)	0.2583 (0.01363)	0.2346** (0.02073)
$AS(III)$	0.2576 (0.0102)	0.2965* (0.02742)	0.2480 (0.013)	0.2504 (0.02061)	–
$AS(IV)$	0.3199 (0.0115)	0.3260 (0.01340)	0.3091 (0.02383)	–	–

Table 12. Values of I^B for AS conditioned on GDP. ** indicates rejection of null hypothesis of equality to the index in the first column at 5% level; * at 10% level

	Non cond.	$GDP(I)$	$GDP(II)$	$GDP(III)$	$GDP(IV)$
$AS(I)$	0.3037 (0.0132)	–	–	0.3133 (0.02310)	0.3103 (0.01724)
$AS(II)$	0.3611 (0.0133)	–	0.4176** (0.02728)	0.3349 (0.01752)	0.3327 (0.02742)
$AS(III)$	0.3364 (0.0136)	0.3954* (0.03469)	0.3213 (0.02367)	0.3350 (0.01851)	–
$AS(IV)$	0.4119 (0.0138)	0.4246 (0.01615)	0.3858 (0.02735)	–	–

Table 13. Values of I^{FL} for AS conditioned on GDP. ** indicates rejection of null hypothesis of equality to the index in the first column at 5% level; * at 10% level

ment. Tables 12 and 13 contain the results when the principal variable is AS and we condition on GDP.

Results are in Tables 12 and 13 (values of the unconditioned indices in the first column are from Table 5). First of all note that many values are not available for lack of data. This was predictable as it is likely to have very few observations for, say, $AS(I)$ and $GDP(I)$. This can be a first hint on the irrelevance of conditioning on GDP in presence of AS. Moreover, the number of statistically significant differences between the unconditioned and conditioned values is particularly low, if compared with the previous cases.

Summing up: we have attempted to identify the relative role of our variables in the explanation of GRV in the approach based on transition matrices. At this stage, we have found a partial confirmation of the hypotheses formulated from the model in Section 2, on the relevance of the dimension of the economy and structural change and on the irrelevance of per capita GDP in the explanation of growth volatility. In addition, in the analysis of conditioned GRV indices, clearer results appear more often at extreme $TGDP$ classes, indicating that in the transition from low to high $TGDP$ levels, the relations are more blurred (as resulted also from the preliminary graphical analysis).

6 Conclusions

This paper investigates the relation between growth volatility and the level of development, structural change and the size of the economy. Growth volatility is measured by a set of indices inspired by the literature on social mobility. Growth volatility appears to be negatively related to total GDP, proxy for the dimension of the economy. In particular it seems appropriate to consider as an additional control for the dimension of the economy the integration in the world markets. Moreover, growth volatility appears to be negatively related to the share of agriculture on GDP, proxy for structural change. Finally, per capita GDP, proxing for the level of development, does not seem to add relevant information when the other variables are considered. A direction for further research may be an assessment of the explanatory power of other factors related to development and to growth volatility, like the growth of a financial sector, in relation to structural change. In [9] we provide a first attempt.

References

1. Acemoglu, D. and F. Zilibotti (1997), "Was Prometheus Unbound by Chance? Risk, Diversification, and Growth", *Journal of Political Economy*, 105, pp. 709-751.
2. Acemoglu, D., S. Johnson, J. Robinson and Y. Thaicharoen (2003), "Institutional Causes, Macroeconomic Symptoms: Volatility, Crises and Growth", *Journal of Monetary Economics*, 50, pp. 49-123.
3. Bartholomew, D.J. (1982), *Stochastic Models for Social Processes*, 3rd edition, John Wiley & Son, New York.
4. Bowman, A.W and A. Azzalini (1997), *Applied Smoothing Techniques for Data Analysis*, Oxford, Clarendon Press.
5. Bowman, A.W. and A. Azzalini. Ported to R by B. D. Ripley, (2004). *sm: kernel smoothing methods: Bowman and Azzalini (1997)*. R package version 2.0-13.
6. Canning, D., L. Amaral, Y. Lee, M. Meyer and H. Stanley (1998) "Scaling the Volatility of GDP Growth Rate", *Economics Letters*, 60, 335-341.
7. Easterly, W., R. Islam and J. E. Stiglitz (2000), "Explaining Growth Volatility", mimeo.
8. Fiaschi, D. and A.M. Lavezzi (2003). "On the Determinants of Growth Volatility: a Nonparametric Approach". Discussion Paper del Dipartimento di Scienze Economiche. Nr. 25. University of Pisa.
9. Fiaschi, D. and A.M. Lavezzi (2005). "On the Determinants of Growth Volatility: a Markov Chain Approach", mimeo: University of Pisa.
10. Grossman, H. and M. Kim (1996), "Predation and Accumulation", *Journal of Economic Growth*, 1(3), 333-350.
11. Horvarth, M. (1998), "Cyclicality and Sectoral Linkages: Aggregate Fluctuations from Independent Sectoral Shocks", *Review of Economic Dynamics*, 1, 781-808.
12. Maddison A. (2001), *The World Economy: a Millenium Prospective*. Paris: OECD.
13. Pasinetti, L. (1981), *Structural Change and Economic Growth : a Theoretical Essay on the Dynamics of the Wealth of Nations*, Cambridge University Press, Cambridge.

14. Pritchett, L. (2000), "Understanding Patters of Economic Growth: Searching for Hills among Plateaus, Mountains, and Plains", *The World Bank Economic Review*, 14, 221-250.
15. Quah D. T. (1993), "Empirical Cross-section Dynamics for Economic Growth", *European Economic Review*, 37, 426-434.
16. Quah D. T. (1996), "Regional Convergence Clusters Across Europe", *European Economic Review*, 40, 951-958.
17. R Development Core Team (2003), *R: A language and environment for statistical computing*. R Foundation for Statistical Computing, Vienna, Austria. ISBN 3-900051-00-3, URL http://www.R-project.org.
18. Ramey, G. and Ramey, V. A. (1995), "Cross-country Evidence on the Link Between Volatility and Growth", *American Economic Review*, 85, 1138-1151.
19. Scheinkman J. and M. Woodford (1994), "Self organized criticality and economic fluctuations". *American Economic Review*, 84, 417-421.
20. Shorrocks, A.F. (1978), "The Measurement of Mobility", *Econometrica*, 46, 1013-1024.

Appendix

A Country List

AFRICA	1 Algeria	2 Angola	3 Benin	4 Botswana
5 Cameroon	6 Cape Verde	7 Cent. Afr. Rep.	8 Chad	9 Comoros
10 Congo	11 Côte d' Ivoire	12 Djibouti	13 Egypt	14 Gabon
15 Gambia	16 Ghana	17 Kenya	18 Liberia	19 Madagascar
20 Mali	21 Mauritania	22 Mauritius	23 Morocco	24 Mozambique
25 Namibia	26 Niger	27 Nigeria	28 Rwanda	29 Senegal
30 Seychelles	31 Sierra Leone	32 Somalia	33 South Africa	34 Sudan
35 Swaziland	36 Tanzania	37 Togo	38 Tunisia	39 Uganda
40 Zambia	41 Zimbabwe	LATIN AMERICA	42 Argentina	43 Brazil
44 Chile	45 Colombia	46 Mexico	47 Peru	48 Uruguay
49 Venezuela	50 Bolivia	51 Costa Rica	52 Cuba	53 Dominican Rep.
54 Ecuador	55 El Salvador	56 Guatemala	57 Haiti	58 Honduras
59 Jamaica	60 Nicaragua	61 Panama	62 Paraguay	63 Puerto Rico
64 Trin. Tobago	OFF WESTERN	65 Australia	66 New Zealand	67 Canada
68 United States	WEST ASIA	69 Bahrain	70 Iran	71 Iraq
72 Israel	73 Jordan	74 Kuwait	75 Lebanon	76 Oman
77 Qatar	78 Saudi Arabia	79 Syria	80 Turkey	81 UAE
82 Yemen	83 W.Bank Gaza	EAST ASIA	84 China	85 India
86 Indonesia	87 Japan	88 Philippines	89 South Korea	90 Thailand
91 Bangladesh	92 Hong Kong	93 Malaysia	94 Nepal	95 Pakistan
96 Singapore	97 Sri Lanka	98 Afghanistan	99 Cambodia	100 Laos
101 Mongolia	102 North Korea	103 Vietnam	EUROPE	104 Austria
105 Belgium	106 Denmark	107 Finland	108 France	109 Germany
110 Italy	111 Netherlands	112 Norway	113 Sweden	114 Switzerland
115 UK	116 Ireland	117 Greece	118 Portugal	119 Spain

Table 14. Country list

Financial Fragility and Scaling Distributions in the Laboratory

Giovanna Devetag[1], Edoardo Gaffeo[2], Mauro Gallegati[3], and Gianfranco Giulioni[4]

[1] DMCS and CEEL, University of Trento, Italy
[2] DE and CEEL, University of Trento, Italy, and SIEC, Politechnic University of Marche, Ancona, Italy
[3] DEA and SIEC, Politechnic University of Marche, Ancona, Italy
[4] DMQTE University of Chieti-Pescara, Italy and SIEC, Politechnic University of Marche, Ancona, Italy

Summary. We present results from human and computer-based experiments aimed at exploring the role of rationality and financial markets in explaining the emergence of some well-known stylized facts regarding industrial and aggregate dynamics. We find that the information conveyed by financial markets helps agents adopt more rational decision processes. Rationality, in turn, is necessary to observe smooth aggregate behaviors.

1 Introduction

In a recent stream of papers, D. Delli Gatti, M. Gallegati and their coauthors (Delli Gatti et al., 2004a, 2004b; Gallegati et al., 2003; hereafter DGG) have emphasized the relationship between industrial dynamics and financial fragility as a key issue in understanding macroeconomic fluctuations. The main insights surfaced from this line of research are that: i) distributions matter; and ii) individuals' level stochasticity and interactions matter. Briefly stated, the message to be conveyed is that, at odds with the mainstream approach to macroeconomic modeling, interactions among heterogeneous agents buffeted with idiosyncratic shocks have a bearing in explaining aggregate fluctuations.

Such a view has emerged after three notable empirical regularities sprung up in the econophysics literature have started to be employed as stylized facts against which the predictive ability of business cycle models can be tested. First, the distribution of firms' size is right skewed and its upper tail follows a power law (Axtell, 2001; Gaffeo et al., 2003). Second, the growth rates of either firms' output and countries' GDP follow a Laplace distribution (Stanley et al., 1996; Amaral et al., 1998). Third, the standard deviation of GDP growth rates and of firms' growth rates scale down with size with an identical exponent (Lee *et al.*, 1998). For standard economic theory, which remains largely rooted on the optimizing behavior of identical agents acting in isolation, these findings are bewildering. In fact, scale invariance and universality calls for critical behavior in self-organizing economic organizations (Bak, 1996; Norrelykke and Bak, 2002), or random multiplicative growth processes in hierarchically organized systems with variable number of components (Amaral *et al.* 1998; Blank and Solomon, 2000). In other terms, scaling and universal features are likely to signal the existence of complex interactions among economic units characterized by heterogeneous sizes, financial positions and possibly behavioral rules.

The 'financial fragility' framework developed in DGG, where heterogeneous firms and a banking sector interact giving rise to complex dynamics, has a case in point. Agent-based simulations have shown this model is capable to replicate an impressive number of stylized facts regarding both industrial and aggregate dynamics (see also Delli Gatti *et al.*, 2004). These results are particularly interesting, given that they are derived in a model which explicitly attempts to address a well known criticism by e.g. Brock (1999) and Durlauf (2003), who warn against the practice of providing interpretative frameworks for scaling laws in economics and finance by simply re-labeling stochastic processes known to generate them, without paying any attention to modeling the interplay of preferences, technology, constraints and beliefs guiding human economic behavior. In fact, DGG show that scaling phenomena can arise endogenously in a model where heterogeneous rational agents choose desired investment in order to maximize a profit function. Individual decision-making processes are buffeted with idiosyncratic *i.i.d.*real (i.e., referred to relative prices) shocks, and the interplay of demand and supply in the credit market gives rise to mean field interactions among agents.

An issue that deserves further research, however, consists in exploring to what degree the assumptions on the environment and individual behavior at the heart of the DGG model can be blamed for its results. In this paper we present results from a controlled experiment that attempts to provide an answer to this question. In particular, our aim is that of investigating whether simulations' findings are confirmed once: *i*) perfectly-rational agents are substituted by less-than-perfectly-rational decision makers; and *ii*) markets, in this case the credit market, does not work properly.

Summarizing our results, we find that some celebrated stylized facts of industrial dynamics, such as the distribution of firms' size scaling down as a power law, are consistent with very low degree of rationality on the parts of economic agents. Interactions on the financial market, in turn, help our experimental subjects to behave in a risk-averse manner, that is a rational decision-making process as soon as the possibility of going bankrupt is taken into account. This last result suggest a possible new role for financial markets, that is fostering rationality.

The remaining of the paper is organized as follows. Section 2 briefly presents the theoretical framework we brought to the laboratory. Section 3 describes the experimental setting, while Section 4 provides details of our findings. Finally, Section 5 summarizes the paper and put forwards its main conclusions.

2 Theoretical Framework

In its basic version, the DGG model consists of only two markets: goods and credit. In order to simplify the analysis as much as possible, we assume that output is supply driven, i.e. firms can sell all the output they produce. In each period, there is a "large" (but, due to the entry-exit process, far from constant) number of firms N, indexed by $i = 1,.., N$, which differ according to their financial conditions. The financial robustness of a firm is proxied by the so-called equity ratio, i.e. the ratio of its equity base or net worth (A_{it}) to the capital stock (K_{it}), $a_{it} = A_{it}/K_{it}$. Since firms sell their output at an uncertain price they may go bankrupt. Bankruptcy occurs if net worth at time t becomes negative, i.e. if the individual price falls below a critical threshold. The probability of bankruptcy turns out to be an increasing function of the interest rate and the capital stock, and a decreasing function of the equity base inherited from the past.

As shown analytically in Greenwald and Stiglitz (1993), in such a setting the problem of the firm consists in maximizing its expected profits net of bankruptcy costs. The optimal capital stock turns out to be a decreasing function of the interest rate and an increasing function of the equity ratio. Output follows the evolution over time of the capital stock, which is in turn determined by investment.

Due to informational imperfections on the equity market, firms can raise funds only on the credit market. For the sake of simplicity we assume that many heterogeneous firms interact with the banking sector (as an aggregate). Investment is financed by means of retained earnings and by new debt. Therefore the demand for credit of each firm is a function of its financial robustness and of the interest rate. In order to determine the supply of credit and its allocation to each firm, we assume that there is a risk coefficient α - i.e. a threshold ratio of bank's equity to credit extended - that the banks try to target, either because of a risk management strategy or as a consequence of prudential regulation on the part of the monetary authorities. Therefore the aggregate supply of credit turns out to be approximately a multiple ($1/\alpha$) of banking sector equity.

Credit has to be allotted to the heterogeneous firms. We assume that each firm obtains a portion of total credit equal to its relative size, i.e. the ratio of the individual capital stock to the aggregate capital stock: highly capitalized (i.e. collateralized, in economics jargon) borrowers have a higher credit supply, and *vice-versa*. The rate of interest charged to each firm is determined in equilibrium when the demand for credit is equal to the credit extended by the bank to that firm. The equilibrium interest rate generally decreases with the firm's and the bank's equity. Investment and output therefore generally increase with the bank's and the firm's net worth.

The bank's equity base, which is the only determinant of total credit supply, increases with the bank's profits, which are affected, among other things, by firms' bankruptcy. When a firm goes bankrupt, i.e. its net worth becomes negative because of a huge loss, the bank has a "bad debt". This is the root of a potential *domino* effect: the bankruptcy of a firm today is the source of the potential bankruptcies of other firms tomorrow via its impact on the bank's equity.

The model has been simulated by means of agent-based techniques, and the initial number of firms has been set equal to 10000. Among the many stylized facts the model can replicate (Delli Gatti *et al.*, 2004c), it deserves to be mentioned that the distribution of firms' size (in terms of capital stock) is characterized by persistent heterogeneity and is well fitted by a power law (Figure 1), while the distribution of firms' growth rates is approximated by a tent-shaped curve, i.e. a Laplace distribution. To test whether the conditional distribution of growth rates has a functional form independent of size, data have been further divided into five bins according to firms' dimension. In fact, the five conditional probability distributions collapse onto a single curve (Figure 2).

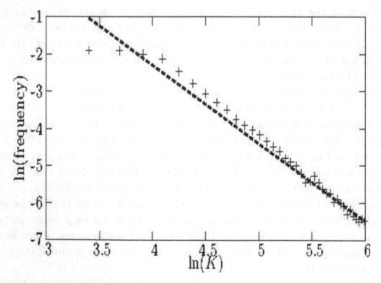

Fig. 1. Log-log plot of firms' size (by capital stock) as derived from simulations of the financial fragility model. The solid line is the OLS regression line through the data, with slope 1.11 ± 0.11.

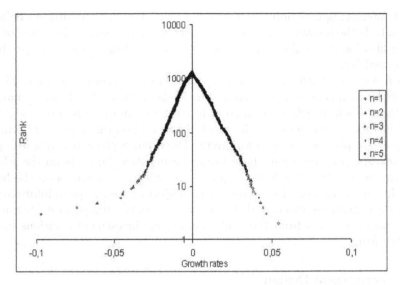

Fig. 2. PDF of simulated firms' growth rates. Rates are grouped in five bins (indicated with n) according to the rank of firms' size. Data collapse onto a single curve, signaling that conditional probabilities are size-independent.

3 Human Subjects and Computational Experiments

Agent-based simulation techniques are powerful methods to derive results, especially when dealing with complex frameworks as the one described above. However, in many cases simulation findings are highly dependent on the width of the parameters' space chosen on the one hand, and the rationale for what is truly going on (i.e., who causes what) during simulations remains somehow hidden on the other one. In other terms, simulations must be confronted with the issues of robustness and causality. As a matter of example, in the DGG model the issue of whether scaling in the firms' size distribution as reported in Figure 1 is merely a result of stochasticity or it is influenced by rational decision processes and institutions (i.e., financial markets) remains unexplored.

A possible method to address this question consists in disentangling the complex interaction of the model's ingredients by analyzing the behavior of real people in a simplified but fully comparable environment, that is by recurring to experiments.

We run a first treatment in which firms (players) are allowed to operate independently, the interest rate is fixed and known to subjects, and firms growth (measured by the equity base dynamics) is therefore only a function of individual independent investment decisions. Fully rational players should take into account the possibility to fail and thus behave in a risk-averse manner, which in this case means that investment is correlated with available equity.

A first interesting question, therefore, is to explore whether subjects behave rationally. If the answer is not, it seems interesting to analyze the shape of resulting firms' size distribution, to assess whether scaling is a mere by-product of stochasticity.

In a second treatment we add the working of a lending market. Firms interact by means of a central bank who lends credit on the basis of market conditions. Each firm faces a different interest rate in every round depending on its financial soundness (i.e., its equity base) and on the difference between aggregate supply and demand for credit. The event represented by a firm going bankrupt (i.e., its equity base becoming negative) impacts on the whole economy by affecting the banks' equity base, with consequences on the level of credit supply and on the interest rate. Subjects have complete information about the conditions affecting their profitability and earnings; strategic uncertainty however stems from the contemporaneous investment decisions made by other firms.

3.1 Experimental Design

Subjects were asked to play a simple investment, repeated game which preserves all the main characteristics of the DGG model. In what follows we describe the two experimental treatments in detail.

Treatment 1: no interaction

In treatment 1 firms are initially endowed with the same amount of equity (A) and debt (L), which together constitute their initial capital endowment $(K = A + L)$. Each firm must decide in every period how much to invest in its production process, i.e., by how much to increase its capital stock K. For simplicity, we assume that investment can only occur by resorting to debt.

Firms make their investment decisions at the beginning of a round, while the final output of that round is determined as follows:

$$Y_{it} = \lambda K_{it}, \; with \; K_{it} = A_{i(t-1)} + L_{i(t-1)} + I_{it} \tag{1}$$

where $I_{it} = \Delta L_{it}$ is the amount of investment in period t, and λ is the productivity parameter, which is the same for all.

The relative price p at which each firm may sell its output is a uniformly distributed random variable in the $[0,2]$ range. A firm's profits in real terms at time t are determined as follows:

$$\pi_{it} = \lambda K_{it} p_{it} - r K_{it} \tag{2}$$

where r is the interest rate, fixed at the same level for all firms. The profit at time t is added to (subtracted from) the firm's equity A, so that $A_{i(t+1)} = A_{it} + \pi_{it}$.

Treatment 2: interaction

In treatment 2, a central bank (played by the computer program) lends to firms on the basis of market conditions. The bank decides in every round the total supply of credit according to the following equation:

$$L_t^S = \frac{E_{(t-1)}}{\alpha} y_t \text{ with } y_t = \frac{N_t - n_{b(t-1)}}{N_t}. \tag{3}$$

The parameter α is the risk coefficient, and $n_{b(t-1)}$ is the number of firms that have left the market in the previous round due to bankruptcy. The term y_t is introduced to adjust for the reduction in the number of active firms overtime, in order to avoid credit oversupply. Each firm faces a different interest rate in every round depending on its financial soundness (i.e., its equity base) and on the difference between aggregate supply and demand for credit:

$$r_{it} = \bar{r} - \phi_1 \ln\left(\frac{A_{i(t-1)}}{\bar{A}_{(t-1)}}\right) - \phi_2 \ln\left(\frac{\frac{K_{i(t-1)}}{K_{(t-1)}^{tot}}}{\sum_{i=1}^{N_{(t-1)}} \frac{K_{i(t-1)}}{K_{(t-1)}^{tot}}}\right) + \\ -\phi_3 \ln\left(\frac{\pi_{i(t-1)}}{\bar{\pi}_{(t-1)}}\right) + \phi_4 \left(\sum_{i=1}^{N_t} I_t - L_t\right) \tag{4}$$

The log-linear functional form (4) approximates the equilibrium equation for the interest rate in the DGG model, which is derived from first order conditions. Such an optimization process does not clearly occur in our experimental setup.

Using an indicator function B_{it} which equals $-A_{it}$ if $A_{it} < 0$ and 0 if $A_{it} \geq 0$ we can finally define the bank's equity base at time t as follows:

$$E_t = E_{(t-1)} + \sum_{i=1}^{N_{(t-1)}} r_{i(t-1)} L_{i(t-1)} - \sum_{i=1}^{N_{(t-1)}} B_{i(t-1)} - \bar{r} E_{(t-1)} \tag{5}$$

The bank's equity base is increased every round by an amount equal to its net profits (i.e., the total amount of interest gained on credit supplied minus financial costs) and it is decreased by an amount equal to "bad debt", i.e., the equity base of firms that went bankrupt in the previous round.

3.2 Implementation

Two experimental sessions were conducted for each treatment, with $N = 20$ subjects participating in each session. Our experimental agents were undergraduate students in the School of Economics of the University of Trento. Subjects were recruited through ads posted at the Computable and Experimental Economics Laboratory (CEEL) building and they had never participated in experiments of this type before. Instructions were distributed at the beginning of the experiment and read aloud by the experimenters. The experiment itself was computerized. It was programmed to last for a minimum of 30 periods, after which the termination was decided on the basis of a random device. The timing structure was communicated to experimental subjects. Finally, the interaction was anonymous.

Subjects made their investment decisions at the beginning of every round and received feedback at the end concerning: 1) their relative price for that round; 2) the profit earned and, consequently, the new updated values of their equity base; 3) their debt and their capital stock; 4) in the case of the interaction treatment, their interest rate. In treatment 2, subjects were also provided feedback about the average payoff obtained, and the average value of the equity base in the economy. The variable I (i.e., the amount to invest) could vary in a range that changed over time in order to calibrate subjects' possible investment decisions on the level of their capital stock, which tended to increase over time. The range was set up as shown in table 1.

Table 1.

periods	1-5	6-15	16-30	31-end
range of I	0-100	100-200	200-400	400-600

The initial show up fee was set at 5 euros, and a subject's final earnings were determined by the final value of her equity base relative to the maximum theoretical value that could be achieved.[5] In addition, a 3 euros penalty had to be paid in case of bankruptcy (subtracted from the show up fee). The initial parameter values were set as follows: $A=20$, $L=40$, $K=60$. The values of the interest rate parameters were $\phi_1 = \phi_2 = \phi_3 = \phi_4 = 0.001$, $\bar{r} = 0.1$. The productivity parameter λ was set equal to 0,20. The parameter α was equal to 1, and finally, the bank's initial equity base E_0 was set at a value of 10.

4 Results

The main goals of our experiment consisted in testing how fully rational investment decisions and feedback from financial markets affects the emergence of some well known regularity in industrial and aggregate dynamics. To accomplish this task, we first run an experiment with human subjects, and subsequently we augmented the number of observations by allowing computerized agents to play strategies identical to those played by real agents. Let us start our discussion from experimental results.

---

[5] In treatment 2, the maximum value of A was determined under the assumption of a constant interest rate of 10%.

4.1 Experimental Results

The choice problem that subjects faced in our experimental setting is undoubtedly complex, hence we did not expect to observe any kind of optimizing behavior, but rather the use of largely sub-optimal and "backward looking" rules of thumb. In fact, several previous experiments have shown that subjects often tend to use simple heuristics that condition the choice in one period to the result obtained in the previous one. As a matter of example, Nagel and Tang (1998) show that subjects playing a centipede game tend to increase the number chosen in the previous round if that round had been successful, and vice-versa if the previous round resulted in a loss. Zwick and Rapoport (2003), in turn, test a simple reinforcement rule in a market entry game to find out that their experimental subjects tended to choose entry more often in a round if their entry had been successful the round before. Finally, Bottazzi and Devetag (2003) report evidence for an analogous one-lag time-dependence in a repeated minority game, while dependence of subjects' choices on more distant rounds seems weak. Accordingly, we investigated whether our subjects' choices in the experiment can be explained by similar patterns.

A natural variable to check for is the payoff obtained in the previous round. Table 2 reports the absolute frequencies (averaged over all groups and treatments) with which a subject's amount of investment was lower than, higher than, and equal to, the investment made in the previous round conditional on the previous round payoff having been positive or negative (a payoff exactly equal to zero was never observed).

Table 2

Investment choices	$\pi_i(t-1) > 0$	$\pi_i(t-1) < 0$
$I(t) > I(t-1)$	0.41	0.30
$I(t) = I(t-1)$	0.30	0.28
$I(t) < I(t-1)$	0.29	0.42

The data show a modest increase in the probability of raising one's investment relative to the previous round if this had been profitable. Overall, the probability that the investment choice in a round will be at least equal to that of the previous round is 71% after a profitable round, while it decreases to 58% after incurring a loss.

A pattern conditioning one's choice to the previous round payoff is very myopic as it does not take into account any measure of risk associated with investment decisions. The risk that subjects should take into account regards the perceived probability of going bankrupt, which occurs when their equity base becomes negative. In order to check for some dependence of choices on parameters signaling "financial solidity", we investigate whether investment in a round depended on a player's equity ratio (i.e., $a = A/K$) in the previous period. Table 3 reports the results of the frequency count, taking as a cutoff level for the equity ratio a value of 0,5:

Table 3

Investment choices	$\pi_i(t-1) > 0$ $a_i \geq 0.5$	$\pi_i(t-1) > 0$ $a_i < 0.5$	$\pi_i(t-1) < 0$ $a_i \geq 0.5$	$\pi_i(t-1) < 0$ $a_i < 0.5$
$I(t) > I(t-1)$	0.28	0.41	0.25	0.28
$I(t) = I(t-1)$	0.47	0.34	0.37	0.36
$I(t) < I(t-1)$	0.24	0.25	0.38	0.36

There seems to be no notable effect of the equity ratio on investment choices. Frequencies are close to those implied by a uniform probability distribution; it has to be noted that the information about the payoff seems more salient than the equity ratio. In particular, after a positive payoff round subjects increase or at least not decrease their investment 75% of the times, while this percentage lowers to 62% after a negative payoff round. The same analysis has been performed taking into consideration the variation of the equity ratio from one period to the next, with analogous results. No differences between groups and treatments were found at this level of analysis.

Next, we tested whether choices in the interaction treatment can be explained by the interest rate that was applied by the bank to each firm. Such interest rate could be lower or higher than a benchmark value of 10% depending on whether the firm has been better than the "average" firm in the economy in terms of profit, equity ratio, and capital stock (see eq. (4)). Subjects were informed of how the interest rate was calculated, and hence they might have taken it into account as a proxy of their relative standing in the economy. Therefore, we conditioned choices of investment in the interaction treatment to both profit and interest rate obtained in the previous period. Instead of testing for variations in investment from one round to the next, we considered whether the amount invested was above or below the 50% of the total amount that subjects could invest. We only considered choices in the first 15 periods, as in the last 15 periods probability of bankruptcy was significantly lower. We used a cutoff level of 0.1 for the interest rate. Table 4 reports the results:

Table 4

Investment choices	$\pi_i(t-1) > 0$ $r_i \geq 0.1$	$\pi_i(t-1) > 0$ $r_i < 0.1$	$\pi_i(t-1) < 0$ $r_i \geq 0.1$	$\pi_i(t-1) < 0$ $r_i < 0.1$
$I(t) > 0.5$	0.28	0.54	0.26	0.38
$I(t) = 0.5$	0.11	0.17	0.04	0.06
$I(t) < 0.5$	0.60	0.29	0.70	0.55

The results show a significant effect of the interest rate. In particular, after a positive payoff and an interest rate below 10%, subjects invest more than half of their amount 54% of the times, while in the worst case of a negative payoff and an interest rate above 10%, 70% of the times subjects invest less than 50% of the total.

From the analysis of individual choice behavior a pair of conclusions can be drawn. First, in spite of having the opportunity to calculate profit maximizing strategies, subjects seem to limit their decisional behavior to rather *naïve* rules and to base their choices on very few pieces of information. When not given explicit information about their risk of bankruptcy, for example, subjects do not seem to take this variable into account when making their choices, acting *de facto* as risk neutral investors. Second, as the introduction of financial markets augments the information space (e.g., the interest rate in the interaction treatment), subjects tend to use more rational decisional processes, for example investing more when their financial position is healthier relative to the average in the economy. These results suggest that financial markets, besides their well-known efficiency-improving role through the coordination and saving and investment, could also play a role in helping economic agents to follow more rational decision processes by disseminating relevant pieces of information about their own. It seems to us that such an issue deserves to be the subjects of further research.

Finally, note that the rate of bankruptcy was equal to 27% overall, with no significant difference between treatments. All bankruptcy events occurred within the first ten periods of the experiment; hence, if a firm was able to survive the first ten periods, the probability of subsequently exiting the market became extremely low. This finding is consistent with empirical evidence on longitudinal firm level data, according to which younger firms have a higher probability of exiting the market (Caves, 1998).

4.2 Simulation results

In order to test the distributional and aggregate features associated with the behavioral rules adopted by our experimental subjects, we run several agent-based simulations by allowing a large number of computerized firms to play the investment game described above over very long horizons. In particular, firms were instructed to use the rules reported in Table 2 and Table 4 for simulations based on the no-interaction and the interaction treatments, respectively. Operationally, the growth rate of capital, comprised in the (-0.01,0.02) interval, has been divided in quantiles of amplitude equal to 0.005. Firms determined their investment rate by choosing a quantile. The effective growth rate of capital was then a random draw from a uniform distribution over the chosen quantile. The number of agents were set at 10000, while simulations spanned over 4200 periods. Firms forced to exit due to bankruptcy were replaced by another firm with common starting conditions for the debt/equity ratio.

The two sets of simulations gave qualitatively different patterns of aggregate behavior. The simple backward-looking decision rule experimental subjects followed when the interest rate was hold constant implies the emergence of a power law for the firms size distribution, but gives rise to an aggregate dynamics, expressed in terms of aggregate output, characterized by recurrent crashes (Figures 3 and 4). In other words, the lack of risk-aversion at an individual level implies the winner-takes-all features typical of a stochastic process composed of random multiplicative shocks combined with an exit threshold (Blank and Solomon, 2000) on the one hand, but returns an unacceptable high degree of aggregate instability on the other one. As far as the random demand allow firms to make profits, they continue to borrow for investment without any regards to their financial soundness. This implies that their equity ratio decreases, with the consequence that the probability to draw a negative demand shock forcing them to bankrupt increases as their size grows. Eventually, the default of very big firms causes huge drops in aggregate output.

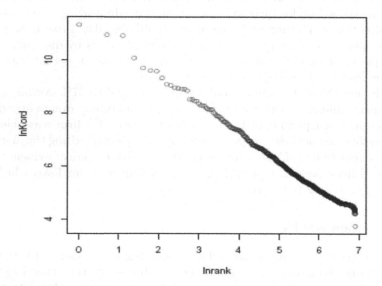

Fig. 3. log-log plot of firms' size distribution in the no interaction treatment.

The introduction of interactions through the credit market translates immediately into interest rates varying with the degree of individual financial fragility. This piece of information was used by experimental subjects to update their investment decisions, as an increasing interest rate could be immediately recognized as a deterioration of their degree of solvability. As suggested before, financial markets can be seen as an institution helping agents to correctly adopt a risk-averse behavior. In fact, simulations for the interaction scenario shows that the average equity ratio stabilized at 0.20 over the whole horizon (Figure 5), while the time series of the aggregate output is much

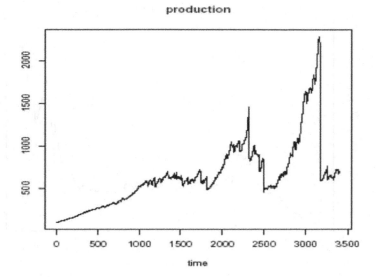

Fig. 4. plot of aggregate output over time in the no interaction treatment.

smoother than before (Figure 6). As regards the firms' size distribution, now we observe a breakdown of scaling: while the size of big firms still follows a power law, the size of small firms seems to obey a log-normal distribution (Figure 7). This pattern is qualitatively similar to the one observed from simulations run with the original model (see Figure 1).

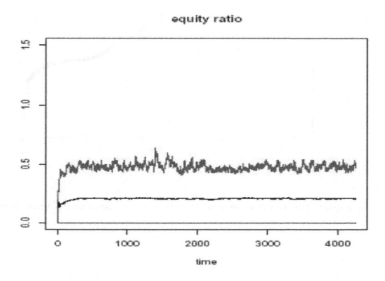

Fig. 5. average equity ratio over time in the interaction treatment.

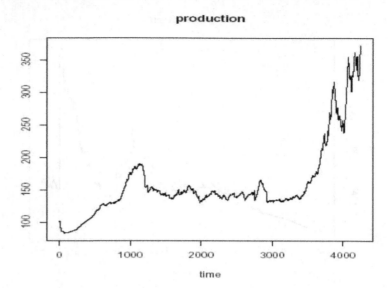

Fig. 6. Plot of aggregate output over time in the interaction treatment.

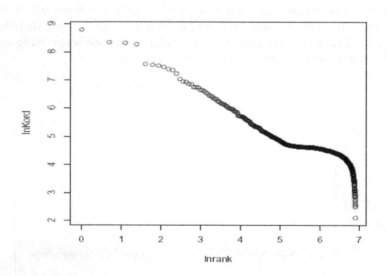

Fig. 7. log-log plot of firms' size distribution in the interaction treatment.

5 Conclusions

The framework recently offered in Delli Gatti *et al.* (2004a; 2004b) and Gallegati *et al.* (2003) has been used as the basis for human and computer-based experiments. Out aim consisted in exploring the role of rationality and financial markets in explaining the emergence of some well-known stylized facts regarding industrial and aggregate dynamics. Two main results have emerged. First, the information conveyed by financial markets seems to help agents to adopt rational decision processes. Second, rationality, on its part, is a necessary condition to observe smooth aggregate behaviors.

References

1. Amaral L, Buldyrev S, Havlin S, Salinger M, Stanley E (1998) Power Law Scaling for a System of Interacting Units with Complex Internal Structure, Physical Review Letters, 80:1385–1388
2. Axtell R (2001) Zipf Distribution of U.S. Firm Sizes, Science, 293:1818–1820
3. Bak P (1997) How Nature Works, Oxford University Press, Oxford
4. Blank A, Solomon S (2000) Power Laws in Cities Population, Financial Markets and Internet Sites, Physica A, 287:279–288
5. Bottazzi, Devetag G (2003) Coordination and Self-Organization in Minority Games: Experimental Evidence, in Kirman A, Gallegati M, Marsili M (Eds.), The Complex Dynamics of Economic Interaction: Essays in Economics and Econophysics, Lecture Notes in Economics and Mathematical Systems, Springer-Verlag, Berlin 283-300
6. Brock W (1999) Scaling in Economics: A Reader's Guide, Industrial and Corporate Change, 8:409–446
7. Caves R (1998) Industrial Organization and New Findings on the Turnover and Mobility of Firms, Journal of Economic Literature, 36:1947–1982
8. Delli Gatti D, Di Guilmi C, Gallegati M, Giulioni G (2004a), Financial Fragility, Industrial Dynamics and Business Fluctuations in an Agent-Based Model, Macroeconomic Dynamics, forthcoming
9. Delli Gatti D, Di Guilmi C, Gaffeo E, Gallegati M, Giulioni G, Palestrini A (2004b) A New Approach to Business Fluctuations: Heterogeneous Interacting Agents, Scaling Laws and Financial Fragility, Journal of Economic Behavior and Organization, forthcoming.
10. Delli Gatti D, Gaffeo E, Gallegati M, Giulioni G, Kirman A, Palestrini A, Russo A (2004c) Complex Dynamics and Empirical Evidence, Information Sciences, forthcoming.
11. Durlauf S (2003) Complexity and Empirical Economics, mimeo
12. Gaffeo E, Gallegati M, Palestrini A (2003) On the Size Distribution of Firms: Additional Evidence from the G7 Countries, Physica A 117-123.
13. Gallegati M, Giulioni G, Kochji N (2003) Complex Dynamics and Financial Fragility in an Agent-Based Model, Advances in Complex Systems 6
14. Greenwald and Stiglitz (1993) Financial Market Imperfections and Business Cycles Quarterly Journal of Economics 108:77–113

15. Lee Y, Amaral L, Canning D, Meyer M, Stanley E (1998) Universal Features in the Growth Dynamics of Complex Organizations, Physical Review Letters, 81:3275–3278
16. Nagel R, Tang FF (1998) Experimental Results on the Centipede Game in Normal Form: an investigation on learning, Journal of mathematical Psychology 42:356–384
17. Nørrelykke S, Bak P (2002), Self-Organized Criticality in a Transient System, Physics Review 68
18. Stanley M, Amaral L, Buldyrev S, Havlin S, Leschorn H, Maas P, Salinger M, Stanley E (1996), Scaling Behavior in the Growth of Nature, 379:804–806
19. Zwick R, Rapoport A (2002) Tacit Coordination in a Decentralized Market Entry Game with Fixed Capacity, Experimental Economics 5(3):253–272

Part II

Complex Economic Network

Heterogeneous Economic Networks

Wataru Souma[1], Yoshi Fujiwara[2], and Hideaki Aoyama[3]

[1] ATR Network Informatics Laboratories, Kyoto 619-0288, Japan. souma@atr.jp
[2] ATR Network Informatics Laboratories, Kyoto 619-0288, Japan. yfujiwar@atr.jp
[3] Department of Physics, Graduate School of Science, Kyoto University, Yoshida, Kyoto 606-8501, Japan. aoyama@phys.h.kyoto-u.ac.jp

Summary. The Japanese shareholding network at the end of March 2002 is studied. To understand the characteristics of this network intuitively, we visualize it as a directed graph and an adjacency matrix. Especially detailed features of networks concerned with the automobile industry sector are discussed by using the visualized networks. The shareholding network is also considered as an undirected graph, because many quantities characterizing networks are defined for undirected cases. For this undirected shareholding network, we show that a degree distribution is well fitted by a power law function with an exponential tail. The exponent in the power law range is $\gamma = 1.8$. We also show that the spectrum of this network follows asymptotically the power law distribution with the exponent $\delta = 2.6$. By comparison with γ and δ, we find a scaling relation $\delta = 2\gamma - 1$. The reason why this relation holds is attributed to the local tree-like structure of networks. To clarify this structure, the correlation between degrees and clustering coefficients is considered. We show that this correlation is negative and fitted by the power law function with the exponent $\alpha = 1.1$. This guarantees the local tree-like structure of the network and suggests the existence of a hierarchical structure. We also show that the degree correlation is negative and follows the power law function with the exponent $\nu = 0.8$. This indicates a degree-nonassortative network, in which hubs are not directly connected with each other. To understand these features of the network from the viewpoint of a company's growth, we consider the correlation between the degree and the company's total assets and age. It is clarified that the degree and the company's total assets correlate strongly, but the degree and the company's age have no correlation.

Keywords. Shareholding network, Visualization, Network analysis, Power law, Company's growth

1 Introduction

The economy is regarded as a set of activities of heterogeneous agents in complex networks. However, many traditional studies in economics are for the activities of homogeneous agents in simple networks, where we call regular networks and random networks simple networks. To overcome such an unrealistic situation many efforts have been made, and a viewpoint of heterogeneous agents has emerged. However, simple networks are adapted in many of the studies of heterogeneous agents. Hence it is important to introduce the true structure of real world networks. Recently, the study of complex networks has revealed the true structure of real world networks: WWW, the Internet, social networks, biological networks, etc. [1][4][5][12]. However the true structure of the economic network is not well known. Hence the purpose of this study is to reveal it.

As is commonly known, if we intend to discuss networks, we must define the nodes and edges. Here, edges represent the relationship between nodes. In business networks, the candidates for the nodes are individuals, companies, industry categories, countries, etc. In this study we consider companies as nodes. Hence, in the next step, we must define the relationship between companies. To define it, we use three viewpoints: ownership, governance, and activity. The ownership is characterized by the shareholding of companies, and the networks constructed by this relationship are considered in this article. The governance is characterized by the interlocking of directors, and networks of this type are frequently represented by a bipartite graph that is constructed with corporate boards and directors. The activity networks are characterized by many relationships: trade, collaboration, etc.

Although we use tree point of view, these have relations with each other. For example, if the owners of a company change, then the directors of that company will change. If the directors of the company change, then the activities of the company will change. If the activities of the company change, then the decisions of the owners and directors will change, and sometimes the owners and the directors will change.

In this article we consider Japanese shareholding networks at the end of March 2002 (see Ref. [9] for shareholding networks in MIB, NYSE, and NASDAQ). In this study we use data which is published by TOYO KEIZAI INC. This data provides lists of shareholders for 2,765 companies that are listed on the stock market or the over-the-counter market. Almost all of the shareholders are non-listed financial institutions (commercial banks, trust banks, and insurance companies) and listed non-financial companies. In this article we ignore shares held by officers and other individuals. The lengths of the shareholder lists vary with the companies. The most comprehensive lists contain information on the top 30 shareholders. Based on this data we construct a shareholding network.

This paper is organized as follows. In Sec. 2, we consider the visualization of the shareholding network as a directed graph and an adjacency matrix. The

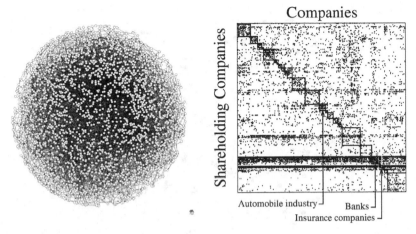

Fig. 1. A visualization of the Japanese shareholding network at the end of March 2002 (left) and a corresponding adjacency matrix (right).

visualization of networks is a primitive, but powerful tool to intuitively understand the characteristics of networks. As an example, we especially consider networks concerned with the automobile industry sector. In Sec. 3, we treat the shareholding network as an undirected network. This is because many useful quantities characterizing networks are defined for undirected cases. We consider the degree distribution, spectrum, degree correlation, and the correlation between the degree and the clustering coefficient. In Sec. 4, we discuss correlations between the degree and the company's total assets and age. Section 5 is devoted to a summary and discussion.

2 Directed Networks

If we draw edges from shareholders to companies, we can obtain shareholding networks as directed graphs. The primitive way to study networks is to use a visualization of them. The visualization of the Japanese shareholding network at the end of March 2002 is shown in the left panel of Fig. 1. This figure is drawn by Pajek, which is a program for analyzing large networks [13]. In this figure, the open circles correspond to companies, and arrows are drawn based on shareholding relationships. The network is constructed from 3,152 nodes and 23,064 arrows. This figure is beautiful, but it is difficult to obtain characteristics of this network.

Frequently, networks are represented by adjacency matrices. Here the adjacency matrix for the directed graph, M_{ij}^d, is chosen based on the shareholding relation: If the i-th company is a shareholder of the j-th company, we assume $M_{ij}^d = 1$; otherwise, $M_{ij}^d = 0$. The size of this matrix (network) is $N = 3,152$. The adjacency matrix is shown in the right panel of Fig. 1. In this figure, the

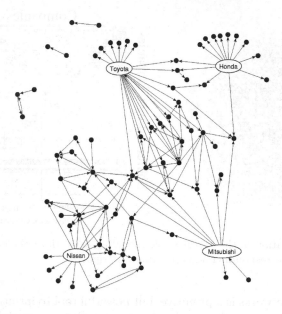

Fig. 2. A shareholding network constructed from companies in the automobile industry sector.

rows correspond to companies and the columns correspond to shareholding companies. The list of companies and the list of shareholding companies are the same. In this figure, the black dots correspond to $M_{ij}^d = 1$, and the others correspond to $M_{ij}^d = 0$. The number of black dots is 23,063.

To define this adjacency matrix we arranged the order of companies according to the company's code number that is defined based on industry categories. The solid lines in this figure indicate industry categories. We make two observations: (i) Dots are distributed in all industry categories in the ranges where financial institutions (banks and insurance companies) are the shareholders; (ii) The density of the black dots is relatively high in each box, except for the financial sector. This indicates that we frequently find companies and shareholders in the same industry category for non-financial firms. Hence this network shows "assortative mixing" on the industry category, except for the financial institutions. The concept of (dis)assortativity is explained in Ref. [11].

2.1 Shareholding Network Constructed from Companies in the Automobile Industry Sector

Here, we consider a network constructed from companies belonging to the automobile industry sector. This sector corresponds to the range shown in the right panel in Fig. 1. The visualization of this network is shown in Fig. 2. In this figure, we include company names only for four major automobile industry companies: Toyota Motor, Nissan Motor, Honda Motor, and Mitsubishi Motor.

If the direction of the arrows is ignored, then Toyota, Honda, and Mitsubishi are connected to each other with two edges through the shortest path. Nissan and Toyota are connected with three edges through the shortest path, and this is also applicable to the case of Nissan and Mitsubishi. However, we need five edges to connect Nissan and Honda through the shortest path. In addition, this path must run through Toyota or Mitsubishi. We presently have no idea how to explain such a network structure, but we believe that a time series analysis of networks is important.

This figure shows that these four major companies have many edges. In the graph theory, the degree of a node is defined by the number of other nodes to which it is attached. The distribution of the degree is an important quantity to study networks [3]. It is well known that the degree distribution of a regular network is represented by the δ-function, because each node has the same degree in the regular network. It is also well known that the degree distribution of random networks, which are proposed by Erdös and Rény, follows the Poisson distribution. A node with a large degree is called a hub. Hence these four major automobile industry companies are hubs in the network. In Sec. 3.1, the details of a study about the degree distribution of the network are explained.

In this figure, we can find that almost no hubs are directly connected. These hubs are mediated by companies, which have a small degree. Networks with such a characteristic are called uncorrelated networks or degree-nonassortative networks [11], and are characterized by a negative correlation between degrees. This is explained in terms of a degree correlation [14] in Sec. 3.2. Intuitively, this nature of the shareholding network is different from that of human networks. This is because, in human networks, for example in the friendship network, the hubs in the network correspond to persons with many friends, and are also friends with each other with a high probability. Networks with such characteristics are called correlated networks or degree-assortative networks, and are characterized by a positive correlation between degrees.

Suppose that a node has k neighbors; then at most $k(k-1)/2$ edges can exist to connect the neighbors with each other. Hence the possible number of triangles (minimum loops) containing this node is also $k(k-1)/2$. The ratio between the possible number of triangles and that of the actually existing triangles defines a clustering coefficient [20]. As we can see in this figure, the clustering coefficient is small for hubs, while it is large for nodes with a small degree. Hence it is expected that degrees and clustering coefficients are negatively correlated. The details are explained in Sec. 3.3.

This figure also shows that the network is constructed from subgraphs: triangles, squares, pentagons, etc. However, not all subgraphs occur with equal frequency. Hence if networks contain some subgraphs as compared to random-

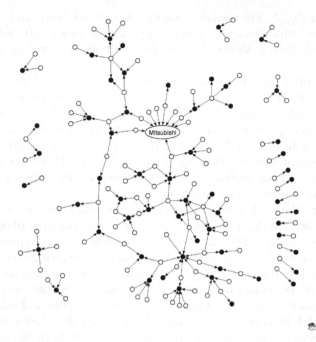

Fig. 3. A shareholding network constructed from edges drawn from the outside of the automobile industry sector to the inside of it. The open circles correspond to non-automobile industry companies, and the filled circles correspond to automobile industry companies. Arrows are drawn from the open circles to the filled circles

ized networks, these subgraphs characterize the networks. These characteristic subgraphs are the building blocks of networks, and are called network motifs [10][17]. Although network motifs are not discussed in this article, the spectrum of the network is considered. This is a primitive way to study subgraphs, and the details are discussed in Sec. 3.4.

2.2 Shareholding Network Constructed from Edges Drawn from the Outside of the Automobile Industry Sector to the Inside of It

Figure 3 is constructed from edges drawn from the outside of the automobile industry sector to the inside of it. To draw this figure, for simplicity, we ignored the arrows connecting financial institutions and automobile industry companies. In this figure the open circles correspond to non-automobile industry companies, and the filled circles correspond to automobile industry companies. Arrows are drawn from the open circles to the filled circles. The network is divided into many subgraphs, but there exists one large lump. This figure contains only Mitsubishi Motors of the four major automobile companies. This means that Mitsubishi Motor is governed by companies outside the automobile industry sector.

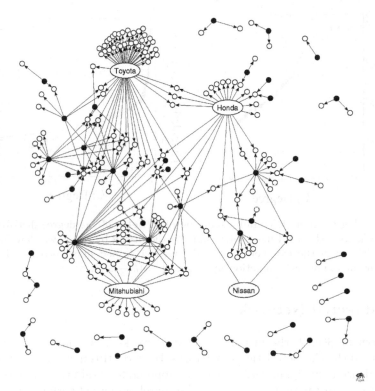

Fig. 4. A shareholding network constructed from edges drawn from the inside of the automobile industry sector to the outside of it. The open circles correspond to non-automobile industry companies, and the filled circles correspond to automobile industry companies. Arrows are drawn from the filled circles to the open circles.

2.3 Shareholding Network Constructed from Edges Drawn from the Inside of the Automobile Industry Sector to the Outside of It

Figure 4 is constructed from edges drawn from the inside of the automobile industry sector to the outside of it. As in the previous case, the open circles correspond to non-automobile industry companies, and the filled circles correspond to automobile industry companies. In this case arrows are drawn from the filled circles to the open circles. The network is divided into many subgraphs, but a large lump exists. We can see that major automobile industry companies are also major shareholders of non-automobile industry companies, except for Nissan Motor. It is expected that such a structure emerged after the year 1999 when Nissan and Renault announced their strategic alliance. Comparing this figure and Figs. 2 and 3 makes us believe that Toyota, Honda and Nissan are leaders in the automobile industry sector, and especially Toyota and Honda are also leaders in the Japanese economy.

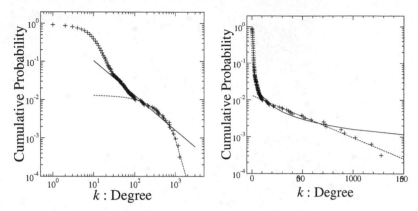

Fig. 5. A log-log plot (left) and a semi-log plot (right) of the degree distribution . In these figures, the solid lines correspond to the fitting by the power law function $p(k) \approx k^{-\gamma}$ with the exponent $\gamma = 1.8$, and the dashed lines correspond to the fitting by the exponential function $p(k) \approx \exp\{-\beta(k - k_0)\}$.

3 Undirected Network

As shown in Sec. 2, the visualization of networks is a powerful method to understand the characteristics of networks. However this method is not always useful. Hence many quantities have been proposed to obtain the characteristics of networks . In the previous section, the shareholding network is represented by a directed graph, but we consider it as an undirected network in this section. This is because many quantities characterizing networks are for undirected cases. In this case, the adjacency matrix M_{ij}^u is chosen: If the i-th company is a shareholder of the j-th company, we assume $M_{ij}^u = M_{ji}^u = 1$; otherwise, $M_{ij}^u = 0$. Hence this matrix is symmetrical.

3.1 Degree Distribution

A degree is the number of edges that attach to a node. In terms of the adjacency matrix, the degree of node i, k_i, is defined by $k_i \equiv \sum_{j=1}^{N} M_{ij}^u$. The log-log plot of the degree distribution is shown in the left panel of Fig. 5, and the semi-log plot is shown in the right panel of Fig. 5. In this figure, the horizontal axis corresponds to the degree and the vertical axis corresponds to the cumulative probability.

In these figures, the solid lines correspond to the linear fitting with the least square method by the power law function, and the dashed lines correspond to that by the exponential function. As we can see, the degree distribution follows the power law distribution with an exponential tail. The probability density function (PDF) $p(k)$ in the power law range is given by

$$p(k) \propto k^{-\gamma},$$

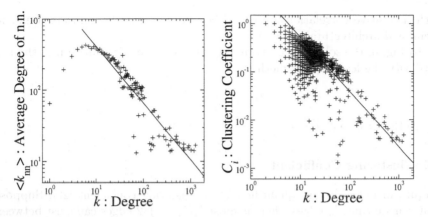

Fig. 6. A log-log plot of the degree correlation (left) and that of the correlation between the degree and the clustering coefficient (right). In the left panel, the solid line corresponds to the fitting by the power law function $\langle k_{nn}\rangle(k) \approx k^{-\nu}$ with the exponent $\nu = 0.8$. In the right panel, the solid line corresponds to the fitting by the power law function $C(k) \approx k^{-\alpha}$ with the exponent $\alpha = 1.1$.

where the exponent $\gamma = 1.8$, and that in the exponential range is given by

$$p(k) \propto e^{-\beta(k-k_0)}.$$

The exponential range is constructed of financial institutions, and the power law range is mainly constructed of non-financial firms. It has also been shown that the degree distributions of different networks in the economy also show power law distributions [18].

3.2 Degree Correlation

To obtain more detailed characteristics of networks, the degree correlation has been considered [14]. The nearest neighbors' average degree of nodes with degree k, $\langle k_{nn}\rangle$, is defined by

$$\langle k_{nn}\rangle = \sum_k k' p_c(k'|k),$$

where $p_c(k'|k)$ is the conditional probability that a link belonging to a node with degree k points to a node with degree k'.

The log-log plot of degree correlation is shown in the left panel of Fig. 6. In this figure, the horizontal axis corresponds to the degree and the vertical axis corresponds to $\langle k_{nn}\rangle$, i.e., the nearest neighbors' average degree of nodes with degree k. We find that the high degree range has a small value of $\langle k_{nn}\rangle$. This means that hubs are not directly connected with each other in this network. Networks with this characteristic are called uncorrelated networks, which are also found in biological networks and WWW. On the other hand, networks with a positive correlation are called correlated networks, and are found in social and architectural networks.

In Fig. 6, the solid line is the fitting by the power law function in the tail part with the least square method:

$$\langle k_{\mathrm{nn}} \rangle(k) \propto k^{-\nu},$$

where the exponent $\nu = 0.8$.

3.3 Clustering Coefficient

Cliques in networks are quantified by a clustering coefficient [20]. Suppose that a node i has k_i edges; then at most $k_i(k_i - 1)/2$ edges can exist between them. The clustering coefficient of node i, C_i, is the fraction of these allowable edges that actually exist e_i:

$$C_i = \frac{2e_i}{k_i(k_i - 1)}.$$

The clustering coefficient is approximately equal to the probability of finding triangles in the network. The triangle is the minimum loop. Hence if node i has a small value of C_i, then the probability of finding loops around this node is low. This means that the network around this node is locally tree-like. The correlation between k_i and C_i is shown in the right panel of Fig. 6. This figure shows that clustering coefficients have a small value in the high degree range. This means that the shareholding network has a local tree-like structure asymptotically.

The solid line in the right panel of Fig. 6 is the linear fitting by the power law function with the least square method:

$$C(k) \propto k^{-\alpha},$$

where the exponent $\alpha = 1.1$. Such a scaling property of the distribution of clustering coefficients is also observed in biological networks, and motivates the concept of hierarchical networks [15][16].

3.4 Spectrum

Here we discuss the spectrum of the network, i.e., the distribution of eigenvalues of the adjacency matrix. The distribution around the origin is shown in the left panel of Fig. 7. In this figure the horizontal axis is an eigenvalue λ_i and the vertical axis is a frequency. As is well known, if the network is completely random the distribution is explained by Wigner's semi-circle law. However, Fig. 7 is apparently different from the semi-circle distribution. We make four observations (see also Ref. [6]): (i) A δ peak at $\lambda_i = 0$, indicating the localized eigenstates that are produced by the dead-end vertices ; (ii) δ peaks

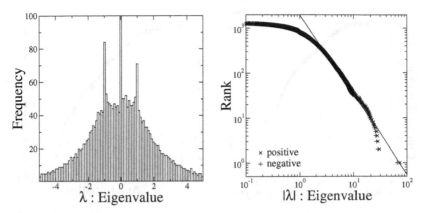

Fig. 7. A distribution of eigenvalues around the origin (left) and a log-log plot of the distribution of the absolute value of eigenvalues (right). In the right panel, the solid line corresponds to the fitting by the power law function $\rho(\lambda) \approx |\lambda|^{-\delta}$ with the exponent $\delta = 2.6$.

at $\lambda_i = \pm 1$, indicating the existence of approximately infinite long chains; (iii) A flat shape in the range $-1 < \lambda_i < 1$ except for $\lambda_i = 0$, indicating the existence of long chains constructed from weakly connected nodes, i.e., nodes with small degrees; and (iv) A fat tail.

The log-log plot of the eigenvalue distribution is shown in the right panel of Fig. 7 (see also Ref. [19]). In this figure the horizontal axis is the absolute value of eigenvalues $|\lambda_i|$, and the vertical axis is the cumulative probability. The plus symbols represent the distribution of the negative eigenvalues, and the cross symbols represent that of the positive eigenvalues. This figure shows that the shape of distribution in the positive eigenvalue range and that in the negative eigenvalue range are almost the same. The linear fitting with the least square method to the tail part is shown by the solid line in the figure. This fitting suggests that the PDF of the eigenvalue $\rho(\lambda)$ is asymptotically given by

$$\rho(\lambda) \propto |\lambda|^{-\delta},$$

where the exponent $\delta = 2.6$. If we compare the values of γ and δ, we can find the scaling law: $\delta = 2\gamma - 1$.

It has recently been shown under an effective medium approximation that the PDF of eigenvalue $\rho(\lambda)$ is asymptotically represented by that of the degree distribution $p(k)$:

$$\rho(\lambda) \simeq 2 |\lambda| p(\lambda^2),$$

if the network has a local tree-like structure [6]. Therefore, if $p(k)$ asymptotically follows the power law distribution, $\rho(\lambda)$ also asymptotically follows the power law distribution, and we can obtain the scaling relation $\delta = 2\gamma - 1$. In addition, the local tree-like structure is guaranteed by the right panel of Fig. 6.

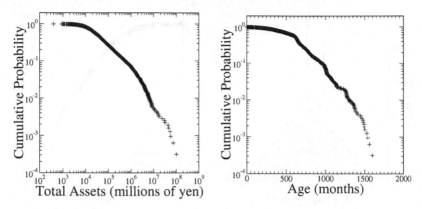

Fig. 8. A log-log plot of the distribution of the company's total assets (left) and a semi-log plot of the company's age (right).

4 Correlation Between Degree and Company's Total Assets and Age

It is interesting to construct models that can explain the topology of share-holding networks. However, in this section, we consider the correlation between the degree and the company's total assets and age. In many complex networks, it is difficult to quantitatively characterize the nature of nodes. However, in the case of economic networks, especially networks constructed of companies, we can obtain the nature of nodes quantitatively. We consider this to be a remarkable characteristic of business networks, and this allows us to understand networks in terms of the company's growth. This is the reason why we consider the correlation between the degree and the company's total assets and age.

The log-log plot of the distribution of the company's asset is shown in the left panel of Fig. 8. In this figure, the horizontal axis is the assets with the unit of millions of yen, and the vertical axis is the cumulative probability. This figure shows that the distribution in the intermediate range follows the power law distribution. In this case, the distribution is for companies listed on the stock market or the over-the-counter market. The completeness of data is a problem in extracting the true nature of the total asset distribution. However, it has been shown that data satisfying this completeness shows more clearly the power law distribution of the total assets [7].

The semi-log plot of the distribution of the company's age is shown in the right panel of Fig. 8. In this figure, the horizontal axis is the age with the unit of months, and the vertical axis is the cumulative probability. This figure shows that the distribution follows approximately the exponential distribution. It is expected that the age of companies has a relation with their lifetime, and it is also clarified that the lifetime of bankrupted companies follows the exponential distribution [8].

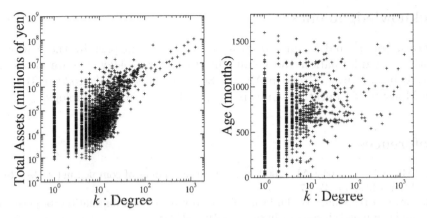

Fig. 9. A log-log plot of the correlation between the degree and the company's total assets (left) and a semi-log plot of the correlation between the degree and the company's age (right).

The log-log plot of the correlation between degrees and total assets is shown in the left panel of Fig. 9. In this figure the horizontal axis is the degree, and the vertical axis is the company's total assets with the unit of millions of yen. This figure shows that the degree and the total assets positively correlate.

The semi-log plot of the correlation between degrees and the company's age is shown in the right panel of Fig. 9. In this figure the horizontal axis is the degree, and the vertical axis is the company's age with the unit of months. This figure shows that the degree and the company's age have no correlation.

These two results suggest that the degree of companies has a relation with their total assets, but no relation with their age. This result means that the size of the company is an important factor to consider with regard to growing economic networks, but the age of the company is not. Old companies are not necessarily big companies. Hence knowing the dynamics of the company's growth is a key concept in considering growing economic networks [2].

5 Summary

In this paper we considered the Japanese shareholding network at the end of March 2002, and found some of the characteristics of this network. However, there are many unknown facts about the characteristics of shareholding networks. For example, these include time series changes of networks, the aspect of weighted networks, flows in networks, and the centrality of networks. Together with these studies, it is also important to study the dynamics of a company's growth. It is expected that the dynamics of economic networks cam be explained in terms of the dynamics of the company's growth.

Acknowledgements

Wataru Souma and Yoshi Fujiwara are supported in part by the National Institute of Information and Communications Technology. We are also supported in part by a Grant-in-Aid for Scientific Research (#15201038) from the Ministry of Education, Culture, Sports, Science and Technology.

References

1. Albert R, Barabási AL (2002) Statistical mechanics of complex networks. Rev. Mod. Phys. 74: 47–97
2. Aoyama H, Souma W, Fujiwara Y (2003) Growth and fluctuations of personal and company's income. Physica A 324: 352–358
3. Barabási AL, Albert R (1999) Emergence of scaling in random networks. Science 286: 509–512
4. Barabási AL (2004) Network Biology: Understanding the cell's functional organization. Nature Rev. Gen. 5: 101–114
5. Dorogovtsev SN, Mendes JFF (2003) Evolution of Networks: From Biological Nets to the Internet and WWW. Oxford University Press, New York
6. Dorogovtsev SN, et al. (2003) Spectra of complex networks. Phys. Rev. E68: 046109
7. Fujiwara Y, et al. (2004) Do Pareto-Zipf and Gibrat laws hold true? An analysis with European firms. Physica A 335: 197–216
8. Fujiwara Y (2004) Zipf law in firms bankruptcy. Physica A 337: 219–230
9. Garlaschelli D, et al. (2003) The scale-free topology of market investments. arXiv:cond-mat/0310503
10. Milo R, et al (2002) Network Motifs: Simple Building Blocks of Complex Networks. Science 298: 824–827
11. Newman MEJ (2002) Assortative Mixing in Networks. Phys. Rev. Lett. 89: 208701
12. Newman MEJ (2003) The Structure and Function of Complex Networks. SIAM Rev. 45: 167–256
13. Pajek: Program for Analysis and Visualization of Large Networks. URL: http://vlado.fmf.uni-lj.si/pub/networks/pajek/
14. Pastor Satorras R, Vázquez A, Vespignani A (2001) Dynamical and Correlation Properties of the Internet. Phys. Rev. Lett. 87: 258701
15. Revasz E, et al. (2002) Hierarchical Organization of Modularity in Metabolic Networks. Science 297: 1551–1555
16. Revasz E, Barabási AL (2003) Hierarchical organization in complex networks. Phys. Rev. E 67: 026112
17. Shen-Orr SS, et al. (2002) Network motifs in the transcriptional regulation network of Escherichia coli. Nature Genet. 31: 64–68
18. Souma W, Fujiwara Y, Aoyama H (2003) Complex networks and economics. Physica A 324: 396–401
19. Souma W, Fujiwara Y, Aoyama H (2004) Random matrix approach to shareholding networks. Physica A, in press
20. Watts DJ, Strogatz SH (1998) Collective dynamics of 'small-world' networks. Nature 393: 440–442.

The Emergence of Paradigm Setters Through Firms' Interaction and Network Formation *

Rainer Andergassen[1], Franco Nardini[2], and Massimo Ricottilli[3]

[1] Rainer Andergassen, Department of Economics, University of Bologna (Italy) anderga@economia.unibo.it
[2] Franco Nardini, Department of Mathematics for the Social Sciences, University of Bologna (Italy) nardini@dm.unibo.it
[3] Massimo Ricottilli, Department of Economics, University of Bologna (Italy) and Centro Interdipartimentale Galvani ricottilli@economia.unibo.it

Summary. Technological innovation requires the gathering of information through a process of searching and learning. We distinguish two different but definitely complementary and overlapping ways through which searching and learning occur. The first exploits the spillover potential that lies in a firm's network and thanks to which gathering innovation-useful information is actually possible. The second is the autonomous capacity that a firm possesses in order to carry out in-house innovative search. We build a model where rationally bounded firms try to increase their innovative capability through endogenous networking. The paper characterizes the emergence of technological paradigm setters in terms of network properties as they result from searching routines, furthermore the corresponding average efficiency of the system in terms of innovative capability is assessed.

Key words: Networks, Bounded Rationality, Technological Change, Innovative Capability, Paradigm Setters.

1 Introduction

Technological innovation requires the gathering of information through a process of searching and learning. In economies at the cutting edge of their frontier the competitive drive compels leading firms to engage in this process lest their advantage be lost to competitors and imitators. This quest for information is largely an adaptive, gradual process in which internal, in-house resources generating innovation-worthy knowledge are woven together with those obtained through technological spillovers proceeding from other firms. Firms, and agents within firms, exhibit bounded rationality and their success

---

* We are very grateful to seminar participants of the WEHIA 04 Kyoto meeting for their very useful comments.

is largely due to but also constrained by their technological capabilities that are, nevertheless, magnified by interaction with other firms. The implication is that the process of search is local and confined within a neighbourhood that is cognitively reachable and that is, therefore, the medium of effective information flows. To this effect, the role of a network in which firms nest is paramount

Because of bounded rationality and radical uncertainty, the process of reciprocal observation and learning occurring through interaction between firms and making the spreading of technological spillovers possible takes place in viable networks. Their importance has been highlighted in recent literature investigating technological and knowledge diffusion, see for example, Cowan and Jonard (2004, 2005), Silverberg and Verspagen (2002), Arenas et Alii, (2001, 2002). Networks, however, are subject to change and growth: connectivity between members, linkages between firms,can be viewed as a process in time. Seminal work in this field has been done by Albert and Barabasi, (2002) for an exhaustive review, Watts and Strogatz (1998), for the emergence of small worlds. When seen as an interactive system (Kirman 1997a, 1997b), the economy generates capabilities that are mutually acquired rendering firms technologically interdependent. It must, nevertheless, be recognised that capability-building also depends on in-house efforts supported by investment in formal and informal research and development. In this paper we treat technological capabilities as a stock of firm-specific knowledge cumulated through time by every single firm in the economy. We, accordingly, propose a model in which it is posited that each firm is endowed with a measurable index of innovation-enabling knowledge that is subject to change as a consequence of an active searching policy. We distinguish two different but definitely complementary and overlapping ways through which searching and learning occur. The first exploits the spillover potential that lies in a firm's network and thanks to which gathering innovation-useful information is actually possible. The second rests with the autonomous capacity that a firm possesses in order to carry out in-house innovative search. While these two searching processes not only coexist but are also reciprocally sustaining, we find it expedient to separate them by integrating a knowledge diffusion mechanism that propagates technological capabilities with an independent stochastic process capturing innovation arrivals due to internal R.&D. A network's evolution depends on how firms assess their performance in terms of innovation-enabling spillovers. In a bounded rationality framework, firms normally explore a limited part of the firms' space and require a protocol to target their information gathering efforts. The paper addresses this issue by designing a routinised behaviour according to which firms periodically reshape the neighbourhood that they observe to glean information by reassessing other firms' contributions to their own capability. The way the specific neighbour-choosing routine is accordingly organised determines in a significant way firms' average innovative capability. This feature is modelled by changing the span of network observation from a very broad setting, the whole economy, to a very narrow one, namely the most proximate neighbourhood membership. As a result of the structure of the model presented in the next section, there are two distinct but to some extent overlapping neighbourhoods which are relevant for firms' interaction. The first is the neighbourhood whose members are observed by each firm and from which capability contributions are obtained. We term this neighbourhood *inward*. The second is the one made up by a firm's observers, i.e. by firms observing and learning from it: it evolves as an active search for new inward members is carried out. We call this neighbourhood *outward*. This process of information interaction leads to the emergence of some firms that are observed by most of the remaining ones. It is they that provide some or much of the overall technological capability and that we term *paradigm setters*. We also assume that the in-house acquired capability is subject to structural shifts by means of periodic random shocks.

To keep the model mathematically tractable, we formalise the features stated above by means of a linear system in which technological capabilities are made to depend on a matrix of interaction with evolving outward neighbours as well as on a vector of in-house generated knowledge. The model is then simulated to determine the emergent properties of neighbourhood formation and stability together with average capability. We aim to identify (i) under what conditions the emergence of technological paradigm setters occurs, (ii) the pattern of neighbourhood formation and (iii) the average relative efficiency in terms of technological capability of the economy as a whole.

The plan of the paper is as follows: section two illustrates the linear model that is implemented to run simulations; section three describes the simulation procedures and the index employed to assess results; section four discusses results obtained and section five draws conclusions and sets an agenda for further research.

2 Firms' Technological Capabilities and Spillover Potential

We view a firm's technological capability as the upshot of an evolutionary process owing to learning, searching and gathering of information ultimately leading to innovation, much of it being explained by the interaction taking place within the system. These three processes are largely overlapping since neither can exist without the other two. In this section, we direct our analysis to investigate firms that are assumed to be technological leaders and whose major interest lies with innovation. We, accordingly, postulate that they possess 'in house' innovative capabilities resulting from past investment and that we distinguish, in a somewhat artificial manner but useful for modelling, from those that are entirely due to spillovers. These capabilities can be viewed and measured in a way akin to the more general category of a firm's knowledge base, cognitive potential or set of skills, know-how and competencies: they can actually be modelled as either a vector arranging different indicators or more simply as a scalar compounding the whole. We choose the latter approach.

Let $V_i(t)$ be the scalar that at time t designates firm i's innovative capability or, to use a term borrowed from biology, its innovative fitness. Then, $V(t)$ is the vector $V(t) = [V_i(t)]$, $i = 1, 2.....J$ arraying the fitness of all firms in the economy. By $C_i(t)$ we further designate the in-house capability cumulated until time t. As mentioned above, the latter, while embodying cumulated knowledge, requires investment to be preserved and eventually improved. Considerable efforts are therefore necessary to remain on the forefront of technological prowess, efforts which need not always prove successful. It is, accordingly, assumed that $C_i(t)$ be stochastically subject to change and $C_i(t) \in (0, 1)$. $C(t)$ is the corresponding vector.

A significant part of total technological capability is explained by firms' interaction with other firms. As mentioned above, this is due to searching activity and to the ease with which each firm is observable by other firms when broadcasting information. For searching to be significant and transmission possible, it is necessary that cognitive proximity generate interaction to let an effective spill over take place. The ensuing ability to broadcast relevant technological information may be measured by a basic index specific to each pair ij of firms in the economy. Accordingly, let a_{ij} indicate this index in terms of the part of each firm j's total innovative fitness that can cognitively be passed on to firm i should the latter be in a position to observe the former. The entire web of interfirm technological spillover can then be designated by a square, $J x J$, matrix A. Clearly, the main diagonal of this matrix is made up by 0's, $a_{ii} = 0$, since no firm broadcasts information to itself. Matrix A is simply an indicator of how well observing firms understand the technology of other firms and, therefore, it states no more than a spill over potential as a structural characteristic of the economy.

Bounded rationality restricts the number of neighgbours that a firm can usefully search to glean technological information and hinders, furthermore, the optimal choice of new ones when a firm gets the chance to adjust its neighbourhood.. Actual observation is restricted to a more or less narrow neighbourhood made up by firms whose informative usefulness has been discovered by a searching process. We accordingly postulate that firms carry out an active search to single out neighbours best suited to pass on innovative capability. It is assumed that each firm i searches among its potential information suppliers j whose broadcasting capacity is $a_i = (a_{ij})$, $j = 1, 2...J$, those that at each point in time it is able to choose as a target for innovative information and actually observe. This choice can be formalised by introducing the proximity matrix $B(t) = [b_{ij}(t)]$ where each $b_{ij}(t) = 1$ or $b_{ij}(t) = 0$ according to whether neighbour j has been or hasn't been identified as a useful contributor. This procedure defines matrix $M(t) = (a_{ij}b_{ij}(t))$. Thus, the innovative capability that is determined by interaction can be formalised by the system $M(t)V(t)$ where actually observed firms are restricted to a limited number of neighbours. The general equation for firm i's innovative capability is[4]

$$V_i(t) = \sum_{j=1}^{J} a_{ij} b_{ij}(t) V_j(t) + C_i(t)$$

and the system for all firms:

$$V(t) = [I - M(t)]^{-1} C(t)$$

where $[I - M(t)]^{-1}$ plays the role of an endogenous matrix multiplier of in-house capabilities.

Since firms are bounded in their rationality, different neighbourhood configurations lead to different multipliers. Attempts to improve networks lead to neighbours' change and thus to change of matrix $M(t)$.

2.1 Neighbourhood Structure

The structure in which we describe firms' innovative capability can be represented by a directed graph of J nodes each of which is connected with other nodes in two different but overlapping ways. The first is the number of connections that each firm establishes when observing other firms to determine its own innovative capability. The number of $k_{i,in} << J$ connections defines for firm i the dimension of its *inward neighbourhood*. This number is substantially smaller than J since searching is costly and observation bounded . This neighbourhood can formally be defined as

$$\Gamma_i(t) = \{\gamma_j; j : j = 1, 2...J \wedge b_{ij}(t) = 1\}$$

This is the set of firms from which at any time t firm i is able to glean innovative capability through observation and learning.

The second kind of neighbourhood, which we term the *outward* , is made up for each firm j by firms that actually observe it. It passively results as a consequence of their networking activity. Let it be defined by:

$$\Psi_j(t) = \{i : i = 1, 2...J \wedge b_{ij}(t) = 1\}$$

Its size determines the impact of an observed firm's technological capability as it propagates throughout the economy contributing to overall performance. For this purpose, we classify the population of firms according to classes of their outward neighbourhood size and then define an impact factor by ranking them.

Definition 1. *Technological paradigm setters emerge when the probability of each impact factor rank is positive.*

---

[4] Absorbing the impact of spillovers is clearly a process that requires an adjustment in time. We simplify this problem by assuming that the time required to complete adjustment is negligible in relation to the system evolution.

2.2 Evolution

Given this neighbourhood structure, evolution owes to two basic determinants: search routines and exogenous changes of individual firms' in-house innovative capabilities. Firms construe their *inward* neighbourhood by an active search aimed to single out members that contribute capability to their own. This search, while bounded by the neighbourhood in which the firm happens to be nested, may take place according to a variety of algorithms. We have chosen one that responds to the criteria of bounded rationality and *satisficing*. We propose two versions that respectively capture a strong and a weak form of bounded rationality. In both, we first conjecture that the cardinality of Γ_i is $|\Gamma_i| = k_{i,in} \ll J$ and generate the choice of neighbours and the evolution of this neighbourhood according to the following *routine*: each firm i assesses the fitness contribution of its existing neighbours and picks out the least contributing one:

$$\gamma_i(t-1) = \arg \min_{j \in \Gamma_i(t)} [a_{ij} b_{ij}(t-1) V_j(t-1)]$$

Secondly, the identified neighbour is substituted with a new one by randomly drawing among the limited number of the latter own neighbours, in the case of weak bounded rationality, or by randomly drawing among the remaining $J - k_{i,in} - 1$ members of the entire economy, in the case of strong but bounded rationality. In either case, to generate a new $\Gamma_i(t)$ it is necessary that this simple condition be satisfied:

$$V_i(t) > V_i(t-1)$$

This procedure redefines at each time step $M(t)$ and the system then generates a new set of solutions.

Next to the dynamics generated by neighbourhood adjustment we introduce in the system the autonomous and independent dynamics involving the in-house capability $C(t)$. This vector is subject to change by a random draw of some $i \in (1, 2...J)$ and by randomly redefining the ith component by a new random value $C_i(t)$ uniformly chosen between 0 and 1. These occurrences are arrivals that take place according to a predetermined mean waiting time μ.

3 Parameters and Benchmarking

The purpose of this model is to study the emergent properties of neighbourhood formation their stability and impact on average technological capabilities. In order to check for the impact of randomness in firms' searching activity we simplify the model by assuming parameters that insure an even field of equal starting points. It is, accordingly, conjectured that all firms have the same skill in broadcasting information and in spilling over their innovative capability. Matrix A will, therefore, be set at $A = (a_{ij} = a)$, $\forall i, j$. Parameter

a measures the strength of interaction and the degree to which each firm's innovative capability depends on other firms', given a matrix $M(t) = [ab_{ij}(t)]$. For the system to yield positive solutions for any given vector $C(t)$ and prevent explosive behaviour in $V(t)$, a must be kept as $a \le \frac{1}{k_{i,in}}$, $\forall i$. We further assume that the number of inward neighbours is the same for all firms and constant in time: $k_{i,in} = const$, $\forall i$.

In order to make a significant assessment of innovative performances as simulations are carried out, it is expedient to set benchmarks against which they can be measured. For this purpose, we propose the following definitions.

Definition 2. *Let the vector $\bar{C}(t)$ be the vector of elements of $C(t)$ rearranged in decreasing order and let the matrix $\overline{B} = [b_{ij}]$ where $b_{ij} = 1, \forall i = 1, 2....k_{in}, \forall i \ne j, j \le k_{in} + 1$ and $b_{ij} = 1, \forall i > k_{in} + 1......J; b_{ij} = 0$ otherwise. Furthermore define $\bar{v}(t) = \frac{1}{J} \sum_{j=1}^{J} \overline{V}_j(t)$, where $\overline{V}(t) = [I - \overline{M}]^{-1}\overline{C}(t)$ and $\overline{M} = a\overline{B}$.*

Definition 3. *Let matrix $\underline{B} = \left\{ b_{ij} = \frac{k_{inn}}{J-1}, \forall i \ne j, b_{ii} = 0 \forall i \right\}$. We then define $\underline{v}(t) = \frac{1}{J} \sum_{j=1}^{J} \underline{V}_j(t)$, $\underline{M} = a\underline{B}$ and $\underline{V}(t) = [I - \underline{M}]^{-1}C(t)$.*

Remark 1. If $\overline{V}(t)$ can be considered as the capability values resulting from an optimally planned network, $\underline{V}(t)$ is the average outcome from a random choice of neighbours.

Definition 4. *Given that $v(t) = \frac{1}{J} \sum_{j=1}^{J} V_j(t)$, an index of average relative capability gains can be construed as*

$$\phi(T) = \frac{1}{T} \sum_{t=t_0}^{T} \frac{v(t) - \underline{v}(t)}{\bar{v}(t) - \underline{v}(t)}$$

where T is the time period considered and t_0 is a sufficiently large transitional period to render initial conditions irrelevant.

The difference $v(t) - \underline{v}(t)$ is the system's average gain resulting from the actual choice over the potential maximum one $\bar{v}(t) - \underline{v}(t)$. Index $\phi(T)$ takes economically meaningful values between 0 and 1 for $T \to \infty$.

In the following section we run simulations with a population of firms $J = 64$, setting the number of inward neighbours $k_{in} = 3$ and the spillover coefficient $a = 0.25$.

4 Simulation Results

The model shown in section two has been simulated by tuning two crucial variables: (i) the mean waiting time of shock arrivals, μ, determining a change of

$C(t)$ and (ii) the neighbour searching routine that we call τ. The latter variable is defined in terms of the population of firms' space spanned when searching for a new neighbour. At one extreme, if a broad capacity to investigate and gather information from all firms is assumed, this space is the whole population of firms in the economy J; otherwise, any intermediate number up to the other extreme where firms are limited to explore only their extant neighbourhood. Thus, by calling π the probability of exploring across the entire firm space at each attempt to change a randomly chosen neighbour, we define it as $\pi \equiv \frac{1}{\tau}$. It follows that when $\tau = 1$, $(\pi = 1)$, the search routine is such that firms always look across the whole economy for a new neighbourhood member but τ tends to infinity, $\tau \to \infty$, when searching is entirely restricted within their neighbourhood ($\pi \to 0$, they never look outside). Intermediate values state the number of times searching occurs inside the firm's neighbourhood and just *once* outside.

4.1 Search Routines and Efficiency

It is in the model's logic that performance be gradually improved by adjusting capability-contributing neighbours whatever the choice of the search routine. This steady improvement, however, is more or less deeply upset by negative as well as positive shocks arriving randomly according to different waiting times, μ. In the following diagram we plot the index defined in the previous section, $\phi(T)$. As it can be seen, it is indeed confirmed that for any average waiting time, a searching strategy confined to one's own neighbourhood with some across-the-board exception brings about a clear relative improvement in the system's average innovative capability but up to a certain point. The differently printed lines correspond to as many shock mean waiting times in ascending order.

That performance improves as the mean waiting time, μ, rises is obviously the case since firms have more time to adjust and absorb random perturbations. What is more interesting, however, is the clearly different outcome of different search routines. It clearly appears that whatever μ turns out to be, performance is noticeably improved towards the theoretical maximum as firms cease to randomly scout the entire economy when they get the chance to substitute the least performing neighbour. As they concentrate on a local rather than a global search, better results are obtained at least up to approximatly τ^*, where τ^* is located in the environs of $\tau = 4 - 5$ in the simulations we have run: this means that it pays to search within one's own neighbourhood and yet to look over the entire economy approximately once every approximately τ^* times. Insisting on an even more localised search, however, invariably worsens performance. The reason why this happens is quite straightforward. Localised search has both an advantage and a disadvantage over the purely random one across the whole economy. The former lies in the fact that if a firm has high performing contributors, their own contributing neighbours are also likely to be high performing ones; thus, excluding the least performing of these high

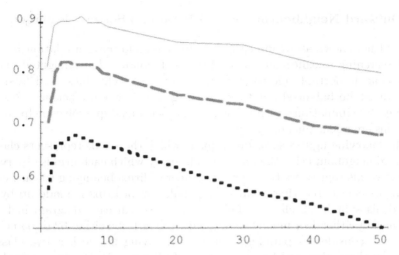

Fig. 1. Efficiency index $\phi(T)$ as a function of τ for mean-waiting time $\mu = 8$, (dots) $\mu = 16$ (dashed line) and $\mu = 32$ (continuous line)

contributors to substitute with one of its own neighbours carries a high probability of chancing on an even more performing contributor. On the contrary, if a firm were to carry its search for substitution over the entire firm space, it would stand an equal probability of finding either a poor or a high performing new neighbour. An improvement in a neighbour's contribution could then come from a comparatively mediocre firm The disadvantage lies in the fact that if a firm gets locked in poor performers, localised search is likely to perpetuate mediocre performance. As a consequence, there is a trade off between localised and across-the-board search. It pays to search locally but to renew the pool of neighbours by searching randomly across the whole economy every now and then. To clarify this point, consider the case of a firm that starts its neighbour adjustment process through local search and by having high performers: it would clearly manage to quickly raise its technological capability. If, however, one of its neighbours is hit by a negative shock sizably lowering its performance, the firm would cease to obtain high positive externalities and its capability growth would consequently plunge. Exclusively localised search would then leave this firm in a poor performing environment until positive shocks arrive to lift it out of this lock in: a further drawback is that positive and negative shocks are equally probable for any given waiting time μ. The only way to break the lock in is to search outside one's own neighbourhood. It follows that a good routine is one that considers local search to exploit the probability of finding high performers but that periodically searches abroad for better ones to avoid a poor performing stalemate.

4.2 Outward Neighbourhoods and Paradigm Setters' Emergence

The evidence shown above indicates that there is an important relationship between searching routines and their efficiency. It follows that farsighted search
still has an important role to play and that short-sightedness and farsightedness must be balanced out to yield the greatest search efficiency. Results
provided by simulations on how technological leadership evolves help to explain this emergent phenomenon.

The following figures show diagrams in which the x-axis represents classes
of inward neighbours (i.e. the number of firms by which each firm is observed)
and the y-axis represents the average number of firms belonging to each class
within the considered time span. The population of firms is made up by 64
individuals split in 32 classes: thus, the first class in each diagram includes
firms having either 1 or nought neighbours, the last one either 62 or 63 neighbours. We consider as paradigm setters firms lying in the last two classes,
that is by those observed by almost 95% of all firms. We begin our analysis
by discussing the case in which $\mu = 8$ and $\tau = 1$.

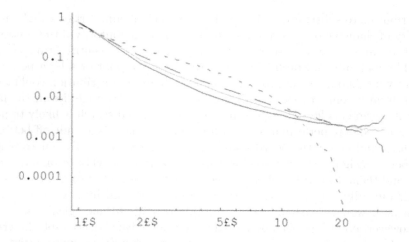

Fig. 2. Average frequency distribution of quantiles per time unit, in $\log - \log$ space;
$\mu = 8$, $\tau = 1$(dots), $\tau = 2$ (dashed line), $\tau = 3$ (gray line) and $\tau = 5$ (black line).

In Figure 2, the dotted line reveals that the last classes (from 20 to 32)
are empty. This means that when the mean waiting time of shock arrivals
is too short and the search routine randomly spans the whole economy, the
probability of paradigm setters' emergence is nil. Furthermore, as we have
argued above, lack of adjustment time and broad randomness of searching
hinder many firms from seeking out high performers and, thus, neighbourhood
configurations remain suboptimal. This pattern retains its main features as
long as the search routine remains defined by $\tau = 1$ even if mean waiting times

lengthen to allow sufficient neighbourhood adjustment (see Figure 3 and 4, dotted lines). Random choice of new neighbours from the entire firm space implies performance mediocrity but, for better or for worse, also avoidance of a lock-in no matter what the shock arrival waiting time is. Firms keep on changing their neighbours so much that no particular firm is allowed to set its technological capability as a paradigm for the rest of the economy: as a result, the last classes remain empty and no paradigm setter emerges. As the routine search becomes more local, see Figure 2, 3 and 4 dashed and continuous gray and black lines for $\tau = 2, 3$ and 5 respectively, paradigm setters begin to emerge and as searching approaches $\tau \to \infty$, that is when the choice of new neighbours is restricted exclusively within the narrow bounds of a firm's own neighbourhood, the same firms tend to persist as paradigm setters. The economy ends up by having a distribution featuring very few paradigm setters and a multitude of firms observed by no one else: firms with zero outward neighbours. This result occurs independently of τ provided that μ is sufficiently high, i.e. if shocks hit the economy sufficiently seldom. The following Figures 2, 3 and 4 illustrates this point for $\mu = 8$, 16 and 32, respectively.

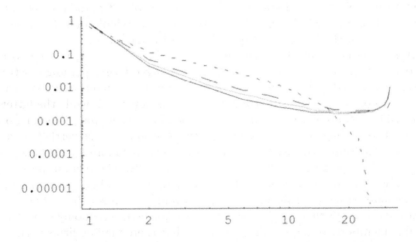

Fig. 3. Average frequency distribution of quantiles per time unit, in the $\log - \log$ space; $\mu =16$, $\tau =1$ (dots), $\tau =2$ (dashed line), $\tau =3$ (gray line) and $\tau =5$ (black line).

In these cases, the probability that there be firms with 62 or 63 technological onlookers is significantly positive (see dashed, gray and black lines). To compound these results, it is important to consider what happens to the two or three firms that temporarily emerge as leaders and eventually become paradigm setters as evolution unfolds, shocks arrive and firms' adjustment takes place. Simulations show that after some time these two or three firms

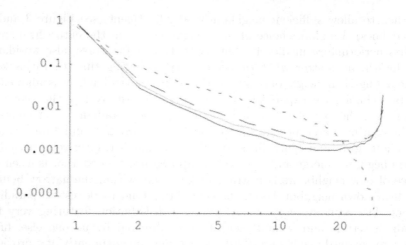

Fig. 4. Average frequency distribution of quantiles per time unit, the $\log - \log$ space; $\mu = 32$, $\tau = 1$ (dots), $\tau = 2$ (dashed line), $\tau = 3$ (gray line) and $\tau = 5$ (black line).

strengthen their acquired status and rise to the role of paradigm setters as the ones that most other firms elect as their inward neighbours: they possess, in other words, the highest number of outward neighbours.

It is apparent that the emergence of paradigm setters is related to the achievement of better performance. Our data suggest that paradigm setters' emergence precedes the attainment of maximum performance, and as such it appears to be its prerequisite: the former emerge for $\tau = 2$, while the latter is achieved for values of τ between 4 and 6. Lock-in and renewal by looking over the whole economy balance out in the region around τ^*. Nevertheless, as τ rises past the observed peak and search becomes ever more local, the lock-in effect becomes pervasive and average capability falls. There is an interesting implication in this statement: past the maximum τ^*, while paradigm setters are the most observed firms in the economy and as such set the technological standards through their spillovers, they need not be the economy's most performing members. It is quite likely, in fact, that there may be firms with very high innovative capability that are left unchosen because never discovered by a very local search routine. In this sense, the system is likely to lock into a progressively more inefficient leadership.

5 Conclusions

The foregoing analysis highlights the importance of searching and networking in fostering the development of technological capabilities to innovate in a context of bounded rationality. Firms obtain information and learn when crucially placed in a cognitive and information providing neighbourhood. Technological spillovers flow and give other firms the opportunity to learn only if networks come into being to give shape to searching and make learning possible. This paper depicts this process as an effort by firms, which do carry out their own in-house innovation capability building, to seek out high performers able to contribute to the latter. Routines differ according to the breadth of this search. Thanks to a simulated, simple, linear system, it is shown that broad, economy wide search routines are inefficient. The system's average innovative capability rises when searching is more local and firms are constrained to seek out new neighbours only among their discarded neighbour's neighbours. Up to a point. While the said ploy implies a technological lock-in, the probability of finding better performing firms increases. This result, however, occurs only if firms still carry out a broad, economy wide search with a frequency that simulations approximately identify. Thus, tuning short-sightedness and farsightedness improves the system's innovative efficiency. Past a given combination of the two, the system slides towards increasing mediocrity but paradigm setters emerge as a permanent and systematic feature of this hypothetical economy. The more local search is, the greater is the likelihood that paradigm setters' emergence occurs. When searching is highly and in the limit exclusively local, a lock-in into mediocre paradigm setters takes place and neighbourhoods are no longer providers of efficient spillovers.

Exogenous shocks set another time scale to the system and upset the learning and neighbour choosing adjustment. As shock arrival waiting time increases the system's technological efficiency rises but there is very little impact on paradigm setters' emergence that, therefore, appears to depend only on the chosen routine pattern.

Further research is required to investigate the effects of a lopsided distribution of spillover coefficients ($a'_{ij}s$) and the possibility that inward neighbourhoods may evolve to produce emergent properties of the system's innovative capabilities.

References

1. Albert R, Barabasi A-L (2002) Statistical mechanics of complex networks. Reviews of Modern Physics 74:47–97.
2. Arenas A, Diaz-Guilera A, Pérez CJ, Vega-Redondo F (2002) Self-organized criticality in evolutionary systems with local interaction. Journal of Economic Dynamics and Control 26:2115–2142.
3. Arenas A, Diaz-Guilera A, Guardiola X, Llas M, Oron G, P érez CJ, Vega-Redondo F (2001) New Results in a Self-organized Model of Technological Evolution. Advances in Complex Systems 1:1–12.
4. Cowan R, Jonard N (2004) Network Structure and the difusion of knowledge. Journal of Economic Dynamics and Control 28:1557–1575.
5. Cowan R, Jonard N (2005) Innovation on a network. Forthcoming in Structural Change and Economic Dynamics.

6. Kirman A (1997a) The Economy as an Evolving Network. Journal of Evolutionary Economics 7:339–353.
7. Kirman A (1997b) The Economy as an Interactive System. In Arthur W B, Durlauf S N, Lane D (eds) The Economy as a Complex Evolving System II. Santa Fe Institute, Santa Fe and Reading, MA, Addison-Wesley.
8. Watts D J, Strogatz S H (1998) Collective Dynamics of 'small world' networks. Nature 393:440–442.

Part III

Economic Dynamics

Statistical Properties of a Heterogeneous Asset Pricing Model with Time-varying Second Moment *

Carl Chiarella[1], Xue-Zhong He[1], Duo Wang[2]

[1] School of Finance and Economics, University of Technology, Sydney, PO Box 123 Broadway, NSW 2007, Australia
[2] LMAM, Department of Financial Mathematics, School of Mathematical Sciences, Peking University, Beijing 100871, People's Republic of China

Summary. Stability and bifurcation analysis of deterministic systems has been widely used in modeling financial markets. However, the impact of such dynamic phenomena on various statistical properties of the corresponding stochastic model, including skewness and excess kurtosis, various autocorrelation (AC) patterns of under and over reactions, and volatility clustering characterised by the long-range dependence of ACs, is not clear and has been very little studied. This paper aims to contribute to this issue. Through a simple behavioural asset pricing model with fundamentalists and chartists, we examine the statistical properties of the model and their connection to the dynamics of the underlying deterministic model. In particular, our analysis leads to some insights into various mechanisms that may generate some of the stylised facts, such as fat tails, skewness, high kurtosis and long memory, observed in high frequency financial data.

JEL classifications: D83; D84; E21; E32, C60

Key words: Fundamentalists, chartists, stability, bifurcation, investors' under- and over-reactions, stylized facts.

1 Introduction

As a result of a growing dissatisfaction with (i) models of asset price dynamics based on the representative agent paradigm, as expressed for example by Kirman [25], and (ii) the extreme informational assumptions of rational expectations, research into the dynamics of financial asset prices resulting from

* Financial support for Duo Wang by NSFC.10271007 and RFDP.20010001042 is acknowledged. This paper was prepared when Wang was visiting the University of Technology, Sydney, whose hospitality is gratefully acknowledged.

the interaction of heterogeneous agents has developed, including for example Arthur et al. [2], Brock and Hommes ([3], [4]), Brock and LeBaron [6], Bullard and Duffy [7], Chen and Yeh ([8], [9]), Chiarella [10], Chiarella et al. [11], Chiarella and He ([12], [13], [14]), Dacorogna et al. [16], Day and Huang [17], Farmer and Joshi [19], Gaunersdorfer [20], Gaunersdorfer et al [21], Hommes ([23], [24]), LeBaron et al. [26], Lux ([27], [28], [29]) and Lux and Marchesi [30]. In particular, Brock and Hommes ([4], [5]) have introduced the concept of an *adaptively rational equilibrium*, where agents adapt their beliefs over time by choosing from different predictors or expectations functions, based upon their fitness functions measured by realized profits. The resulting dynamical system is capable of generating the entire *zoo* of complex behaviour from local stability to high order cycles and chaos as various key parameters of the model change. Brock and Hommes's framework has been extended further in Gaunersdorfer [20] and Chiarella and He ([12], [13], [14]) to incorporate heterogeneous variance, risk and learning under both Walrasian auctioneer and market maker scenarios. It is found that the relative risk attitudes, different learning mechanisms and different market clearing scenarios affect asset price dynamics in a very complicated way. It has been shown (e.g. Hommes [24]) that such simple nonlinear adaptive models are capable of explaining important stylized facts, including fat tails, clustered volatility and long memory, of real financial time series.

There are two goals this literature is trying to achieve, first to explain various market behaviour and, second to replicate the econometric properties and stylized facts of financial time series. These heterogeneous agent models, in particular the theoretically oriented ones, have shown great potential in achieving the first goal. However, we still have some distance in achieving the second goal. On the one hand, it is well known that most of the stylized facts can only be observed for high frequency data, such as weekly, daily or intraday data. However, most of the heterogeneous agent models have difficulty in replicating realistic time series at high frequency. On the other hand, stability and bifurcation analysis have been widely accepted tools in investigating such models, but it is not clear yet how various types of bifurcations of the deterministic system can be used to characterize various price volatility and time series properties (such as distribution density and autocorrelations of returns) of the corresponding stochastic asset pricing models. These issues have been tackled recently in He [22] through a simple fixed fraction heterogeneous asset pricing model. This paper is largely motivated by these issues, in particular by He [22], and seeks to contribute to both of the aforementioned goals.

Gaunersdorfer et al. [21] consider a simple asset pricing model of fundamentalists and chartists. The fundamentalists believe that tomorrow's price will move in the direction of the fundamental price, while chartists derive their beliefs from a simple technical trading rule using only the latest observed price and extrapolation of the latest observed price change. By assuming a constant conditional variance for both types of traders and imposing a penalty function in the fitness function of the chartists (to ensure that speculative bubbles cannot last forever), they show that volatility clustering can be characterized by a coexistence of a stable steady state and a stable limit cycle, which arises as a consequence of a so-called Chenciner bifurcation. Simple economic intuition of this result is also provided. This paper is a significant contribution in relating asset price volatility to the bifurcation nature of the underlying deterministic system. However, it is not clear if such a characterization of volatility clustering still holds for the corresponding stochastic system when the fundamental price follows a random process, in particular when the trading frequency is high.

Given the variety of technical trading rules and differing risk aversion of various investors, this paper introduces a risk adjusted demand function for the chartists by assuming that they use a weighted process of past prices to estimate both conditional mean and variance. Therefore their demand function is a nonlinear function of the conditional mean and variance, instead of a linear function of the conditional mean only as in Gaunersdorfer et al. [21]. It is found that the mechanism of variance adjusted demand function of the chartists provides a natural way to prevent the price from *getting stuck* in a speculative bubble[3]. Similar to Brock and Hommes ([4], [5]), an adaptive model based on the fitness function is then obtained. We examine how the price dynamics of the risky asset is affected by the reactions of investors, the switching intensity of the fitness function, and the weighting process and risk adjustment of the chartists.

The plan of the paper is as follows. We first develop a simple adaptive asset pricing model of fundamentalists, chartists and noise traders. A statistical analysis is then conducted when parameters are near the bifurcation boundaries of the underlying deterministic model and a connection between the Hopf (flip) bifurcation and under(over)-reaction AC patterns is established. The paper finishes with a discussion on the capability of the model to generate the stylised facts and the long memory of high frequency financial time series.

2 The Model

Following the framework of Brock and Hommes [5], consider an asset pricing model with one risky asset and one risk free asset. It is assumed that the risk free asset is perfectly elastically supplied at the risk-free rate r (per annum). Let p_t be the price (ex dividend) per share of the risky asset at time t and $\{y_t\}$ be the stochastic dividend process of the risky asset. Then the wealth of investor h at $t+1$ is given by

$$W_{h,t+1} = RW_{h,t} + R_{t+1}z_{h,t}, \tag{1}$$

where

---

[3] In Gaunersdorfer et al. [21], this is ensured by artificially adding a penalty function in the fitness function of the chartists, which is unnecessary in our model.

$$R_{t+1} = p_{t+1} + y_{t+1} - Rp_t \qquad (2)$$

is the excess capital gain/loss, $R = 1 + r/K$, K is the trading frequency per annum[4], $W_{h,t}$ is the wealth at time t and $z_{h,t}$ is the number of shares of the risky asset purchased at t. Denote by $F_t = \{p_{t-1}, \cdots; y_{t-1}, \cdots\}$ the information set formed at time t. Let $E_{h,t}, V_{h,t}$ be the *beliefs* of investor type h about the conditional expectation and variance, based on F_t. Then it follows from (1) and (2) that

$$E_{h,t}(W_{t+1}) = RW_t + E_{h,t}(R_{t+1})z_{h,t}, \qquad V_{h,t}(W_{t+1}) = z_{h,t}^2 V_{h,t}(R_{t+1}). \quad (3)$$

Assume investor has a CARA (constant absolute risk aversion) utility function $u(W) = -e^{-a_h W}$ with risk aversion coefficient $a_h > 0$, and maximises his/her expected utility of wealth, leading the optimal demand

$$z_{h,t} = \frac{E_{h,t}(R_{t+1})}{a_h V_{h,t}(R_{t+1})}. \qquad (4)$$

In the following discussion, we adopt the popular fundamentalist/chartist model by assuming that all investors can be grouped into either fundamentalists (type 1) or chartists (type 2).

Fundamentalists—The fundamentalists are assumed to believe that the expected market price p_t is mean reverting to their perceived fundamental value p_t^* and the conditional variance of the market price is constant. That is,

$$\begin{cases} E_{1,t}(p_{t+1}) = p_{t-1}^* + v(p_{t-1} - p_{t-1}^*), & 0 \le v \le 1 \\ V_{1,t}(p_{t+1}) = \sigma_1^2, \end{cases} \qquad (5)$$

where p_t^* is the fundamental price of the risky asset estimated by the fundamentalists at some cost, v is the speed of mean reversion estimated by the fundamentalists, and $\sigma_1 > 0$ is a constant. In particular, $E_{1,t}(p_{t+1}) = p_{t-1}^*$ for $v = 0$ and $E_{1,t}(p_{t+1}) = p_{t-1}$ for $v = 1$. The conditional expectation of the fundamentalists (5) can also be written as $E_{1,t}(p_{t+1}) = (1 - v)p_{t-1}^* + vp_{t-1}, 0 \le v \le 1$, which is a weighted average of the fundamental price and latest price. Hence small (large) values of v indicate that the fundamentalists give more (less) weight to the fundamental price and less (more) weight to the latest price, believing that price moves quickly (slowly) towards its fundamental value p_t^*. For convenience of discussion, we say the fundamentalists *over(under)-react* (to the market price) when more (less) weight v is given to the market price.

Chartists—Unlike the fundamentalists who are able to obtain a good estimate of the fundamental value, chartists base their trading strategy on signals generated from the costless information contained in recent prices. The signal may be generated by comparing the latest price p_{t-1} with some reference

---

[4] Typically, annually, quarterly, monthly, weekly and daily trading periods correspond to $K = 1, 4, 12, 52$ and 250, respectively.

price trends $\tilde{p}_{t-1}$, such as a moving average process. Specifically, the chartists consider the realizations of $\tilde{p}_{t-1}$ as random drawn from some distribution. The distribution can be conditional on past realized values. For simplicity, we assume that $\tilde{p}_{t-1}$ is conditionally distributed on prices p_{t-2} and p_{t-3} with weighting probabilities w and $1 - w$, respectively. Then the conditional mean and variance of the trend can be estimated, respectively, as

$$\begin{cases} \bar{p}_{t-2} \equiv wp_{t-2} + (1 - w)p_{t-3}, & 0 \leq w \leq 1, \\ \bar{\sigma}_{t-2}^2 \equiv w[p_{t-2} - \bar{p}_{t-2}]^2 + (1 - w)[p_{t-3} - \bar{p}_{t-2}]^2. \end{cases} \tag{6}$$

Based on the trading signals $p_{t-1} - \tilde{p}_{t-1}$ and the conditional mean and variance estimates (6), we make the following assumptions for the chartists:

$$\begin{cases} E_{2,t}(p_{t+1}) = p_{t-1} + g(p_{t-1} - \bar{p}_{t-2}), & g \in \mathbb{R}, \\ V_{2,t}(p_{t+1}) = \sigma_1^2[1 + b\bar{\sigma}_{t-2}^2], & b \geq 0, \end{cases} \tag{7}$$

where $g \in \mathbb{R}$ is the estimated extrapolation rate of the chartists. That is, the chartists' beliefs are based on the latest price and their extrapolation of the trading signals generated from the trend. In particular, chartists are called *trend followers* when $g > 0$ and are *contrarians* when $g < 0$. For $w = 1$, $E_{2,t}(p_{t+1}) = p_{t-1} + g(p_{t-1} - p_{t-2})$ which is the case discussed in Gaunersdorfer et al. [21] and, for $w = 0$, $E_{2,t}(p_{t+1}) = p_{t-1} + g(p_{t-1} - p_{t-3})$. Similarly, for convenience of discussion, we say the chartists *over(under)-react* when they extrapolate strongly (weakly), that is when $|g|$ is large (small). With regard to the chartists' estimate of the variance, they use the historical variance to scale up the fundamental variance through the parameter b. High b reflecting a greater sensitivity to variance risk.

For the dividend process, we assume $E(y_t) = \bar{y}, V(y_t) = \sigma_y^2$. In particular, when $p_t^* = p^*$ is constant, $\bar{y} = rp^*$. Based on the above assumptions, the optimal demands for the fundamentalists and chartists are given, respectively, by

$$\begin{cases} z_{1,t} = [p_{t-1}^* + v(p_{t-1} - p_{t-1}^*) + \bar{y} - Rp_t]/A_1, \\ z_{2,t} = [p_{t-1} + g(p_{t-1} - \bar{p}_{t-2}) + \bar{y} - Rp_t]/A_{2,t}. \end{cases} \tag{8}$$

where

$$A_1 = a_1(\sigma_1^2 + \sigma_y^2), \qquad A_{2,t} = a_2[\sigma_y^2 + \sigma_1^2(1 + b\bar{\sigma}_{t-2}^2)] \tag{9}$$

and $\bar{p}_t$ and $\bar{\sigma}_t^2$ are defined by (6).

Let $U_{1,t}$ and $U_{2,t}$ be the realized profit of the fundamentalists and chartists, respectively, defined by $U_{i,t} = R_t z_{i,t-1} - C_i, i = 1, 2$, where $C_i \geq 0$ measures the total cost incurred by the agents in generating information to form their estimates. Let $n_{i,t}$ be the fractions of agents of type i at time t. Assume the fractions are formed on the basis of discrete choice probability (see Manski and McFadden [31], Anderson, de Palma and Thisse [1], Brock and Hommes ([4], [5])), namely

$$n_{i,t} = exp[\beta U_{i,t-1}]/Z_{t-1}, \qquad (i = 1, 2), \qquad Z_{t-1} = \sum_{i=1}^{2} exp[\beta U_{i,t-1}], \tag{10}$$

where $\beta(> 0)$ is the *intensity of choice* measuring how fast agents switch among different prediction strategies. Let $m_t = n_{1,t} - n_{2,t}$. Then $n_{1,t} = (1 + m_t)/2, n_{2,t} = (1 - m_t)/2$ and

$$m_t = \tanh\left[\frac{\beta}{2}R_{t-1}(z_{1,t-2} - z_{2,t-2}) - \frac{\beta}{2}C\right], \tag{11}$$

where $C = C_1 - C_2$. It is assumed that the fundamentalists expend more on generating relevant information to generate their estimates so that the constant $C \geq 0$ in general.

To clear the market through a Walrasian scenario, we introduce noise traders whose supply of the risky asset is denoted by $z_{n,t} \sim \mathcal{N}(0, \sigma_n^2)$. The market is viewed as finding the price p_t that equates the sum of these demand from both fundamentalist and chartist schedules to the supply of the noise traders via

$$(1 + m_t)z_{1,t} + (1 - m_t)z_{2,t} = z_{n,t}. \tag{12}$$

Substituting (8) and (11) into (12), one obtains the market cleaning price p_t

$$p_t = \frac{F(p_{t-1}, p_{t-1}^*, m_t, \bar{p}_{t-2}, \bar{\sigma}_{t-2}^2, z_{n,t}, y_t)}{G(p_{t-1}, p_{t-1}^*, m_t, \bar{p}_{t-2}, \bar{\sigma}_{t-2}^2, z_{n,t}, y_t)}, \tag{13}$$

where

$$F(p_{t-1}, p_{t-1}^*, m_t, \bar{p}_{t-2}, \bar{\sigma}_{t-2}^2, z_{n,t}, y_t) = A_{2,t}[1 + m_t][p_{t-1}^* + v(p_{t-1} - p_{t-1}^*) + \bar{y}]$$
$$+ A_1[1 - m_t][p_{t-1} + g(p_{t-1} - \bar{p}_{t-2}) + \bar{y}] - z_{n,t}A_1A_{2,t},$$
$$G(p_{t-1}, p_{t-1}^*, m_t, \bar{p}_{t-2}, \bar{\sigma}_{t-2}^2, z_{n,t}, y_t) = R[(1 + m_t)A_{2,t} + A_1(1 - m_t)],$$

$\bar{p}_t$, A_1 and $A_{2,t}$ are defined by (6) and (9). Because of the stochastic nature of the dividend process y_t, the fundamental value p_t^* and the noise supply, the equilibrium price equation (13) is equivalent to a high-order nonlinear stochastic discrete system in general.

When the stochastic terms in equation (13) take their mean levels, one obtain the corresponding deterministic system. For the deterministic system, it is found from Chiarella et al. [15] that the constant fundamental price p^* is the unique steady state price of the system. In addition, the steady state price becomes unstable through a Hopf bifurcation only when the chartists are trend followers (that is when $g > 0$). However, when the chartists are contrarians (that is when $g < 0$), the local stability region is bounded by both Hopf and flip bifurcations. However, depending on the weighting parameter w, one of the boundaries may be unbinding. In particular, the steady state price can become unstable through a Hopf bifurcation when w is small, or a flip bifurcation when w is large, or both bifurcations when w is near 0.5. We refer to Chiarella et al. [15] for more detailed analysis on the dynamics of the deterministic system. In the following discussion, we examine the statistical properties of the market price of the stochastic model (13). In particular we are interested in how statistical properties exhibit different behaviour near different types of bifurcation boundaries of the deterministic system. In so doing, we obtain some insights into the interaction of the dynamics of the deterministic system and the noise processes.

In the following, we choose the fundamental price $p_o^* = \$100$, annual risk-free rate $r = 5\%$, annual volatility of the fundamental price $\sigma = 20\%$. Trading frequency $K = 250$, which corresponds to a daily trading period. Correspondingly, the total risk-free return per trading period $R = 1 + r/K = 1.0002$, daily price volatility $\sigma_1^2 = (p^*\sigma)^2/K = 8/5$ and daily dividend volatility $\sigma_y^2 = r^2\sigma_1^2 = 1/250$. We also choose the risk aversion coefficients for both types of investor as $a_1 = a_2 = 0.8$, the cost difference $C = 0$, the coefficient of endogenous variance component $b = 1$ and the switch intensity $\beta = 0.1$.

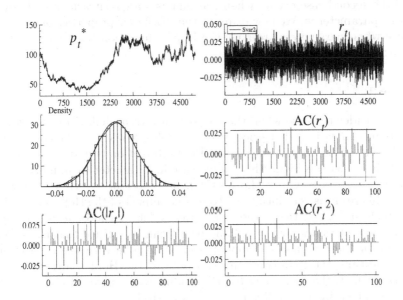

Fig. 1. Time series plots of the prices p_t^*, returns r_t and autocorrelations of r_t, $|r_t|$ and r_t^2.

We assume that the fundamental price p_t^* follows a stationary random walk process

$$p_t^* = p_{t-1}^*[1 + \epsilon_t], \qquad \epsilon_t \sim N(0, \sigma_p^2), \qquad p_0^* = \$100.00. \tag{14}$$

Fig.1 displays a realisation of the fundamental prices, p_t^* and the corresponding relative return $r_t = p_t^*/p_{t-1}^* - 1$. This fundamental price is used for all of the following simulations. The simulation contains 5,000 observations, corresponding to about 20 years of daily prices. The corresponding statistical properties, including distribution density of the return r_t and the autocorrelations (ACs) of $r_t, |r_t|$ and r_t^2, are also included in Fig. 1 One can see that the ACs of r_t, $|r_t|$ and r_t^2 are not significant at the 95% level, which is consistent with the stationary random walk process (14).

3 Hopf Bifurcations and Under-Reaction AC Patterns

Chiarella *et al.* [15] have extensively analysed the dynamical properties of the deterministic skeleton of the price dynamics (13). It is found that the fundamental steady state price of the underlying deterministic system becomes unstable through a Hopf bifurcation only in two situations. The first case occurs when the chartists are trend followers (with any weighting parameter w) and the second case occurs when the chartists are contrarians with lower weighting parameter w. We now examine the stochastic properties of the market price when parameters are near the Hopf bifurcation boundary in these two different situations.

3.1 Case 1

We first consider the case when the chartists are trend followers. It is found from various simulations that prices become explosive when (i) the parameters v, g and w are either near or outside the Hopf bifurcation boundary, or (ii) the switching intensity β is high. Also, an increase in parameter b can make prices become less explosive. Given the destabilizing effect of the trend following strategy, this result is very intuitive. To see the impact of the Hopf bifurcation on the stochastic properties of the stochastic model, we choose $w = 0.5, g = 1$ and two different values for $v = 0.3$ and 0.9. It can be verified that the steady state price of the deterministic model is locally stable with these selections and, in addition, the parameter moves closer to the Hopf bifurcation as v increases. In fact, the corresponding Hopf bifurcation values of g for $v = 0.3$ and 0.9 are given by $g^* = 1.458$ and 1.3, respectively.

To see the impact of the stochastic fundamental price and noise traders on the price dynamics, we consider two different situations: $\sigma_n = 0$ and $\sigma_n = 1$. For $\sigma_n = 0$, the market price is influenced by the random fundamental price p_t^* only. When $\sigma_n = 1$, the market price is influenced by both the random fundamental price p_t^* and the noisy supply. Fig. 2 illustrates the time series of the market price p_t, the densities of the market fraction m_t and the return r_t, and the ACs of $r_t, |r_t|$ and r_t^2 for four combinations of $(v, \sigma_n) = (0.3, 0), (0.3, 1), (0.9, 0)$ and $(0.9, 1)$. Our numerical simulations show that the features illustrated in Fig. 2 are robust, from which we arrive at the following observations.

(i) The market prices p_t follow closely the random fundamental price p_t^* in all cases. This may partially be due to the local stability of the steady state

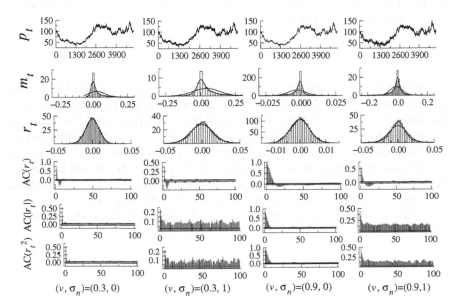

Fig. 2. Time series of the market price p_t, distribution density of the population fraction difference m_t and return r_t (compared with the normal distribution) and ACs of r_t, $|r_t|$ and r_t^2 for $w = 0.5$, $\beta = 0.1$, $a = 0.8$, $g = 1$, $b = 1$, $C = 0$, two different values of $v = 0.3$ and 0.9 and without, with the noisy supply $\sigma_n = 0, 1$.

price of the underlying deterministic model, partially due to the mean reverting expectation scheme of the fundamentalists.

(ii) The returns are not normally distributed as indicated by the significant ACs in all cases. However, including noise traders makes the returns even less normal, leading to skewness and high kurtosis.

(iii) The ACs of returns show a strong under-reaction pattern, characterised by the oscillating ACs with positive ACs for small lags, as the parameters move closer to the Hopf bifurcation (the cases $(v, \sigma_n) = (0.3, 0)$ and $(0.9, 0)$). Such strong patterns are be washed out by introducing the noise traders (the case $(v, \sigma_n) = (0.3, 1)$ and $(0.9, 1)$).

(iv) The noise traders make a significant contribution to the volatility clustering effect characterised by the significant ACs of the $|r_t|$ and r_t^2 in cases of $(v, \sigma_n) = (0.3, 1)$ and $(0.9, 1)$.

(v) The distribution of the population difference m_t shows that the investors tend to switch toward the fundamental (trend following) strategy as the parameters move away from (closer to) the Hopf bifurcation boundary, though the volatility of m_t is increased by the introduction of the noise traders.

Overall, we can see that the Hopf bifurcation can be used to explain the under-reaction of AC patterns. Also the stylised facts, including skewness, kurtosis, volatility clustering and significant ACs of $|r_t|$ and r_t^2, are observed when the noise traders act as liquidity traders.

3.2 Case 2

We now consider the case when the chartists are contrarians and the weight parameter w is small. In this case, the flip bifurcation boundary becomes non-binding. It is found from simulations that, different from the previous case, prices do not become explosive on both sides of the Hopf bifurcation. An intuitive explanation is the fact that both fundamental and contrarian strategies are stabilizing strategies. This may also be due to the unbinding flip bifurcation. To see how the market price behavior near the Hopf bifurcation, we choose $v = 0.5, w = 0.3$ and $g = -1$ and -3, which are located on both sides of the Hopf bifurcation value $g^* = -2.65$ of the underlying deterministic system.

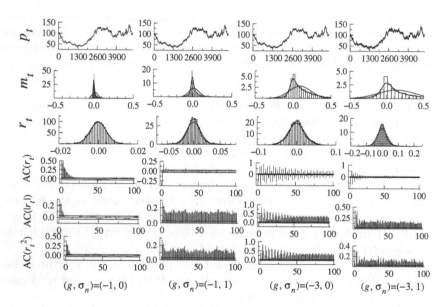

Fig. 3. Time series of the market price p_t, distribution density of the population fraction difference m_t and return r_t (compared with the normal distribution) and ACs of $r_t, |r_t|$ and r_t^2 for $w = 0.3, \beta = 0.1, a = 0.8, v = 0.5, b = 1, C = 0$, two different values of $g = -1$ and -3 and without, with the noise supply $\sigma_n = 0, 1$.

Fig. 3 illustrates the time series of the market price p_t, the densities of the market fraction m_t and the return r_t, and the ACs of $r_t, |r_t|$ and r_t^2 for four combinations of $(g, \sigma_n) = (-1, 0), (-1, 1), (-3, 0)$ and $(-3, 1)$. Similar statistical properties are found when the parameters are located inside the local stability region, but near the Hopf bifurcation boundary, of the underlying deterministic system (that is the cases $(g, \sigma_n) = (-1, 0)$ and $(-1, 1)$). However, when the parameters are located outside the local stability region, the steady state price of the deterministic system becomes unstable through a Hopf bifurcation and the ACs of returns of the stochastic model display over-reaction patterns characterised by oscillating and decaying ACs which are negative for odd lags and positive for even lags, though the patterns are washed out significantly by the presence of the noise traders.

4 Flip Bifurcations and Under/Over-Reaction AC Patterns

This section examines the impact of the flip bifurcation (of the deterministic system) on the stochastic properties of the market price. As we know that the fundamental steady state price of the underlying deterministic system becomes unstable through a flip bifurcation when the chartists are contrarians and the weighting parameter w is higher (in this case the Hopf bifurcation of the underlying deterministic system becomes unbinding). Because of the stablizing effect of both fundamental and contrarian strategies, prices become less explosive on both sides of the flip bifurcation boundary (but more sensitive to the switching intensity β). To see the market price behavior near the flip bifurcation, we choose $v = 0.5, w = 0.8$ and $g = -1$ and -3, which are located on both sides of the flip bifurcation value $g^* = -2.18773$ of the underlying deterministic system.

Fig. 4 illustrates the time series of the market price p_t, the densities of the market fraction m_t and the return r_t, and the ACs of $r_t, |r_t|$ and r_t^2 for four combinations of $(g, \sigma_n) = (-1, 0), (-1, 1), (-3, 0)$ and $(-3, 1)$. Similar statistical properties are found when the parameters are located inside the local stability region, but near the flip bifurcation boundary, of the underlying deterministic system (that is the cases $(g, \sigma_n) = (-1, 0)$ and $(-1, 1)$). In particular, the under-reaction AC pattern is observed when there are no noise traders. However, for $g = -3$, the steady state price of the underlying deterministic system becomes unstable through a flip bifurcation and both the stochastic fundamental price and the noise traders have different impacts on the stochastic properties of the market price which are summarized as follows:

(vi) Both the population fraction difference m_t and the return r_t have bimodal distributions when there are no noise traders present. Correspondingly, the AC displays a very strong over-reaction pattern characterised by slowly decaying oscillating $AC(r_t)$ which are negative for odd lags and

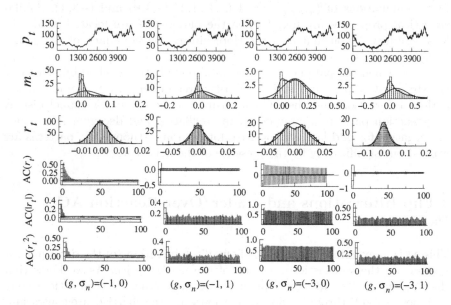

Fig. 4. Time series of the market price p_t, distribution density of the population fraction difference m_t and return r_t (compared with the normal distribution) and ACs of r_t, $|r_t|$ and r_t^2 for $w = 0.8, \beta = 0.1, a = 0.8, v = 0.5, b = 1, C = 0$, two different values of $g = -1$ and -3 and without, with the noisy supply $\sigma_n = 0, 1$.

positive for even lags. This strong AC pattern is also preserved for the ACs of $|r_t|$ and r_t^2. All these features are underpinned by the flip bifurcation of the underlying deterministic system.

(vii) When the noise traders are present, most of the statistical features in (vi) disappear. The bi-modal distributions of m_t and r_t disappear, showing skewness and high kurtosis. In addition, the strong over-reaction AC patterns on returns are washed out and, apart from the first two lags, become insignificant. However, the ACs of both $|r_t|$ and r_t^2 are significant.

Overall, we can see that, when the parameters cross the flip bifurcation of the deterministic system, the ACs pattern of the stochastic returns changes from under-reaction to over-reaction and this is also consistent with trader behaviour. When the noise traders are present, the model generates most of the stylised facts that we discuss in the following section.

5 The Stylised Facts and Long Memory

The above analysis has led us to our first goal—to gain some insights into various types of market price behaviour and to understand the possible mechanism generating such behaviour. We have shown that various statistical aspects of the stochastic system, including distributions and under/over-reaction AC patterns are closely related to the stability and bifurcation analysis of the deterministic system.

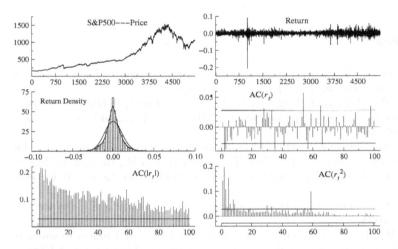

Fig. 5. Time series of the prices and (relative) returns of S&P500 from 18/11/1983 to 19/11/2003, distribution density of the return (compared with the normal distribution) and ACs of returns, absolute returns and squared returns for lags up to 100.

We now try to address our second goal, which is the most important and difficult part of asset price modeling, to replicate the econometric properties and the stylized facts of financial time series. As a benchmark of such financial time series, we consider the daily accumulated index of the S&P500 from 18/11/1983 to 19/11/2003. Fig. 5 illustrates the price and (relative) return time series for the last 20 years, the return distribution and the ACs of returns, absolute returns and squared returns for lags up to 100. For a comprehensive discussion of stylized facts and the so-called long memory of returns of high frequency financial time series, we refer to Ding et al. [18] and Pagan [32]. The stylised facts include excess volatility, volatility clustering, skewness and excess kurtosis, while the long memory refers to the insignificant ACs on returns, but significant and decayed ACs on the absolute and squared returns, as illustrated by Fig. 5 for the S&P500 daily index. Given the simplicity of our model, it is not easy to reproduce all those features. However, based on our analysis in the previous sections, one can see that the model is indeed able to produce most of those features when the noise traders act as liquidity traders. In particular, for the cases when the chartists are contrarians and the parameters are near either the Hopf or flip bifurcation (but from inside the local stability region) of the underlying deterministic system $(w, g, \sigma_n) = (0.3, -1, 1), (0.8, -1, 1)$, the return distributions show skewness and excess kurtosis, the distributions of the population fraction difference show some herding behaviour, while the ACs of returns display similar patterns to the S&P500, apart from significant ACs of returns for the first few lags and persistent ACs of the absolute and squared returns.

References

1. Anderson, S. and A. de Palma, J. Thisse (1993) Discrete Choice Theory of Product Differentiation, MIT Press, Cambridge, MA
2. Arthur, W., Holland, J., LeBaron, B., Palmer, R., Tayler, P. (1997) Asset pricing under endogeneous expectations in an artifical stock market Economic Notes 26(2):297–330
3. Brock, W., Hommes, C. (1997) Models of Complexity in Economics and Finance. chapter 1, pp. 3–44 In: Heij, C., Schumacher, J.M., Hanzon, B., Praagman, C. (eds) Systems Dynamics in Economic and Finance Models, Wiley, New York
4. Brock, W.,Hommes, C. (1997) A rational route to randomness, Econometrica 65:1059–1095
5. Brock, W., Hommes, C. (1998) Heterogeneous beliefs and routes to chaos in a simple asset pricing model, Journal of Economic Dynamics and Control 22:1235–1274
6. Brock, W., LeBaron, B. (1996) A structural model for stock return volatility and trading volume, Review of Economics and Statistics 78:94–110
7. Bullard, J., Duffy, J. (1999) Using Genetic Algorithms to Model the Evolution of Heterogeneous Beliefs, Computational Economics 13:41–60
8. Chen, S.-H., Yeh, C.-H. (1997) Toward a computable approach to the efficient market hypothesis: an application of genetic programming, Journal of Economic Dynamics and Control 21:1043–1063
9. Chen, S.-H., Yeh, C.-H. (2002) On the emergent properties of artificial stock markets: the efficient market hypothesis and the rational expectations hypothesis, Journal of Economic Behavior and Organization 49:217–239
10. Chiarella, C. (1992) The dynamics of speculative behaviour, Annals of Operations Research 37:101–123
11. Chiarella, C., Dieci, R., Gardini, L. (2002) Speculative behaviour and complex asset price dynamics, Journal of Economic Behavior and Organization 49:173–197
12. Chiarella, C., He, X. (2001) Asset pricing and wealth dynamics under heterogeneous expectations, Quantitative Finance 1:509–526

13. Chiarella, C., He, X. (2002) Heterogeneous beliefs, risk and learning in a simple asset pricing model, Computational Economics 19:95–132
14. Chiarella, C., He, X. (2003) Heterogeneous beliefs, risk and learning in a simple asset pricing model with a market maker, Macroeconomic Dynamics 7:503–536
15. Chiarella, C., He, X., Wang, D. (2004) Asset price dynamics with time-varying second moment, Technical report, University of Technology, Sydney
16. Dacorogna, M., Muller, U., Jost, C., Pictet, O., Olsen, R., Ward, J. (1995) Heterogeneous real-time trading strategies in the foreign exchange market, European Journal of Finance 1:383–403
17. Day, R., Huang, W. (1990) Bulls, bears and market sheep, Journal of Economic Behavior and Organization 14:299–329
18. Ding, Z., Granger, C., Engle, R. (1993) A long memory property of stock market returns and a new model, Journal of Empirical Finance 1:83–106
19. Farmer, J. D., Joshi, S. (2002) The price dynamics of common trading strategies, Journal of Economic Behavior and Organization 49:149–171
20. Gaunersdorfer, A. (2000) Endogenous fluctuations in a simple asset pricing model with heterogeneous agents, Journal of Economic Dynamics and Control 24:799–831
21. Gaunersdorfer, A., Hommes, C., Wagener, F. (2003) Bifurcation routes to volatility clustering under evolutionary learning, Technical Report 03-03, CeN-DEF, University of Amsterdam, Working Paper
22. He, X. (2003) Asset pricing, volatility and market behaviour-a market fraction approach, Technical Report 95, Quantitative Finance Research Center, University of Techonology, Sydney
23. Hommes, C. (2001) Financial markets as nonlinear adaptive evolutionary systems, Quantitative Finance 1:149–167
24. Hommes, C. (2002) Modeling the stylized facts in finance through simple nonlinear adaptive systems, Proceedings of National Academy of Science of the United States of America 99:7221–7228
25. Kirman, A. (1992) Whom or what does the representative agent represent?, Journal of Economic Perspectives 6:117–136
26. LeBaron, B., Arthur, W., Palmer, R. (1999) Time series properties of an artifical stock market, Journal of Economic Dynamics and Control 23:1487–1516
27. Lux, T. (1995) Herd behaviour, bubbles and crashes, Economic Journal 105:881–896
28. Lux, T. (1997) Time variation of second moments from a noise trader/infection model, Journal of Economic Dynamics and Control 22:1–38
29. Lux, T. (1998) The socio-economic dynamics of speculative markets: Interacting agents, chaos, and the fat tails of return distributions, Journal of Economic Behavior and Organization 33:143–165
30. Lux, T., Marchesi, M. (1999) Scaling and criticality in a stochastic multi-agent model of a financial market, Nature 397(11):498–500
31. Manski, C., McFadden, D. (1981) Structural Analysis of Discrete Data with Econometric Applications, MIT Press
32. Pagan, A. (1996) The econometrics of financial markets, Journal of Empirical Finance 3:15–102

Deflationary Recessions in a General Equilibrium Framework

Luca Colombo[1] and Gerd Weinrich[2]

[1] Università Cattolica, Istituto di Economia e Finanza, Largo Gemelli, I-20123 Milano. Tel. +39-02-7234.2637, fax +39-02-7234.2781
lucava.colombo@unicatt.it
[2] Università Cattolica, Largo Gemelli 1, I-20123 Milano. Tel. +39-02-7234.2728, fax +39-02-7234.2671 gerd.weinrich@unicatt.it.

Summary. This paper investigates the role of fiscal and monetary shocks in the occurrence of deflationary recessions. Our model is based on a temporary equilibrium approach with stochastic rationing, where inventory dynamics is explicitly taken into account, amplifying spillover effects between markets. This setting allows us to study the driving forces behind disequilibrium phenomena, and to investigate the efficacy of alternative policies in overcoming them. In particular, we provide for an application of our approach to the study of the Japanese deflationary recession.

JEL Classification: D45, D50, E32, E37.

Key words: Inventories, non-tâtonnement, price adjustment, deflationary recessions

1 Introduction

The motivation for this paper has come from the observation that current mainstream economic theory has difficulties to convincingly explain deflationary recessions in a general-equilibrium framework. Since in a recession many factors of production are underemployed, their prices should decrease. In a standard general-equilibrium model these decreases would be assumed to occur quickly. However, the recent experience of the Japanese economy as well as that of the Great Depression indicate that a deflationary recession may last over a long time. In fact, recession and deflation are a problem in Japan since the mid-1990s.

To analyze deflationary recessions in a general-equilibrium framework we choose a non-tâtonnement approach and build a macroeconomic model with temporary equilibrium with rationing in each period and price adjustment from period to period. This makes it possible to study the dynamic functioning of an economy in which disequilibrium phenomena like underemployment, deflation and excess productive capacities/inventories are allowed to occur. In addition we put special emphasis on the occurrence of inventories of produced goods because they reinforce the importance of spillover effects between markets and therefore favour the persistence of aggregate-demand deficiencies.

We find that restrictive fiscal and monetary policies can cause a permanent recession. Contrary to conventional wisdom, this is favored by sufficient downward flexibility of the nominal wage. Vice versa, to best avoid or overcome long-term unemployment, wages should be rigid downwards. We further obtain that the stationary Walrasian equilibrium is unstable whereas there exist recessionary Keynesian underemployment equilibria which are quasi-stationary (i.e. with real variables constant but nominal variables deflating) and locally stable. Finally, by using our framework to analyze the recent deflationary Japanese recession, we argue that a mix of fiscal and monetary measures should help in overcoming it.

2 The Model

There are overlapping–generations consumers, producers and a government. They exchange between them a composite consumption good, labor and money. Time is divided in periods of equal length and prices are fixed at the beginning of each period. At these prices trades take place, and the fact that prices are typically not Walrasian implies that there is some rationing; in particular there is the possibility of involuntary unemployment. To obtain a feasible allocation we employ the concept of temporary equilibrium with quantity rationing. Once goods have been exchanged, the economy moves to the next period and new prices are quoted. At these prices a new allocation obtains and so on.

The adjustment of prices at the beginning of a new period is linked to the size of imbalance on the various markets that materialized in the equilibrium with rationing of the previous period. In each period t these imbalances will be measured by the ratios of aggregate demands and supplies

$$\gamma_t = \frac{Y_t^d}{Y_t^s} \, , \, \lambda_t = \frac{L_t^d}{L_t^s}$$

on the goods and labor market, respectively.

Consumers and firms hold expectations γ_t^e and λ_t^e of these values. In an equilibrium with rationing these expectations will be correct. More precisely, consumers are subject to stochastic rationing in case they are on the long side of a market.[3] That rationing will be such that the expected transaction

---

[3] Stochastic rationing - as opposed to deterministic rationing - is compatible with manipulability of the rationing mechanism and therefore provides an incentive for rationed agents to express demands that exceed their expected trades, as argued by Green (1980), Svensson (1980), Douglas Gale (1979, 1981) and Weinrich (1982,

is proportional to a consumer's effective demand (or supply), with the factors of proportionality on each market being the rationing coefficients γ_t and λ_t. Given also the price p_t of the good, the nominal wage, w_t, and previous period's net aggregate profits, $(1 - tax)\Pi_{t-1}$, that are distributed to young consumers, the aggregate effective demand of young consumers, after resolving an optimization problem that takes account of stochastic rationing, is a function[4]

$$X_t = X(\gamma_t^e, \lambda_t^e; p_t, w_t, (1 - tax)\Pi_{t-1}).$$

To this we have to add old consumers' aggregate demand which is just M_t/p_t - the real value of the money stock held entirely by the old consumers - and G which denotes real government spending. Thus, setting $\alpha_t = w_t/p_t$, $m_t = M_t/p_t$ and $\pi_t = \Pi_{t-1}/p_t$, the aggregate demand on the goods market is

$$Y_t^d = X^d \left(\gamma_t^e, \lambda_t^e; \alpha_t, (1 - tax)\, \pi_t\right) + m_t + G \equiv Y^d(\gamma_t^e, \lambda_t^e; \alpha_t, m_t, (1 - tax)\, \pi_t, G).$$

Aggregate supply of labor is fixed and denoted by L^s.

Producers derive their effective labor demands $\ell_t^d = \ell^d(\gamma_t^e; \alpha_t)$ and production decisions $y_t^p = f(\ell_t^d)$ from maximizing expected profits subject to a stochastic rationing mechanism, i.e.

$$\max_\ell [\min\{\gamma_t^e, 1\}\, (f(\ell) + s_t) - \alpha_t \ell]$$

where s_t is the amount of stored goods carried over from the previous period. In case of demand rationing on the labor market, only the share λ_t of firms' aggregate labor demand will be realized. Taking account of this fact, aggregating over all firms yields thus an aggregate labor demand and an aggregate goods supply function

$$L^d \left(\gamma_t^e; \alpha_t\right)\ ,\ \ Y^s \left(\gamma_t^e, \lambda_t^e; \alpha_t, S_t\right) = Y^p \left(\gamma_t^e, \lambda_t^e; \alpha_t\right) + S_t \tag{1}$$

where S_t denotes aggregate inventories held by firms at the beginning of period t.[5]

3 Equilibrium

Given $\alpha_t, m_t, (1 - tax)\,\pi_t, S_t$ and G, a *temporary equilibrium with rationing* is given by a fixpoint $(\gamma_t^*, \lambda_t^*)$ of the mapping

---

1984, 1988). This renders possible to have imbalances of aggregate demand and supply in an equilibrium with rationing and use the size of these imbalances for a determination of price adjustment. For a definition of manipulability see for example Böhm (1989) or Weinrich (1988).

[4] For the formal derivation of this function see Colombo and Weinrich (2003).

[5] For the precise derivation of these functions see Colombo and Weinrich (2003).

$$\chi : (\gamma_t, \lambda_t) \mapsto (\frac{Y^d(\gamma_t, \lambda_t; \alpha_t, m_t, (1 - tax)\, \pi_t, G)}{Y^s\,(\gamma_t, \lambda_t; \alpha_t, S_t)}, \frac{L^d\,(\gamma_t; \alpha_t)}{L^s})$$

and by expectations $(\gamma_t^e, \lambda_t^e) = (\gamma_t^*, \lambda_t^*)$. In that case all agents have perfect foresight of the rationing coefficients which they reproduce by their optimal actions. Thus no agent has an incentive to change something. In Colombo and Weinrich (2003) we show that such an equilibrium exists and is unique, i.e. the mapping

$$(\alpha_t, m_t, (1 - tax)\, \pi_t, S_t, G) \mapsto (\gamma_t^*, \lambda_t^*)$$

is well defined. This yields an aggregate allocation

$$\overline{Y}_t = \min\left\{Y^d(\gamma_t^*, \lambda_t^*; \alpha_t, m_t, (1 - tax)\, \pi_t, G), Y^s\,(\gamma_t^*, \lambda_t^*; \alpha_t, S_t)\right\}$$
$$\equiv \mathcal{Y}\,(\alpha_t, \pi_t, m_t, S_t, G, tax)$$

$$\overline{L}_t = \min\left\{L^d\,(\gamma_t; \alpha_t), L^s\right\} \equiv \mathcal{L}\,(\alpha_t, \pi_t, m_t, S_t, G, tax).$$

According to the whether γ_t^* and λ_t^* are larger or smaller than one there are different types of equilibrium. More precisely we can summarize the various possibilities as follows:

γ_t^*	λ_t^*	*Type*	*T*	S_t	*Unemployment*
< 1	< 1	*Keynesian unemployment*	*K*	+	+
> 1	> 1	*Repressed inflation*	*I*	0	0
> 1	< 1	*Classical unemployment*	*C*	0	+
< 1	> 1	*Underconsumption*	*U*	+	0

Moreover there are intermediate cases the most important of which is the Walrasian equilibrium which corresponds to $\gamma_t^* = \lambda_t^* = 1$.

4 Dynamics and Numerical Analysis

The model discussed above gives rise to a four dimensional non-linear dynamic system, with state variables α_t, m_t, π_t and S_t, which given by the sequence $\{(\alpha_t, m_t, \pi_t, S_t)\}_{t=1}^{\infty}$ where

$$\alpha_{t+1} = \frac{w_{t+1}/w_t}{p_{t+1}/p_t}\alpha_t,$$

$$m_{t+1} = \frac{1}{p_{t+1}/p_t}\left[m_t + G + (1 - tax)\, \pi_t\right] - \pi_{t+1},$$

$$\pi_{t+1} = \frac{\Pi_t/p_t}{p_{t+1}/p_t} = \frac{\mathcal{Y}\,(\alpha_t, \pi_t, m_t, S_t) - \alpha_t \mathcal{L}\,(\alpha_t, \pi_t, m_t, S_t)}{p_{t+1}/p_t},$$

$$S_{t+1} = Y^s\,(\lambda_t^*, \gamma_t^*) - \mathcal{Y}\,(\alpha_t, \pi_t, m_t, S_t) + S_t.$$

Prices and wages are assumed to rise (fall) whenever an excess of demand (supply) is observed, i.e.

$$\frac{p_{t+1}}{p_t} = \begin{cases} 1 - \mu_1 \left(1 - \gamma_t^*\right) \text{ if } \gamma_t^* \leq 1 \\ 1 - \mu_2 \left(1 - \gamma_t^*\right) \text{ if } \gamma_t^* \geq 1 \end{cases} ; \quad \mu_1, \ \mu_2 \ \in [0, 1]$$

and

$$\frac{w_{t+1}}{w_t} = \begin{cases} 1 - \nu_1 \left(1 - \lambda_t^*\right) \text{ if } \lambda_t^* \leq 1 \\ 1 - \nu_2 \left(1 - \lambda_t^*\right) \text{ if } \lambda_t^* \geq 1 \end{cases} ; \quad \nu_1, \ \nu_2 \ \in [0, 1],$$

where μ_1 and μ_2 (ν_1 and ν_2) denote downward and upward price (wage) flexibility, respectively.

As there are four nondegenerate equilibrium regimes, the dynamic system can be viewed as the union of four subsystems each of which may become effective through endogenous regime switching. Being of such a complex nature, we cannot be study it by means of analytical tools only. Therefore we resort to numerical simulations to investigate the dynamic behavior of the model.[6] Note that, under suitable parameter specifications (see Colombo and Weinrich, 2003), as a benchmark a stationary Walrasian equilibrium is obtained for

$$\alpha^* = 0.85, \ m^* = 46.25, \ \pi^* = 15, \ S^* = 0, \ G^* = 7.5, \ tax^* = 0.5, \qquad (2)$$

with trading levels $L^* = Y^* = 100$. Furthermore, in the following we set $\mu_1 = \mu_2 = \nu_2 = 0.1$, whereas ν_1 is varied between 0 and 0.1.

4.1 Fiscal and Monetary Shocks

We first consider a change in public expenditure G. Starting from the Walrasian equilibrium, $(\alpha_0, m_0, \pi_0, S_0) = (\alpha^*, m^*, m^*, S^*)$, the bifurcation diagram in Figure 1 shows, for $\nu_1 = 0$, the stationary values of employment to which the system converges in dependence of values of G between 0 and 15. From this it is evident that $\overline{L} < L^*$ for $G < G^*$ and $\overline{L} = L^*$ for $G = G^*$. When $G < G^*$, aggregate demand Y^d is diminished which creates an excess supply on the goods market. Consequently firms reduce their production and cut back on employment. The result is an excess supply on the labor market, too, and the economy enters in a state of Keynesian unemployment. The imbalance on the goods market gives rise to a price decrease whereas on the labor market the nominal wage cannot decrease as $\nu_1 = 0$. As a result the real wage increases. This is illustrated in Figure 2 which shows the time series for employment L, inventories S, the real money stock m and the real wage α for the first 200 periods where $G = 7$. The real wage is rising until approximately period 30 when it has become large enough so that the system enters into a regime of Classical unemployment. Here the goods price decreases and

---

[6] See Colombo and Weinrich (2003) for a more detailed analysis characterizing the dynamic system associated to each regime.

the real wage falls until at around period 50 it settles at a stationary value $\overline{\alpha} > \alpha^*$. Since the nominal wage rate does not change, the constant real wage implies that the goods price does not change either beyond period 50, and the economy has reached a stationary state at the frontier between Keynesian and Classical unemployment. In that state there is market clearing on the goods market but excess supply on the labor market.

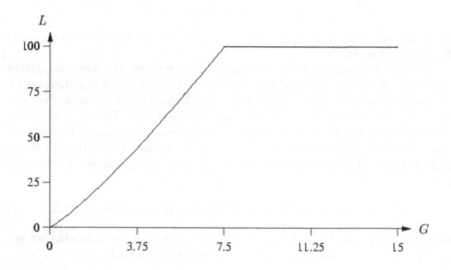

Fig. 1. Stationary employment values when $\nu_1 = 0$.

Next consider, again for $G < G^*$, what happens when $\nu_1 > 0$. The charts in Figure 3 show, analogously to Figure 1, stationary values of employment for various values of G. The two charts in the top row refer to downward wage flexibilities of $\nu_1 = 0.025$ and $\nu_1 = 0.1$, respectively. The striking result is that a little downward wage flexibility has an enormous effect on the impact of fiscal restraint as is documented by the discontinuity of the graphs at $G = G^*$. Note that this does not happen in the model without inventories, as is shown by the lower chart in Figure 3 where $\nu_1 = 0.1$ but S_t is exogenously set to zero at the beginning of each period.

Why inventories have such a dramatic effect is easily explained. When aggregate demand is diminished due to a decrease in G, inventories become positive and rise further as excess supply on the goods market builds up. Since S_t influences Y^s positively (see Equation (1)) and $\gamma_t = \overline{Y}_t / Y^s (\lambda_t, \gamma_t; \alpha_t, S_t)$ by definition, an increase in S_t reduces γ_t which diminishes the labor demand of firms, and thus increases further the excess supply on the labor market. Therefore, the downward flexible wage rate decreases more than would be the case without inventories. If the decrease in the wage rate is larger than

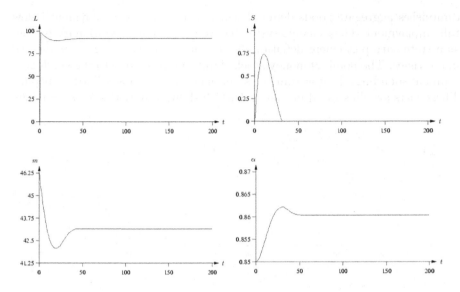

Fig. 2. Time series when $\nu_1 = 0$ and $G = 7$.

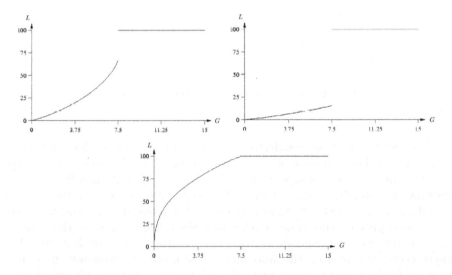

Fig. 3. Stationary employment values for $\nu_1 = 0.025$, $\nu_1 = 0.1$ (top charts), and $\nu_1 = 0.1$ and $S_t \equiv 0$.

the decrease in the goods price, the real wage decreases, and it may continue to decrease permanently approaching a limit level below the Walrasian real wage. The lower real wage diminishes labor income of workers which diminishes aggregate goods demand which in turn keeps employment below full employment. The dynamic system converges to a quasi-stationary Keynesian state with permanent deflation of all nominal variables but constant real magnitudes. The nominal money stock shrinks because, due to the small government spending, the government is permanently realizing a budget surplus. These facts are illustrated in Figure 4 which shows time series for $\nu_1 = 0.025$.

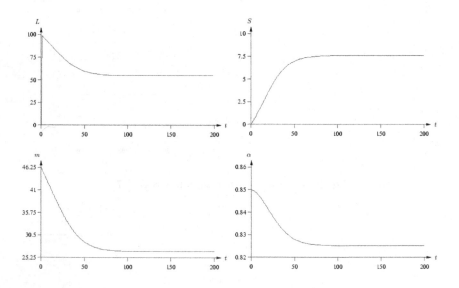

Fig. 4. Time series when $\nu_1 = 0.025$ and $G = 7$.

It is easy to show that restrictive monetary shocks can have the same effects as those associated to fiscal shocks investigated above. Consider for instance a reduction in the initial money stock from m^* to $m_0 = 40$, keeping all other parameters and initial values at their Walrasian levels. Since m_0 is the demand of old agents at time $t = 0$, aggregate demand is reduced. Consequently there is excess supply on the goods market and, since firms adjust to the reduced transaction level on the goods market, they reduce their labor demand. Thus there is excess supply on the labor market, too, and the economy enters in a state of Keynesian unemployment. What happens next depends on whether the nominal wage is flexible downwards.[7]

As shown in Figure 5, where $\nu_1 = 0.025$, when downward wage flexibility is allowed, employment, real wage and real money all converge to values lower than the respective Walrasian values. The reason is similar to that already

---

[7] With nominal wage rigid downwards, a restrictive money shock has a temporary but not lasting effect on economic activity, as it is shown in Colombo and Weinrich (2003)].

discussed in the context of fiscal shocks: the presence of inventories increases the fall of labor demand by firms which in turn depresses labor income and aggregate demand. The system tends to a quasi-stationary Keynesian state with permanent deflation of nominal variables. Note that, in our framework, money is not neutral in the long run: the restrictive monetary shock has caused a permanent decrease in employment and output.[8]

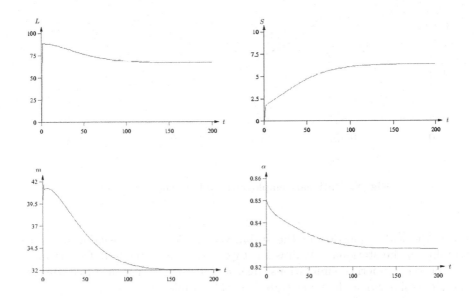

Fig. 5. Time series when $\nu_1 = 0.025$ and $m_0 = 40$.

At this point the natural question is which downward wage flexibility is needed to drive the economy into a permanent recession or even depression. The answer is given in the bifurcation diagram of Figure 6. From that it can be seen that approximately until $\nu_1 = 0.02$ the economy is capable of returning to the full employment after the monetary shock, whereas for speeds of wage adjustment larger than this the economy gets trapped in underemployment.

5 Stability

The fact that a restrictive monetary shock may lead to a quasi-stationary Keynesian unemployment state as limit of the dynamic system's trajectory

---

[8] As for fiscal shocks, this is not the case when $S_t \equiv 0$. In fact, the real wage decreases initially but then the decrease in the goods price dominates that in the nominal wage, and the real wage moves back to its Walrasian level, as do all the other variables.

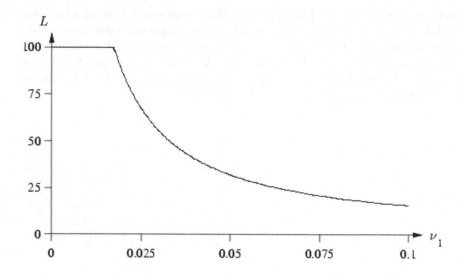

Fig. 6. Stationary employment values when $m_0 = 40$.

raises the wider question of the stability of such a state. Analogously, the stability of the stationary Walrasian state may be investigated. To anticipate the answer, numerical simulations suggest that the latter is unstable whereas the Keynesian unemployment state is locally stable.

Let us first look at the stationary Keynesian state. The limit values of the state variables of the simulation shown in Figure 5, where we have an initial reduction of the money stock to $m_0 = 40$ at a downward wage flexibility of $\nu_1 = 0.025$ and policy parameters $G^* = 7.5$ and $tax^* = 0.5$, are (approximately)

$$\overline{\alpha} = 0.8281, \overline{m} = 31.9263, \overline{\pi} = 15.7889, \overline{S} = 6.4060, \tag{3}$$

with a stationary employment level $\overline{L} = 66.9342$. We now take these stationary values as new initial values, i.e. set $(\alpha_0, m_0, \pi_0, S_0) = (\overline{\alpha}, \overline{m}, \overline{\pi}, \overline{S})$, and perform a bifurcation analysis with respect to each of the state variables around these initial values and dependent variable the employment level $\overline{L}$. The results are shown in Figure 7. The diagrams show that local deviations in directions of any of the state variables do not change the fact that the system has this state as an attracting long-run rest point. Although this is not a strict proof of local stability, it is highly indicative for the behavior of the dynamic system.

Moreover, in order to return from the stationary recessionary state to permanent full employment by changing *only* the money stock, an increase in that variable to its Walrasian value (46.25) is not sufficient, but a higher increase, to approximately 51, would be needed. The case is still worse for a

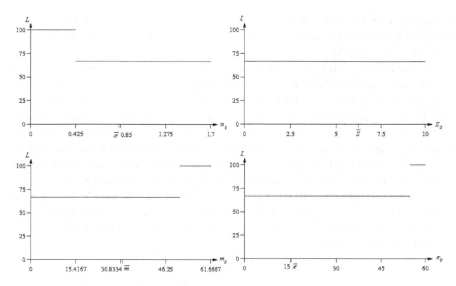

Fig. 7. Local stability of the Keynesian unemployment state.

unilateral variation of real profits which would have to be more than tripled, to around 52, to get permanently out of unemployment. In both cases the reason is intuitively clear: only a substantial increase in purchasing power of consumers, old or young, can succeed to change the situation of insufficient aggregate demand causing Keynesian unemployment. A detailed analysis of the respective time series shows that, after an initial shock to $m_0 > 51$ or $\pi_0 > 52$, the system enters in the regime of repressed inflation from where it converges to the Walrasian equilibrium.

The panel of Figure 7 showing the bifurcation diagram over inventories indicates that no change in the initial value of S is by itself sufficient to overcome the recessionary state. Even a momentaneous decrease of inventories to zero is not capable to lead the dynamic system out of the Keynesian unemployment regime because the insufficient demand very quickly builds up inventories again and restores the old situation.

A less immediate intuition is available for the effects of a change in the initial value of the real wage α. An increase of its value from the stationary state value typically results in a state of classical unemployment, because the high real wage not only creates excess demand on the goods market but also excess supply on the labor market. This implies a decrease in the nominal wage, an increase in the goods price and a decrease in the real wage. This process continues until the real wage is low enough again to make the system enter into the Keynesian regime. There inventories are built up and the system converges to the quasi-stationary Keynesian state. On the other hand, a decrease of α_0 from $\bar{\alpha}$ typically leads the dynamic system to enter into the regime of repressed inflation. From there it may return to the Keynesian regime (for values of α above 0.425) or converge to the Walrasian state (for α smaller than 0.425).

By performing bifurcation analysis analogous to the ones above, one can also show that the Walrasian stationary equilibrium $(\alpha^*, m^*, \pi^*, S^*)$ is locally unstable in three of the four directions defined by the state variables, namely m, π and S. Only in direction α is the system locally stable. Whenever $m_0 < m^*$, $\pi_0 < \pi^*$ or $S_0 > S^*$, the dynamic system diverges from the Walrasian state and converges to the quasi-stationary Keynesian unemployment state considered above.[9] Note that the above results are obtained for specific parameter values. One particularly significant parameter is the downward wage flexibility ν_1. Here this value has been set to 0.025, but similar results hold whenever ν_1 is above the benchmark value seen in Figure 6 (approximately 0.018) that separates the stationary values of employment from full to below full employment.

6 An Application to the Japanese Deflationary Recession

As shown in the previous sections, our framework permits to describe and to discuss the outcome of alternative economic policies for an economy facing prolonged deflationary and recessionary situations. The behavior of the Japanese economy since the Nineties, having been stuck in a quasi-stationary Keynesian unemployment state for over a decade, provides for a clear example and allows for natural experiments on the outcome of alternative economic policies.

In particular, mimicking a proposal to overcome the Japanese recession by Ben Bernanke (see *The Economist*, June 21st 2003, p.74), we focus here on a balanced-budget tax reduction combined with a simultaneous increase in the money stock. Provided it is of a large enough magnitude this policy proves indeed a remedy to recession in the long run. Consider, for instance, a fiscal-monetary policy combination calling for a reduction in the tax rate (from $tax^* = 0.5$ to $tax = tax^* - \Delta tax$), accompanied by a decrease in G by $\Delta G = \Delta tax \cdot \pi^*$ (from $G^* = tax^* \cdot \pi^*$ to $G = tax \cdot \pi^* = G^* - \Delta G$) and an increase in m_0 by $\Delta m = \Delta G$. Starting from the quasi-stationary Keynesian state illustrated in (3) and setting government spending more precisely to $G = tax \cdot \pi^*$, with $\pi^* = 15$ the Walrasian value of real profits so that the government's budget is balanced at any Walrasian equilibrium, one can show that a decrease in the tax rate monotonically increase the long-run value of employment, allowing the economy to reach the full employment stationary Walrasian equilibrium.[10]

Note that there are two important ingredients of the policy outlined above. On the one hand, there is a balanced-budget fiscal measure in the form of a

---

[9] See Colombo and Weinrich (2003) for the bifurcation diagrams.
[10] See Colombo and Weinrich (2003) for a more detailed analysis.

reduction in the tax rate, that should reduce both output and employment. This textbook economics implication, however, impinges on the fact that the balanced-budget multiplier refers to a static situation with prices and wages fixed, whereas here we have a dynamic analysis with flexible prices, wages and money stock. In our case, the fall in the employment is just a short-run phenomenon. Nonetheless, it is an all but welcome occurrence.[11] To avoid this, a simultaneous increase in the money stock m_0 can be used. Specifically, set $m_0 = \overline{m} + G^* - G$, together with $(\alpha_0, \pi_0, S_0) = (\overline{\alpha}, \overline{\pi}, \overline{S})$. If $(G, tax) = (0, 0)$, then $m_0 = 31.9263 + 7.5 = 39.4263$ and $\overline{L}_1 = 73.8248$. Similarly, for $(G, tax) = (2.55, 0.17)$, $m_0 = 36.8763$ and $\overline{L}_1 = 71.4806$. In both cases subsequent employment values increase monotonically to full employment. This shows that the combined measure of tax reduction and expansive monetary policy works well in our model economy.

7 Concluding Remarks

Our framework has described consistent allocations in every period, even when prices are not at their market clearing levels, obeying at the same time a well defined dynamics. This has rendered possible to study, in a general equilibrium framework, the dynamic behavior of an economy in which disequilibrium phenomena are allowed to occur. In particular, we have shown how, starting from a Walrasian equilibrium, the economy may end up in a permanent recession (i.e. in a locally stable deflationary quasi-stationary state with unemployment) following fiscal or monetary shocks. Being able to characterize the driving forces behind disequilibrium phenomena, our setting has allowed us to investigate the performance of alternative policy measures aimed at solving them. In this respect, we have provided for an application to the Japanese deflationary recessions, testing the efficacy of a mix of fiscal and monetary policies.

References

1. Böhm, V., 1989. *Disequilibrium and Macroeconomics*, New York, Basil Blackwell.

---

[11] To better understand what is going on, consider the two extreme cases of $(G, tax) = (0, 0)$ and $(G, tax) = (15, 1)$. Looking at the respective time series, after one iteration we have $\overline{L}_1 = 60.7438$ in the first case and $\overline{L}_1 = 73.1357$ in the second one. Comparing with the initial level $\overline{L} = 66.9342$ (stationary at $(G, tax) = (G^*, tax^*) = (7.5, 0.5))$ these values fit quite well with the textbook prediction. Subsequently $\overline{L}_{t+1} > \overline{L}_t$ in the case $(G, tax) = (0, 0)$ and $\overline{L}_{t+1} < \overline{L}_t$ in the case $(G, tax) = (15, 1)$. The first time employment is larger in the first case than in the second is in period 8 when $\overline{L}_8 = 65.9933$ in case $(G, tax) = (0, 0)$ and $\overline{L}_8 = 65.9735$ in case $(G, tax) = (15, 1)$.

2. Colombo L. and G. Weinrich, 2003. "Unemployment and Inventories in the Business Cycle", http://www.unicatt.it/docenti/weinrich/invent.pdf.
3. Gale, Douglas, 1979. "Large Economies with Trading Uncertainty", *The Review of Economic Studies* 46, 319-338. ¡
4. Gale, Douglas, 1981. "Large Economies with Trading Uncertainty: a Correction", *The Review of Economic Studies* 48, 363-364.
5. Green, J., 1980. "On the Theory of Effective Demand", *Economic Journal* 90, 341-353.
6. Svensson, L.E.O., "1980. Effective Demand and Stochastic Rationing", *The Review of Economic Studies* 47, 339-355.
7. Weinrich, G., 1982. "On the Theory of Effective Demand", *Economic Journal* 92, 174-175.
8. Weinrich, G., 1984. "On the Theory of Effective Demand under Stochastic Rationing", *Journal of Economic Theory* 34, 95-115.
9. Weinrich, G., 1988. "On the Foundations of Stochastic
10. Non-Price Rationing and the Adjustment of prices", *Rivista di matematica per le scienze economiche e sociali*, 107-131.

Concepts of Thermodynamics in Economic Growth

Jürgen Mimkes

Physics Department, University of Paderborn, Germany mimkes@physik.upb.de

Summary. The Solow-Swan model of economic growth is reviewed on the basis of natural production. Natural growth is a biochemical process based on the laws of thermodynamics. Economic production - like work in thermodynamics - is a non exact differential. The production function $Y(a, b)$ as a function of laborers (a) and (b) depends on the path of integration. The production function may be calculated for the special processes like constant mean capital per labor (T), (which corresponds to the Carnot process in thermodynamics): $Y(a, b) = L\{ak + bl + T(\ln\{a^a b^b\})\}$. The elasticity coefficients or exponents a, b with $a + b = 1$ are determined by the production factors! The production function $Y(a, b)$ has been applied to optimizing production processes in farming and leads to a Boltzmann distribution of production factors. The main source of economic growth is entropy, the chance of diversification, the variety of know how and ideas. The results lead to a new model of economic growth for interdependent economic systems like Japan and the US, East and West Germany, North and South America, and agrees well with data for these economies.

Key words: production function, laws of thermodynamics, economic growth, entropy of mixing. *Classification codes:* C1, E2, I5.

1 Introduction

In the last ten years new interdisciplinary approaches to economics have developed in social and natural science, especially by W. Weidlich 1972, D. K. Foley 1994, J. Mimkes 1995, H. G. Stanley 1999, Y. Aruka 2001, and others. In the present paper the mechanism of economic production is discussed on the basis of natural production.

Why should concepts of thermodynamics be applied to economic growth? The appropriate mathematics field for (infinitesimal) growth is calculus of differential forms. Calculus in one dimension is generally part of mathematical courses in economics. With two production factors, calculus in two dimensions is the proper mathematical tool for economic growth, and differs from calculus in one dimension: In one dimension there are only exact differentials that can be integrated independently of the path of integration, in two dimensions there are also non exact differential that depend on the path of integration. This is familiar in natural sciences, but, unfortunately, not in general to students of social and economic sciences. Thermodynamics is the prominent application of calculus in two dimensions in natural science. Accordingly, we may expect the concepts of thermodynamics to be valuable to economic growth.

The neoclassical growth theory is based on the Solow-Swan model (Barro,1995). The production function $Y = f(K, L)$ is determined by the production factors capital (K) and labor (L). The difference of production (Y) and consumption (C) leads to savings, (S). In economic sciences the equilibrium of production is generally calculated from an exact differential form:

$$dY = \frac{\partial f}{\partial K}dK + \frac{\partial f}{\partial L}dL = 0. \tag{1}$$

The integral of dY in Eq. (1) leads to the production function Y: Solutions for Y have been given by Gossen (2a) and Cobb Douglas (2b):

$$Y = C\ln(L^a K^b) \tag{2a}$$
$$Y = AL^a K^b \tag{2b}$$

with the elasticity constants $a + b + c = 1$.

However, many questions are left open in the Solow Swan model, e. g.:

1. Does a general production function Y(K, L) in two dimensions always exist?
2. How are the elasticity constants defined?
3. Are production (Y) and consumption (C) related, $Y = Y(C)$ or $C = C(Y)$?

Agriculture like many other production processes relies on closed production cycles. But, if the production of one cycle - like one year - was given by the closed exact integral of Eq.(1), this would result in zero output:

$$\oint dY = \oint (\frac{\partial f}{\partial K}dK + \frac{\partial f}{\partial L}dL) = 0 \tag{3}$$

The integral of an exact differential form (dY) depends on the borders of the integral, only. A closed cyclic process starts and ends at the same point and the corresponding integral, Eq. (3), will be zero. This indicates that cyclic economic production may indeed only be represented by a non exact differential form.

2 Non Exact Differential Forms

We now look into the calculus of non exact differential forms (Kaplan, 2003). The integral depends on the border and on the path of integration. A simple example of a non exact form is given by:

$$\oint \delta W = \oint \{a(K,L)dK + b(L,K)dL\} \neq 0 \qquad (4a)$$

Non exact differential forms are marked by "δ" and the non exact differential form δW of production, Eq.(4a) replaces Eqs.(1) and (3). The closed integral of non exact forms will generally not be zero, the output of Eq.(4a) will depend on the path of integration or the production process. Splitting the closed path of Eq.(4a) into two parts, one (δW) from A to B and the other ($\delta W'$) from B to A, we obtain

$$\oint \delta W = \oint_A^B \delta W + \oint_B^A \delta W' = \oint_A^B \delta W - \oint_A^B \delta W' = Y - C = S \neq 0 \quad (4b)$$

If we call the first path Y and the returning path C, the output of cyclic production (4b) leads to non zero profit, which may be invested or saved (S). The concept of non exact differential forms leads to the proper relation of neoclassical theory, $Y - C = S$.

3 Law of Economic Production

According to the laws of calculus (Kaplan, 2003) a non exact form (δW) may be turned into an exact form ($d\sigma$) by an integrating factor (T). Two equivalent non exact forms δW and $-Td\sigma$ may differ by an exact form (dK), which will vanish at closed integration:

$$\delta W - dK - T d\sigma. \qquad (5)$$

In thermodynamics this is called the (first and second) law of work (W). K is called energy, T is called mean energy per particle or temperature, σ is the entropy.

In economics we may call Eq.(5) the law of economic production, K is the capital, T is the mean capital per person, a mean price level or the standard of living. σ is again the entropy, which is related to probability (P),

$$\sigma = \ln P \qquad (6)$$

Probability is determined by combinatorial calculations, as will be discussed, below. Entropy is not restricted to natural science, but is a function of mathematical statistics and has already been introduced to economics by Roegen (Georgescu-Roegen, 1974). Eq.(5) is the answer to the first question of the introduction: Does a general production function exist? The answer is: No! The non exact differential form of production δW may not be integrated in general, there is no general production function (W). But for a given production process, for a specific path of integration the production function (W) may be calculated. This will be discussed in the next section.

4 Production at Constant Price Level T

For one season a number of skilled laborers (L) are working on a farm at constant wages and prices. At constant price level $T = T_0$ we may integrate δW in Eq.(5) from 0 to Y to obtain the income of one season:

$$Y = K - T_0 \sigma \tag{7}$$

L_1 people are specialized in raising cattle, L_2 people are specialized in producing grain, L_3 people are skilled in repairs. During a season each group of laborers produces a certain amount of capital (K), which may differ for each group due to different number of laborers (L_i) and different productivity (k_i) in each section:

$$K = \sum L_i k_i = L \sum (L_i/L) k_i \tag{8}$$

The exact number of possibilities (P) to put (L) people to work in (i) different areas of production is given by the number of combinations,

$$P = \frac{L!}{L_1! L_2! ... L_n!}. \tag{9}$$

Entropy (σ) is given by

$$\sigma = \ln P = \ln \frac{L!}{L_1! L_2! ... L_n!} = -L \sum \frac{L_n}{L} \ln(\frac{L_n}{L}) \tag{10}$$

(The calculation has been performed by applying Stirling's formula for $L!$). The ratios L_i/L in Eqs.(8) and (10) may be renamed by $x_i = L_i/L$ with

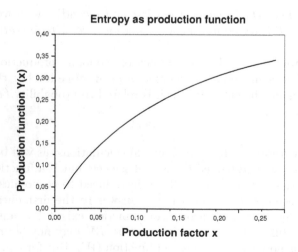

Fig. 1. The production function $Y(x) = L\{kx + Tx \ln x\}$ according to Eq.(11).

$\sum x_i = 1$. The production output of a farm with L laborers after one season at constant prices and wages, Eq.(7) now is:

$$Y(K, L) = L\{\sum x_i k_i + T \sum x_i \ln x_i\}. \tag{11}$$

The production function (11) is plotted in fig. 1, a growing production factor x leads to saturation of productivity.

The results are very close to Gossen's equation (2a) and his law of saturation. Apparently the Gossen equation corresponds to the entropy of production at constant T, with only one small modification, which also answers the second question in the introduction: the constants of elasticity, a, b, c are given by the production factors x_i.

5 Optimization of Production at Constant Price Level T

Production may be optimized by maximizing the production function (11):
1. How many laborers are needed in each field of production? This question may be answered by the deriving Eq.(11) by x:

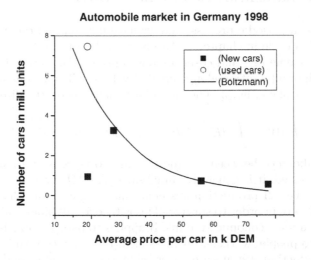

Fig. 2. Distribution of cars x_i in four price classes (i) sold in Germany 1998, data points according to Statistisches Bundesamt [2000]. Only the complete market (new and used cars) follows a Boltzmann distribution.

$$\frac{\partial Y}{\partial x_i} = L\{k_1 + T(\ln x_i + 1)\} = 0 \tag{12a}$$

$$\ln x_i + 1 = -k_1/T \tag{12b}$$

$$x_i = Ae^{\frac{-k_i}{T}} \tag{12c}$$

The optimal number of laborers (12c) is a Boltzmann distribution and depends exponentially on the productivity k_i of the group (i) in relation to the standard of living (T).

2. The Boltzmann distribution has been investigated in a similar problem: How many cars of four different price categories (P_i) have been sold in Germany in 1998 at constant standard of living T? The answer is again a Boltzmann distribution,

$$x_i = Ae^{\frac{-P_i}{T}} \tag{12d}$$

Fig. 2 shows the automobile market in Germany in 1998. The distribution of sold cars in the four price categories according to the official data follows a Boltzmann distribution. Similar distributions have been found for wealth [Willis, 2004], however, some authors have proposed other distributions [Levy, 1996].

6 The Mechanism of Economic Growth

Agriculture is a periodic process, and we have not finished the calculation of the production cycle in chapter 4. According to Eq.(4b) only the closed cycle will lead to income (Y) and costs or consumption (C). Carnot has proposed an ideal path of integration along constant values of T and constant values of σ in the $T - \sigma$ plane, fig. 3. The sign of δW depends on the direction of the cycle,

$$\oint \delta W = \oint (dK - Td\sigma) = \Delta T \Delta \sigma = \Delta Q = Y - C \tag{13a}$$

ΔQ is the difference between income (Y) and costs/consumption (C). If the profit ΔQ is saved, it is generally called savings, $(\Delta Q = S)$.

A farm - like all production systems - has to gain more than it spends, $Y - C = \Delta Q > 0$, in order to survive. This is true for all biological, thermodynamic, human and economic systems. It applies to plants, animals, to electric generators, to people, workers, business men, to farms, companies, countries. All these production systems are governed by the first and second law of thermodynamics (5) and the Carnot process. In the Carnot process - at constant T - we may separate the two functions Y and C. The production function Y of the Carnot process has already been calculated in Eq.(11), consumption C is obtained, accordingly,

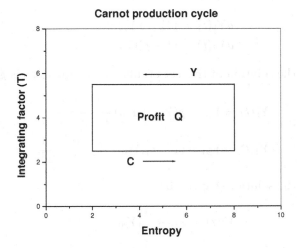

Fig. 3. The Carnot process of a farm in a $T - \sigma$ diagram. A farm is a generator of capital and operates like a generator of (electrical) energy. Y: The farmer collects a high amount of energy/capital (Y) like grain, cattle etc. from the laborers. C: The farmer pays a lower amount of energy/capital (C) for food, costs, wages etc. to the laborers. The profit is $\Delta Q = Y - C$. The efficiency (r) of the system is given by the ratio of profit ΔQ to invested capital $K, r = \Delta Q / K$.

$$Y(K, L) = K + T_2 \Delta \sigma = L\{\sum x_1 k_i + T_2 \sum x_i \ln x_i\} \qquad (13b)$$

$$C(K, L) - K + T_1 \Delta \sigma - L\{\sum x_1 k_i + T_i \sum x_i \ln x_i\} \qquad (13c)$$

The functions Y and C have been calculated for a given process or path of integration. This answers the third question stated in the introduction: Y and C are both defined only by the specific production process.

7 Economic Growth of Interdependent Systems

Farms, companies, business firms contain at least two groups of people: farmer and laborers, owner and workers, capital and labor, first and third world, Y and C. Both groups together form the economic system. Accordingly, both groups will discuss, how to divide the profit $\Delta Q = Y - C$ of each cycle. This is negotiated periodically by workers and employers, by unions and industry, by world trade conferences.

As a result of the negotiations the worker will obtain the percentage "p" of the profit, the employer will take the fraction $(1 - p)$. If both groups reinvest their shares $p(Y_2 - Y_1)$ and $(1 - p)(Y_2 - Y_1)$, they will grow in time (t) and we obtain a set of two equations, $C = Y_1(t)$ and $Y = Y_2(t)$:

$$dY_1(t) = p(Y_2 - Y_1)dt \tag{14a}$$
$$dY_2(t) = (1 - p)(Y_2 - Y_1)dt \tag{14b}$$

For $p \neq \frac{1}{2}$ the solution of this set of differential equations is given by:

$$Y_1(t) = Y_{10} + p(Y_{20} - Y_{10})\frac{e^{pt} - 1}{1 - 2p} \tag{15a}$$

$$Y_2(t) = Y_{20} + (1 - p)(Y_{20} - Y_{10})\frac{e^{pt} - 1}{1 - 2p} \tag{15b}$$

For $p = \frac{1}{2}$ the solution is given by

$$Y_1(t) = Y_{10} + \frac{1}{2}(Y_{20} - Y_{10})t \tag{16a}$$

$$Y_2(t) = Y_{20} + \frac{1}{2}(Y_{20} - Y_{10})t \tag{16b}$$

8 Results: Economic Growth of Interdependent Countries

The equations (15) and (16) may be applied to all interdependent systems, to workers and employers, unions and industry, or interdependent countries. The results are compared to data of interdependent countries in figs. 4 to 11.

8.1 Growing Economies ($0 < p \leq 0,5$)

$p = 0$, fig. 4: If all profit goes to the richer party, the income (Y_2) of group (2) will grow exponentially, the income of the first party stays constant, (Y_{10}). The efficiency of the system, $r = (Y_2 - Y_1)/K$ grows with time.

$p = 0,10$, fig. 4: at 10% of the profit for the poorer party (Y_1) and 90% for the rich party (Y_2) both parties grow exponentially. Examples are Japan and Germany after World War II, both economies were depending on the US and were growing exponentially, as indicated in fig. 5.

$p = 0,25$, fig. 4: at 75% for the rich party Y_2 still grows exponentially, but less than with 90% of the profit. Y_1 gets 25% of the profit, but as it is linked to a less growing Y_2 the poorer party Y_1 is growing less for longer times than at 10% of the profit.

This indicates clearly, that a high rise in wages will weaken the economy and lead to lower wages in the long run. Low increase of wages will lead to more exponential growth for industry and workers. Workers as well as their managers will have to be more patient with pay raises, like in Germany or Japan after world war II, fig. 5. The same is now observed in the rising economies in East Asia.

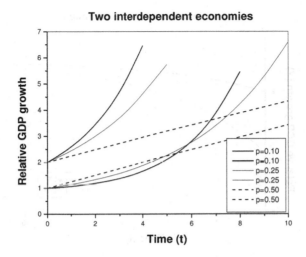

Fig. 4. The development of income of two interdependent economic systems starting at $Y_1 = 1$ and $Y_2 = 2$. The profit for the poor side varies from $p = 0, 10$ to $p = 0, 50$. In the long run the income of workers (Y_1) will grow faster with lower pay raise p !!

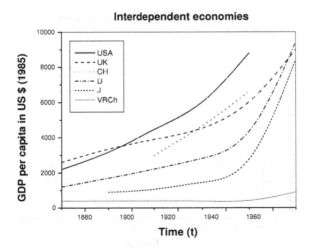

Fig. 5. Economic growth of US, UK, Switzerland, Japan, Germany, China between 1870 and 1990 (after Barro and iMartin, 1995). The victorious allies USA and UK have grown exponentially. Japan and Germany only started to grow exponentially after World War II by international trade at low wages. China was excluded and did not take part in economic growth, at that time.

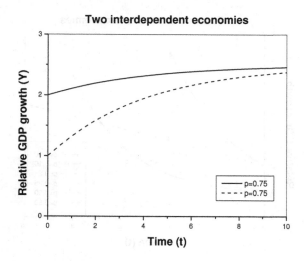

Fig. 6. The development of income of two interdependent economic systems stating at $Y_1 = 1$ and $Y_2 = 2$. At high values of profit for the poor side, $p = 0,75$, economic growth is reaching a boundary with time.

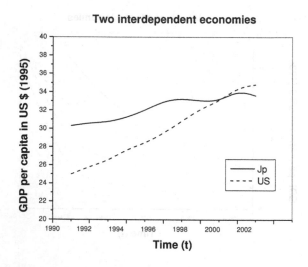

Fig. 7. The development of mean income (GDP / person) of the USA and Japan in quarters between 1980 and 2000. The efficiency of trade between the two countries is declining with time [World Factbook, USA, 2004].

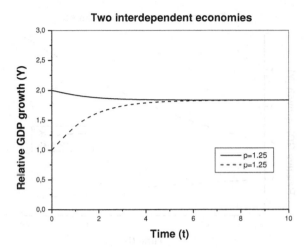

Fig. 8. The development of standard of living of two interdependent economic systems stating at $Y_1 = 1$ and $Y_2 = 2$. At high values of p for the poor side, $p > 1$ both economies will converge below $Y_2 = 2$.

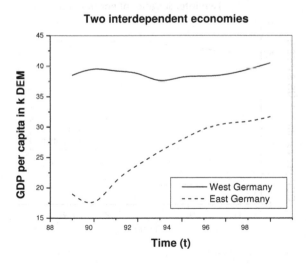

Fig. 9. Mean income in West and Germany between 1989 and 1998 due to productivity and capital transfer. In 1998 East Germany reached about 80% of the mean income in West Germany (Fründ 2002).

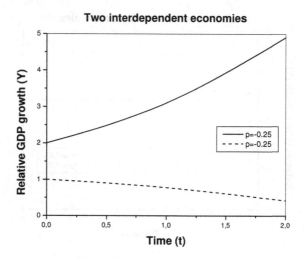

Fig. 10. The development of mean income of two interdependent economic systems stating at $Y_1 = 1$ and $Y_2 = 2$. $p = -0,25$ means the poor side is exploited. The income of the dominating party grows exponentially, the exploited party falters.

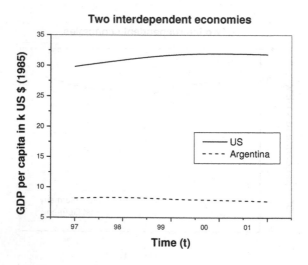

Fig. 11. The development of mean income of USA and Argentina between 1997 and 2001. Due to the large amounts of capital transferred to the US, the mean income in Argentina was reduced, considerably.[World Factbook, USA, 2004 : Ministerio de Economía, Argentina, 2004].

$p = 0, 50$; fig.4: An even split between the two parties seems to be a fair deal. But this "fair" deal leads only to a linear growth of both parties. The efficiency of the system, $r = (Y_2 - Y_1)/K$ stays constant with time.

8.2 Stagnation: USA - Japan ($0, 5 < p \leq 1$)

$p = 0, 75$; fig. 6: A high factor $p > 0, 5$ leads to decreasing efficiency of the system, $r = (Y_2 - Y_1)/K$ declines with time. Y_2 and Y_1 are reaching a boundary. (Y_1) is trailing a decreasing (Y_2). After Japan and Germany have acquired many production plants, the factor p has grown and the efficiency of the exports to the US started to decrease. The slowly growing economic levels (Y_2) of Japan and USA (Y_1) are trailing each other with low efficiency, fig. 7.

$p = 1, 00$; If all profit goes to the poor side, the income of the poor party soon reaches the constant income of the rich party, the efficiency of the system is zero!

8.3 Converging Economies, West and East Germany ($p > 1$)

$p = 1, 25$; fig. 8: If more than 100% of the profit goes to the poor party, (Y_2) will decrease, and (Y_1) will catch up with (Y_2). This has been observed in the relationship of West and East Germany after reunion in 1990, fig. 9.

8.4 Exploitation ($p < 0$)

$p = -0, 25$; fig. 10: If the poor side (Y_1) makes only losses, $p < 0$, it and will go bankrupt. The richer party (Y_2) will grow exponentially.

This has been observed in Argentina in 2000, when many rich people transferred there assets to the banks in the US, see data in fig. 11. In the last 500 years many European countries have exploited their colonies. The European countries have grown exponentially, the colonies faltered.

References

1. ARUKA, Y, ED.: Evolutionary Controversies in Economics. Springer, Tokyo 2001.
2. BARRO R, AND SALA I MARTIN, X,: Economic Growth. McGraw-Hill (1995).
3. CIA WORLD FACTBOOK: USA, 2004.
4. FOLEY, D. K.: A Statistical Equilibrium Theory of Markets. Journal of Economic Theory Vol. 62, No. 2, April 1994.
5. FOWLER, R. AND GUGGENHEIM, E. A.: Statistical Thermodynamics. Cambridge University Press (1960).
6. FRÜND, TH.: Diploma Thesis. Physics Department, Paderborn 2002.
7. GEORGESCU-ROEGEN, N.: The entropy law and the economic process. Cambridge, Mass. Harvard Univ. Press, 1974.

152   Jürgen Mimkes

8. STATISTISCHES BUNDESAMT: Statistisches Jahrbuch für das Vereinte Deutschland. Wiesbaden (1991) and (2000).
9. KAPLAN, W.: Advanced Calculus. Addison-Wesley (2003).
10. LEVY, M. AND SOLOMON, S.: Power Laws are Logarithmic Boltzmann Laws. International Journal of Modern Physics C , Vol. 7, No. 4 (1996) 595.
11. MIMKES, J.: Society as a many-particle System. J. Thermal Analysis 60 (2000) 1055 - 1069.
12. MIMKES, J.: Binary Alloys as a Model for Multicultural Society . J. Thermal Anal. 43 (1995) 521-537.
13. MINISTERIO DE ECONOMÍA Y OBRAS Y SERVICIOS PÚBLICOS, SECRETARÍA DE PROGRAMACIÓN ECONÓMICA, ARGENTINA: 2004.
14. Report Deutsches Institut für Wirtschaftsforschung, DIW, Berlin 1993.
15. STATISTISCHES BUNDESAMT: Statistisches Jahrbuch 2000. Wiesbaden 2000.
16. STANLEY, H. E., AMARALA, L. A. N., CANNING, D., GOPIKRISHNANA, P., LEEA Y. AND LIUA, Y.: Econophysics: Can physicists contribute to the science of economics?. Physica A, 269 (1999) 156 - 169.
17. WEIDLICH, W.: Sociodynamics. Amsterdam : Harwood Acad. Publ., 2000.
18. WEIDLICH, W.: The use of statistical models in sociology. Collective Phenomena 1, (1972) 51.
19. WILLIS, G., MIMKES, J.: Economics Working Paper Archive. WUSTL, 2004.

Firm Dynamics Simulation Using Game-theoretic Stochastic Agents

Yuichi Ikeda[1], Osamu Kubo[2], and Yasuhiro Kobayashi[3]

[1] Author to whom correspondence should be made. Hitachi Research Laboratory, Hitachi Ltd. ,Omika 7-1-1, Hitachi, Ibaraki 319-1292, Japan
yikeda@gm.hrl.hitachi.co.jp
[2] Hitachi Research Laboratory, Hitachi Ltd. ,Omika 7-1-1, Hitachi, Ibaraki 319-1292, Japan okubo@gm.hrl.hitachi.co.jp
[3] Hitachi Research Laboratory, Hitachi Ltd. ,Omika 7-1-1, Hitachi, Ibaraki 319-1292, Japan kobayash@gm.hrl.hitachi.co.jp

Summary. Decision-making is a crucial task for the business planning of industrial firms, in order to cope with uncertainties in the business environment. A method of firm dynamics simulation, i.e. the game-theoretic stochastic agent, was developed by applying game theory to a stochastic agent model in order to analyze the uncertain business environment. Each stochastic agent is described by a Langevin-type equation with an additional term for rational decision-making. In this paper, the dynamics of firms in computer related industries, which consist of three industrial sectors, i.e. the large scale integrated circuit sector, the personal computer sector, and the liquid crystal display sector, are simulated using the game-theoretic stochastic agents model. Then, the importance of the herding behavior of firms is demonstrated to reproduce the formation and collapse of the bubble in the Japanese computer related industry markets during the late 90s.

Key words: [83.10.Pp] Particle dynamics, [02.50.Le] Decision theory and game theory, [05.10.Gg] Stochastic analysis methods (Fokker-Planck, Langevin, etc.), [05.10.Ln] Monte Carlo methods, [05.20.-y] Classical statistical mechanics, [05.40.Jc] Brownian motion

1 Introduction

Decision-making is a crucial task for the business planning of industrial firms, in order to cope with uncertainties in the business environment. The valuation result using the net-present value for a given business plan is significantly changed by three factors, i.e. discount rate, profit volatility, and business plan feasibility [1]. Business planning and demand forecasting can seldom be estimated reliably and it is almost hopeless to tackle the problem by standard economics and financial engineering. In a previous paper [2] applying game theory to a stochastic agent model develops a method of firm dynamics simulation, i.e. the game-theoretic stochastic agent. It was shown that the model can be reduced to the real option valuation method using approximations. A decision-making process with an enhancement option and an abandonment option was investigated using the real option version of the model. In this paper, the dynamics of firms in computer related industries, which consist of three industrial sectors, i.e. the large scale integrated circuit (LSI) sector, the personal computer (PC) sector, and the liquid crystal display (LCD) sector, are simulated using the game-theoretic stochastic agent. In Section 2, the model of the game-theoretic stochastic agent is described. Each stochastic agent is described by a Langevin-type equation [3] [4] [5] with an additional term for rational decision-making [6] and an interaction term between agents. A transition probability for each vertex, which corresponds to a Nash equilibrium, is evaluated using backwards induction, after calculating the payoff for each strategy without volatility. In Section 3, the procedure of parameter calibration is described in detail. In Section 4, the herding behavior of firms is investigated and its characteristic features are discussed. The importance of the herding behavior of firms is demonstrated to reproduce the formation and collapse of the bubble in the computer related industry market in Japan during the late 90s.

2 Game-theoretic Stochastic Agent Model

A group of firm agents, in which each agent is characterized by revenue $R_i(t)(i = 1, \cdots, N)$, is discussed using non-equilibrium statistical mechanics. The revenue $R_i(t)$ for the i-th agent is assumed to be described by the Langevin-type equation,

$$\frac{dR_i}{dt} = \sum_k D_i w_{ik} \delta(t - t_k) - \gamma_i R_i - \frac{\partial U}{\partial R_i} + \sigma_i \xi_i(t) + \eta(t) R_{c(i)}, \qquad (1)$$

where γ_i, U, σ_i, and ξ_i are a friction coefficient, an interaction energy, volatility, and the Gaussian white noise, respectively. The rest of economic system is considered by the stochastic term . The agent described by Eq. (1) is called the game-theoretic stochastic agent.

Each agent behaves rationally to maximize its payoff PV_i. Rational decision-making of the i-th agent is made by the first term of the RHS of Eq. (1). $D_i w_{ik} = \pm D_i (k = 1, \cdots, K)$ is the planned revenue of the i-th agent. D_i is the mean absolute difference of revenue, $D_i = |R_i(t) - R_i(t - \Delta t)|$. $w_{ik} = w_{ik}(PV_i)$ is a transition probability at a decision-making point in time k, and depends on the payoff PV_i of the i-th agent. The strategy u of the i-th agent is $u_i = (w_{i1}, \cdots, w_{ik}, \cdots, w_{iK})$. The total number of strategies summed over all agents is equal to 2^{NK}. A payoff PV_i is the summed discounted cash flow over $l = t/\Delta t$,

$$PV_i = (1 - T) \sum_l \{R_i(l\Delta t) - C_i((l\Delta t)\}/(1 + r)^l, \tag{2}$$

where T is a tax rate, and the cost is assumed to be linear and quadratic proportional to the revenue,

$$C_i(t) = \alpha R_i(t) + \beta R_i(t)^2. \tag{3}$$

The second term of the RHS of Eq. (3) corresponds to the fact that larger firms are less efficient. A transition probability w_{ik}, which corresponds to a Nash equilibrium, is evaluated using backward induction after calculating the payoff $(PV_1, \cdots, PV_i, \cdots, PV_N)$ for each strategy without volatility.

The third term of the RHS of Eq. (1) represents an interaction acting on the i-th agent, which is proportional to the difference of the revenue from the other agents, $\frac{\partial U}{\partial R_i} = \sum_{j \neq i}^N F_{i-j}(R_i - R_j - \overline{R_i - R_j})$, where $\overline{R_i - R_j}$ is the average difference of revenue between the i-th agent and the j-th agent.

The last term of the RHS of Eq. (1) represents the irrational herding behavior [6]. $\eta(t)$ and $c(i)$ are the time-varying strength of the herding behavior and the competitor of the i-th agent, respectively. The time-varying strength of the herding behavior $\eta(t)$ is assumed to be externally given in this model.

3 Parameter Calibration

Parameters are calibrated in order to reproduce the market data of the computer related industries in Japan as a Nash equilibrium. The market data of the computer related industries in Japan are shown in Fig. 1. Revenue for the LSI, PC, and LCD sectors, shown in Fig. 1 (a), were divided into periods I to III, according to the stock price of Yahoo Japan, shown in Fig.1 (b). Here, Yahoo Japan was chosen as a representative of Japanese information technology firms, though stock prices showed a similar pattern for many Japanese industrial firms from 1999 to 2000. Period I which starts at Q3-97 and ends at Q4-98, is named the normal period, because the stock price of Yahoo Japan is relatively stable. Period II, which starts at Q1-99 and ends at Q1-00, is named the bubble period. The stock price of Yahoo Japan rose sharply during 1999 and felt significantly in early 2000. Period III, which starts at Q2-00 and ends at Q2-01, is named the post-bubble period.

It is regarded that the market data during Period I was formed as a consequence of rational decision-making of firms. Calibration of the parameters was made using the market data during Period I. The time-varying strength of the herding behavior was assumed to be small enough to neglect, i.e. for Period I. Rational behavior of each agent makes the trend of the market . Thus, the

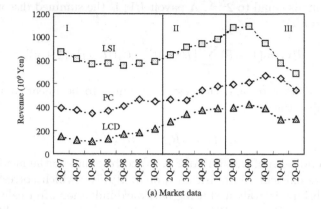

(a) Market data

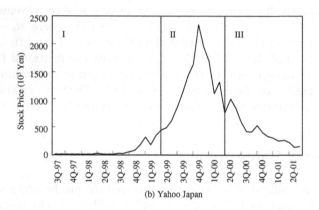

(b) Yahoo Japan

Fig. 1. The market data of the computer related industries in Japan is shown. Market data for the LSI, PC, and LCD sectors, shown in Fig. 1 (a), were divided into Periods I to III, according to the stock price of Yahoo Japan, shown in Fig.1 (b). Here, Yahoo Japan was chosen as a representative of Japanese information technology firms, though stock prices showed similar price changes for many Japanese industrial firms from 1999 to 2000.

first term of the RHS of Eq.(1) is replaced by the trend of the market μ_i for Period I,

$$\sum_k D_i w_{ik} \delta(t - t_k) \to \mu_i. \qquad (4)$$

Eq. (1) was applied to the parameter calibration for the three agents system,

$$R_{PC}(t + \Delta t) - R_{PC}(t) = \xi_{PC} - \gamma_{PC} R_{PC}(t)$$
$$+ F_{PC-LSI}\{R_{PC}(t) - R_{LSI}(t)\} + F_{PC-LCD}\{R_{PC}(t) - R_{LCD}(t)\}, \tag{5}$$

$$R_{LSI}(t + \Delta t) - R_{LSI}(t) = \xi_{LSI} - \gamma_{LSI} R_{LSI}(t)$$
$$+ F_{LSI-PC}\{R_{LSI}(t) - R_{PC}(t)\} + F_{LSI-LCD}\{R_{LSI}(t) - R_{LCD}(t)\}, and \tag{6}$$

$$R_{LCD}(t + \Delta t) - R_{LCD}(t) = \xi_{LCD} - \gamma_{LCD} R_{LCD}(t)$$
$$+ F_{LCD-PC}\{R_{LCD}(t) - R_{PC}(t)\} + F_{LCD-LSI}\{R_{LCD}(t) - R_{LSI}(t)\}, \tag{7}$$

where the first term of the RHS of Eq. (5) to (7) includes the trend of the market μ_i and other terms,

$$\xi_i = \mu_i + \sum_{j \neq i}^{N} F_{i-j} \overline{R_i - R_j}. \tag{8}$$

With the following constraints on the interaction parameters ,

$$F_{PC-LSI} = F_{LSI-PC}, \tag{9}$$

$$F_{PC-LCD} = F_{LCD-PC}, and \tag{10}$$

$$F_{LSI-LCD} = F_{LCD-LSI}, \tag{11}$$

the parameter calibration was made for the quarterly market data of the period I by the least square fitting, and the results obtained are summarized in Table 1.

Table 1. Model Parameters

i	ξ_i	γ_i	F_{i-j}	F_{i-j}	σ_i	D_i
PC	199243	0.145198	-0.00294063	-0.546054	27357.7	34894.4
			$(j = \text{LSI})$	$(j = \text{LCD})$		
LSI	207888	-0.0207392	-0.00294063	-0.345892	22701	59239.6
			$(j = \text{PC})$	$(j = \text{LCD})$		
LCD	-437974	-0.67084	-0.546054	-0.345892	33007.2	31380.0
			$(j = \text{PC})$	$(j = \text{LSI})$		

The parameters α and β in Eq. (3) were determined in order to reproduce the market data as a Nash equilibrium,

$$\chi^2 = \sum_l \left\{ \left(\frac{R_{PC}(l\Delta t) - R_{PC}^{(m)}(l\Delta t)}{\overline{R}_{PC}} \right)^2 + \right.$$

$$\left. \left(\frac{R_{LSI}(l\Delta t) - R_{LSI}^{(m)}(l\Delta t)}{\overline{R}_{LSI}} \right)^2 + \left(\frac{R_{LCD}(l\Delta t) - R_{LCD}^{(m)}(l\Delta t)}{\overline{R}_{LCD}} \right)^2 \right\},$$

$$(12)$$

where $R_i^{(m)}(l\Delta t)$ was calculated revenue using the game-theoretic stochastic agent model. The dependencies of χ^2 on the parameters α and β are shown in Fig. 2. Fitted curves are $\chi^2 = 1 \times 10^{15}\beta^2 - 9 \times 10^7\beta + 2.52(R^2 = 0.7189)$ and $\chi^2 = 1178.5\alpha^2 - 2241.4\alpha + 1066.1(R^2 = 0.9998)$ for Fig.2 (a) and (b), respectively. The optimal value for the parameters were chosen to be $\alpha = 0.96$ and $\beta = 4 \times 10^{-8}$.

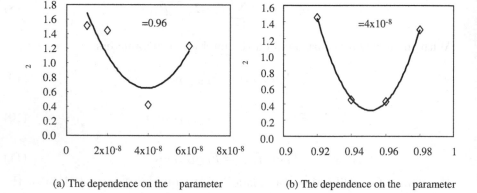

(a) The dependence on the   parameter   (b) The dependence on the   parameter

Fig. 2. The dependencies of χ^2 on the parameters α and β are show. The parameters α and β in Eq. (3) were determined in order to reproduce the market data as a Nash equilibrium. Fitted curves are $\chi^2 = 1 \times 10^{15}\beta^2 - 9 \times 10^7\beta + 2.52(R^2 = 0.7189)$ and $\chi^2 = 1178.5\alpha^2 - 2241.4\alpha + 1066.1(R^2 = 0.9998)$ for Fig.2 (a) and (b), respectively. The optimal value for the parameters were chosen to be $\alpha = 0.96$ and $\beta = 4 \times 10^{-8}$.

Revenue for Period I was calculated using Eq. (5) to (7) with the obtained parameters and $\eta_i(t) = 0$. The first terms of RHS of Eq. (5) to (7) were replaced to incorporate rational decision-making,

$$\xi_i \rightarrow \xi_i - D_i/2 + \sum_k D_i w_{ik} \delta(t - t_k), \tag{13}$$

where the factor 2 of the second term of RHS is due to the absolute value in the definition of D_i. The initial values of revenue were taken at 3Q-97. A transition probability w_{ik} was evaluated using backward induction after calculating the payoff $(PV_1, \cdots, PV_i, \cdots, PV_N)$ for each strategy without volatility. The calculated revenue for Period I is shown in Fig. 3. The agreement of the calculated revenue and the market data of revenue are fairly good, even though only three agents, i.e. the LSI agent, the PC agent, and the LCD agent, were used to model the computer related industry.

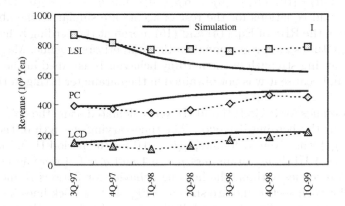

Fig. 3. The calculated revenues for Period I using Eq. (5) to (7) with the obtained parameters are shown. The first terms of RHS of Eq. (5) to (7) were replaced to incorporate rational decision-making, as Eq. (13). The initial values of revenue were taken at Q3-97. A transition probability was evaluated using backward induction after calculating the payoff for each strategy without stochastic force. The agreement of the calculated revenue and the market data of revenue are fairly good, even though only three agents, i.e. the LSI agent, the PC agent, and the LCD agent, were used to model the computer related industry.

4 Simulations

Revenue for Periods II and III was calculated using the six agents system, i.e. the LSI1 agent, the LSI2 agent, the PC1 agent, the PC2 agent, the LCD1 agent, and the LCD2 agent,

$$
R_{PC1}(t + \Delta t) - R_{PC1}(t)
$$
$$
= p\xi_{PC} - pD_{PC}/2 + \sum_k pD_{PC}w_{PC1k}\delta(t - t_k) - \gamma_{PC}R_{PC1}(t) \tag{14}
$$
$$
+ F_{PC-LSI}\{R_{PC1}(t) - R_{LSI1}(t) - R_{LSI2}(t)\}
$$
$$
+ F_{PC-LCD}\{R_{PC1}(t) - R_{LCD1}(t) - R_{LCD2}(t)\} + \eta(t)R_{PC2}(t), and
$$

$$R_{PC2}(t + \Delta t) - R_{PC2}(t)$$
$$= q\xi_{PC} - qD_{PC}/2 + \sum_k qD_{PC}w_{PC2k}\delta(t - t_k) - \gamma_{PC}R_{PC2}(t)$$
$$+ F_{PC-LSI}\{R_{PC2}(t) - R_{LSI1}(t) - R_{LSI2}(t)\}$$
$$+ F_{PC-LCD}\{R_{PC2}(t) - R_{LCD1}(t) - R_{LCD2}(t)\} + \eta(t)R_{PC1}(t),$$

(15)

where $p = R_{PC1}(0)/(R_{PC1}(0) + R_{PC2}(0))$ and $q = R_{PC2}(0)/(R_{PC1}(0) + R_{PC2}(0))$, and equations for the rest of agents are similar to the above. The last terms in the RHS of Eq. (14) and (15) represent the herding behavior, and those magnitudes are proportional to the competitors revenue. Magnitude of the time-varying strength of the herding behavior is assumed in the following calculation, because it was not obtained in the parameter fitting in the period I.

The revenues for Periods II and III were calculated using the six agents system without the herding behavior. The time-varying strength of the herding behavior $\eta(t)$ was given as follows, $\eta(t) = 0.0$ for the period II and $\eta(t) = 0.0$ for the period III. The parameters $p = 0.4$ and $q = 0.6$ were assumed. The calculated revenues without the herding behavior for Periods II and III were summed for each sector and are shown in Fig. 4. The thick lines indicate the calculated revenues. The rise and fall of revenue observed in the market data in Periods II and III were not reproduced. For the LSI sector and the PC sector, the calculated revenues were smaller than the market data during Period II and the first half of Period III. For all sectors, the calculated revenues were larger than the market data during the latter half of the period III.

Next, the revenues for Periods II and III were calculated using the six agents system with the herding behavior. The time-varying strength of the herding behavior $\eta(t)$ was given as follows, $\eta(t) = +0.1$ for Period II and $\eta(t) = -0.1$ for Period III. The parameters $p = 0.4$ and $q = 0.6$ were assumed. The calculated revenues with the herding behavior for Periods II and III were summed for each sector and are shown in Fig. 5. The thick lines indicate the calculated revenues. The rise and fall of revenues observed in the market data in Periods II and III were reproduced fairly well. Fig. 5 shows that the bubble formation during Period II and the bubble collapse during Period III were reproduced for all sectors. It is, however, noted that the calculated revenue for the LCD sector was larger than the market data during Period III. The importance of the herding behavior of firms was demonstrated to reproduce the formation and collapse of the bubble in the computer related industry market in Japan during the late 90s.

5 Conclusions

Decision-making is a crucial task for the business planning of industrial firms in order to cope with uncertainties in the business environment. A method of firm dynamics simulation, i.e. the game-theoretic stochastic agent, was

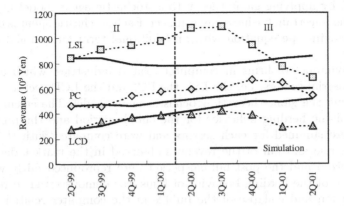

Fig. 4. The revenues for Periods II and III were calculated using the six agents system without the herding behavior. The time-varying strength of the herding behavior $\eta(t)$ was given as follows, $\eta(t) = 0.0$ for the period II and $\eta(t) = 0.0$ for the period III. The calculated revenues without the herding behavior for Periods II and III were summed for each sector and are shown. The thick lines indicate the calculated revenues. The rise and fall of the revenues observed in the market data in the period II and III were not reproduced.

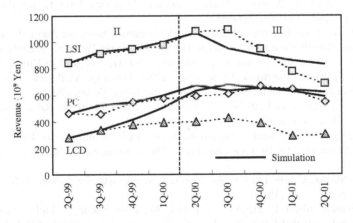

Fig. 5. The revenues for Periods II and III were calculated using the six agents system with the herding behavior. The time-varying strength of the herding behavior $\eta(t)$ was given as follows, $\eta(t) = +0.1$ for Period II and $\eta(t) = -0.1$ for Period III. The calculated revenues with the herding behavior for Periods II and III were summed for each sectors and is shown. The thick lines indicate the calculated revenues. The rise and fall of revenues observed in the market data in Periods II and III were reproduced fairly well. The figure shows that the bubble formation during Period II and the bubble collapse during Period III were reproduced for all sectors.

developed by applying game theory to a stochastic agent model in order to analyze the uncertain business environment. Each stochastic agent is described by a Langevin-type equation with an additional term for rational decision-making.

The dynamics of firms in computer related industries, which consist of three industrial sectors, i.e. the LSI, the PC, and the LCD sectors, are simulated using the game-theoretic stochastic agent model. The calculated revenues with the herding behavior for the bubble period and the post-bubble period were summed for each sectors and were compared with the market data. The rise and fall of the revenues observed in the market data in the bubble period and the post-bubble period were reproduced fairly well. The importance of the herding behavior of firms was demonstrated to reproduce the formation and collapse of the bubble in the computer related industry market in Japan during the late 90s. An appropriate implementation of the game-theoretic stochastic agent model will be valuable to analyze the uncertain business environment and to provide feasible decision-making for business planning.

References

1. Y. Ikeda, et al., A Simple Risk Model for Corporate Valuation, submitted to Asia-Pacific Financial Markets.
2. Y. Ikeda et al., Forecast of Business Performance using an Agent-based Model and Its Application to a Decision Tree Monte Carlo Business Valuation, Applications of Physics in Financial Analysis 4 (2003), accepted in Physica A
3. F. Schweitzer, Brownian Agents and Active Particles, Collective Dynamics in the Natural and Social Sciences, Berlin, Springer-Verlag (2003)
4. P. Richmond, and L. Sabatelli, Langevin processes, agent models and socio-economic systems, Physica A 336 (2004) 27-38
5. M. Levy, H. Levy, and S. Solomon, Microscopic Simulation of Financial Markets, From investor Behavior to Market Phenomena, San Diego, Academic Press (2000)
6. R. Gibbons, Game Theory For Applied Economists, Princeton, NJ, Princeton University Press (1992)
7. S. Bakhchandani, D. Hirshleifer, and I. Welch, A Theory of Fad, Fashion, Custom and Cultural Change as Information Cascades, Journal of Political Economy, 100 (1992) 992-1026

Part IV

Agent-based Modeling

A Functional Modularity Approach to Agent-based Modeling of the Evolution of Technology *

Shu-Heng Chen[1] and Bin-Tzong Chie[2]

[1] AI-ECON Research Center, Department of Economics, National Chengchi University, Taipei, 116, Taiwan chchen@nccu.edu.tw
[2] AI-ECON Research Center, Department of Economics, National Chengchi University, Taipei, 116, Taiwan chie@aiecon.org

Summary. No matter how commonly the term *innovation* has been used in economics, a concrete analytical or computational model of innovation is not yet available. This paper argues that a breakthrough can be made with *genetic programming*, and proposes a functional-modularity approach to an agent-based computational economic model of innovation.

Key words: Agent-Based Computational Economics; Innovation; Functional Modularity; Genetic Programming.

1 Motivation and Introduction

No matter how commonly the term *"innovation"* or *"technological progress"* has been used in economics, or more generally, in the social sciences, a concrete analytical or computational model of innovation is not yet available. Studies addressing specific technology advancements in different scientific and engineering fields are, of course, not lacking; however, the *general representation* of technology, based on which innovation can be defined and its evolutionary process studied, does not exist.

While direct modeling of innovation is difficult, economists' dissatisfaction with the neo-classical economic research paradigm is increasing, partially due to its incompetence in terms of producing novelties (or the so-called *emergent property*). We cannot assume in advance that we know all new goods and new technology that will be invented in the future. Therefore, in our model, we must leave space to anticipate the unexpected. Recently, [2, 3] introduced Zabell's notion of *unanticipated knowledge* to economists [11]. This notion is

* NSC research grant No. 92-2415-H-004-005 is gratefully acknowledged.

motivated by *population genetics*. In probability and statistics it is referred to as the *law of succession*, i.e. how to specify the conditional probability that the next sample is never seen, given available sets of observations up to now. However, the Ewens-Pitman-Zabell induction method proposed by Aoki is still rather limited. Basically, the nature of diversity of species and the nature of human creativity should not be treated equally [5].

This paper proposes *genetic programming* as a possible approach leading to simulating the evolution of technology. Our argument is based on two essential standpoints. First of all, as regards the innovation process, we consider it to be a *continuous process* (*evolution*), rather than a *discontinuous process* (*revolution*). According to the continuity hypothesis, novel artifacts can only arise from antecedent artifacts. Second, the evolution can be regarded as a *growing process* by combining low-level building blocks or features to achieve a certain kind of high-level functionality. In plain English, new ideas come from the use (the combination) of the old ideas (building blocks). New ideas, once invented, will become building blocks for other more advanced new ideas. This feature, known as *functional modularity*, can be demonstrated by GP, and that will be shown in this paper.

2 Background

The idea of functional modularity is not new to economists. For example, Paul Romer has already mentioned that "Our physical world presents us with a relatively small number of building blocks–the elements of the periodic table– that can be arranged in an inconceivably large number of ways." (Romer, 1998). That GP can deliver this feature has already been well evidenced in a series of promising applications to the scientific, engineering, and financial domains.

A decade ago, financial economists started to apply the functional-modularity approach with GP to discover *new* trading rules. [9] and [1] took *moving average rules* and *trading range break-out rules* as the building blocks (primitives). GP was employed to grow new trading rules from these primitives. Hence, GP has already demonstrated the evolution of trading technology: combining low-level building blocks (MA, filter, or break-out rules) to achieve a certain kind of high-level functionality (profitable performance).

John Koza's application of GP to *Kepler's law* is another striking example. Here, not only did GP rediscover the law, but also, as the system climbed up the fitness scale, one of its interim solutions corresponded to an earlier conjecture by Kepler, published ten years before the great mathematician finally perfected the equation [8, 4]. A further application of GP by John Koza to analog circuits shows that GP-evolved solutions can actually compete with human ingenuity: the results have closely matched ideas contrived by humans. Koza's GP has produced circuit designs that infringe 21 patents in all, and duplicate the functionality of several others in novel ways [10].

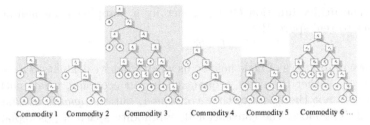

Commodity 1   Commodity 2   Commodity 3   Commodity 4   Commodity 5   Commodity 6 ...

Fig. 1. A Functional-Modularity Representation of Commodities.

Commodities are associated with their respective production processes which, when written in LIST programming language, can be depicted as parse trees as shown here.

3 Commodities and Production

Commodities in economic theory are essentially empty in terms of content. Little attention has been paid to their size, shape, topology, and inner structure. A general representation of commodities simply does not exist in current economic theory. In this paper, each commodity is associated with its *production process*. Each production process is described by a sequence of processors and the materials employed. In general, each sequence may be further divided into many parallel subsequences. Different sequences (or subsequences) define different commodities. The commodity with the associated processor itself is also a processor whose output (i.e. the commodity) can be taken as a material used by an even higher level of production. With this structure, we can ascertain the two major elements of GP, namely, the *function set* and the *terminal set*. The former naturally refers to a set of *primitive processors*, whereas the latter refers to a set of *raw materials*. They are denoted respectively by the following,

$$Function \ \ Set : \Xi = \{F_1, F_2, ..., F_k\}, \tag{1}$$
$$Terminal \ \ Set : \Sigma = \{X_1, X_2, ..., X_\kappa\}. \tag{2}$$

Each sequence (commodity, processor) can then be represented by a *LISP S-expression* or, simply, *a parse tree* (Fig. 1). The evolution of production processes (commodities) can then be simulated by using standard GP. The *knowledge capital* of the society at a point in time can then be measured by the complexity and the diversity of its existing production processes.

4 Commodity Space

Before introducing the functional-modularity approach to preferences, let us start with a brief review of the utility function used in conventional economic theory. The utility function $U(.)$ is generally a mapping from non-negative real space to real space $\mathcal{R}$.

$$U : \mathcal{R}_+^n \to \mathcal{R} \tag{3}$$

This above mapping is of little help to us when what we evaluate is a sequence of processors rather than just a quantity. In our economy, what matters to consumers is not the *quantity* they consumed, but the *quality* of what they consumed. Therefore, the conventional commodity space $\mathcal{R}_+^n$ is replaced by a new commodity space which is a collection of sequences of processors. We shall call the space $\mathcal{Y}$. The representation of the commodity space $\mathcal{Y}$ can be constructed by using the *theory of formal language*, for example, the *Backus-Nauer form* (**BNF**) of grammar. So $\mathcal{Y}$ is to be seen simply as the set of all expressions which can be produced from a start symbol Λ under an application of *substitution rules* (*grammar*) and a finite set of primitive processors (Σ) and materials (Ξ). That is, $\mathcal{Y}$ represents the set of all commodities which can be produced from the symbols Σ and Ξ.

$$\mathcal{Y} = \{Y \mid \Lambda \Rightarrow Y\} \tag{4}$$

While, as we saw in Fig. 1, each Y ($Y \in \mathcal{Y}$) can be represented by the language of expression trees (**ETs**), a more effective representation can be established by using *Gene Expression Programming* (**GEP**) developed by [7]. In GEP, the individuals are encoded as *linear strings of fixed length* (the genome or set of chromosomes) which are afterwards expressed as nonlinear entities of different sizes and shapes, i.e. different expression trees. As [7] showed, the interplay of chromosomes and expression trees in GEP implies an unequivocal translation system for translating the language of chromosomes into the language of ETs. By using GEP, the commodity space can then be defined as a subset of the *Kleene star*, namely,

$$\mathcal{Y} = \{Y_n \mid Y_n \in (\Sigma \cup \Xi)^* \cap GEP\}, \tag{5}$$

where Y_n is a string of length n,

$$Y_n = y_1 y_2 ... y_n, \quad y_i \in (\Sigma \cup \Xi), \forall i = 1, ..., n. \tag{6}$$

We have to emphasize that, in order to satisfy the syntactic validity, $\mathcal{Y}$ is only a subset of the Kleene star $(\Sigma \cup \Xi)^*$. To make this distinction, the $\mathcal{Y}$ described in (5) is referred to as the *strongly-typed Kleene star*. Each Y_n can then be translated into the familiar parse tree by using GEP. This ends our description of the commodity space.

5 Preferences

Unlike a commodity space, a preference space cannot be a collection of finite-length strings, since they are not satisfied by the *non-saturation* assumption.

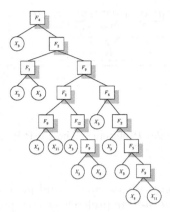

Fig. 2. Preference: The Parse-Tree Representation
What is shown here is only part of the potentially infinitely large parse tree, i.e.
only U^l of $[U^l]$.

Economic theory assumes that consumers always prefer more to less, i.e. the marginal utility can never be negative. Even though we emphasize the *quality* dimension instead of the *quantity* dimension, a similar vein should equally hold: *you will never do enough to satisfy any consumer*. If consumers' preferences are represented by finite-length strings, then, at a point, they may come to a state of complete happiness, known as the *bliss point* in economic theory. From there, no matter how hard the producers try to upgrade their existing commodities, it is always impossible to make consumers feel happier. This is certainly not consistent with our observation of human behavior. As a result, the idea of a commodity space cannot be directly extended to a preference space.

To satisfy the non-saturation assumption, a preference must be a string of infinite length, something like

$$...u_1u_2...u_l... = ...U^l... \tag{7}$$

However, by introducing the symbol ∞, we can regain the finite-length representation of the preference, i.e.

$$\infty u_1u_2...u_l\infty = \infty U^l\infty = [U^l]. \tag{8}$$

First of all, as we mentioned earlier, consumers may not necessarily know what their preferences look like, and may not even care to know. However, from Samuelson's *revealed preference theory*, we know that consumers' preferences *implicitly* exist. Equation (8) is just another way of saying that consumers' preferences are *implicit*. It would be pointless to write down the consumers' preferences of the 30th century, even though we may know that these are much richer than what has been revealed today. To approximate the feedback

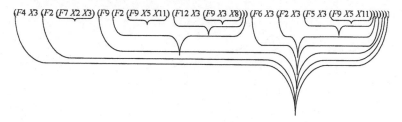

Fig. 3. Modular Preference: The LISP Representation

relation between technology advancements and preferences, it would be good enough to work with *local-in-time* preferences (temporal preferences).

Secondly, Equation (8) enables us to see the possibility that preference is adaptive, evolving and growing. What will appear in those ∞ portions may crucially depend on the commodities available today, the commodities consumed by the consumer before, the consumption habits of other consumers, and other social, institutional and scientific considerations.

6 Utility Function

Given the preference $[U^l]$, let $U \mid [U^l]$ be the utility function derived from $[U^l]$. $U \mid [U^l]$ is a mapping from the *strongly-typed Kleene Star* to $\mathcal{R}_+$.

$$U \mid [U^l] : \mathcal{Y} \to \mathcal{R}_+. \tag{9}$$

Hereafter, we shall simply use U instead of $U \mid [U^l]$ as long as it causes no confusion.

The modular approach to preference regards each preference as a hierarchy of modular preferences. Each of these modular preferences is characterized by a parse tree or the so-called building block. For example, the preference shown in Fig. 2 can be decomposed into modular preferences of different depths. They are all explicitly indicated in Fig. 3. Consider S_i to be the set of all modular preferences of *depth i*. Then Table 1 lists all modular preferences by means of these S_i. From both Fig. 3 and Table 1, it is clear that each subtree at a lower level, say S_j, can always find its parent tree, of which it is a part, at a higher level, say S_i where $i > j$. This subsequence relation can be represented as follows:

$$S_i \sqsupset S_j. \tag{10}$$

A commodity Y_n is said to *match* a modular preference S_i of U^l if they are exactly the same, i.e. they share the same the LISP expression and the same tree representation. Now, we are ready to postulate the first regularity condition regarding a well-behaved utility function, which is referred to as the *monotonicity condition*.

Table 1. Modular Preferences Sorted by Depth

D (d)	Subtrees or terminals	
1	$X_2, X_3, X_5, X_8, X_9, X_{11}$	1
2	$S_{2,1} = (F_7 X_2 X_3)$	2
	$S_{2,2} = (F_9 X_5 X_{11})$	
	$S_{2,3} = (F_9 X_3 X_8)$	
	$S_{2,4} = (F_9 X_5 X_{11})$	
3	$S_{3,1} = (F_{12} X_3 (F_9 X_3 X_8))$	4
	$S_{3,2} = (F_5 X_3 (F_9 X_5 X_{11}))$	
4	$S_{4,1} = (F_2 (F_9 X_5 X_{11})(F_{12} X_3 (F_9 X_3 X_8)))$	8
	$S_{4,2} = (F_2 X_3 (F_5 X_3 (F_9 X_5 X_{11})))$	
5	$S_5 = (F_6 X_3 (F_2 X_3 (F_5 X_3 (F_9 X_5 X_{11}))))$	16
6	$S_6 = (F_9 (F_2 (F_9 X_5 X_{11})(F_{12} X_3 (F_9 X_3$	32
	$X_8)))(F_6 X_3 (F_2 X_3 (F_5 X_3 (F_9 X_5 X_{11})))))$	
7	$S_7 = (F_2 (F_7 X_2 X_3)(F_9 (F_2 (F_9 X_5 X_{11})(F_{12}$	
	$X_3 (F_9 X_3 X_8)))(F_6 X_3 (F_2 X_3 (F_5 X_3$	64
	$(F_9 X_5 X_{11}))))))$	
8	$S_8 = (F_4 X_3 (F_2 (F_7 X_2 X_3)(F_9 F_2 (F_9 X_5 X_{11}))$	128
	$(F_{12} X_3 (F_9 X_3 X_8)))(F_6 X_3 (F_2 X_3 (F_5 X_3$	
	$(F9 X_5 X_{11}))))))$	

Given a preference $[U^l]$, the associated utility function is said to satisfy the *monotonicity condition* iff

$$U(Y_{n_i}) > U(Y_{n_j}) \qquad (11)$$

where Y_{n_i} and Y_{n_j} are the commodities matching the corresponding modular preferences S_i and S_j of U^l and S_i and S_j satisfy Equation (10).

The *monotonicity* condition can be restated in a more general way. Given a preference $[U^l]$ and by letting $\{h_1, h_2, ...h_j\}$ be an increasing subsequence of $\mathcal{N}_+$, then the associated utility function is said to satisfy the *monotonicity condition* iff

$$U(Y_{n_j}) > U(Y_{n_{j-1}}) > ... > U(Y_{n_2}) > U(Y_{n_1}) \qquad (12)$$

where $Y_{n_1}, ..., Y_{n_j}$ are the commodities matching the corresponding modular preferences $S_{h_1}, ..., S_{h_j}$ of U^l, and

$$S_{h_i} \sqsupset S_{h_{i-1}} \sqsupset ... \sqsupset S_{h_2} \sqsupset S_{h_1}. \qquad (13)$$

If S_k is a subtree of S_i as in Equation (10), then S_k is called the *largest subtree* of S_i if S_k is a *branch* (descendant) of S_i. We shall use "$S_i \lhd S_k$" to indicate this largest-member relation. Depending on the grammar which we use, the largest subtree of S_i may not be unique. For example, each modular preference in Fig. 2 has two largest subtrees. In general, let $S_{h_1}, S_{h_2}, ...S_{h_j}$ be all the largest subtrees of S_i, denoted as follows:

$$S_i = \sqcup_{h_1}^{h_j} S_k \lhd \{S_{h_1}, S_{h_2}, ...S_{h_j}\}, \tag{14}$$

where $\{h_1, h_2, ...h_j\}$ is a non-decreasing subsequence of $\mathcal{N}_+$. Notice these largest trees may not have sub-relationships (10) among each other. However, they may have different depths, and the sequence $\{h_1, h_2, ...h_j\}$ ranks them by depth in an ascending order so that S_{h_1} is the largest subtree with minimum depth, and S_{h_j} is the one with maximum depth.

The second postulate of the well-behaved utility function is the property known as *synergy*. Given a preference $[U^l]$, the associated utility function is said to satisfy the *synergy condition* iff

$$U(Y_{n_i}) \geq \sum_{k=1}^{j} U(Y_{n_k}), \tag{15}$$

where Y_{n_i} and $\{Y_{n_k}; k = 1, ..., j\}$ are the commodities matching the corresponding modular preferences S_i and $\{S_{h_k}; k = 1, ..., j\}$ of $[U^l]$, and S_i and $\{S_{h_k}; k = 1, ..., j\}$ satisfy Equation (14).

For convenience, we shall also use the notation $\sqcup_{k=1}^{j} Y_{n_k}$ as the synergy of the set of commodities $\{Y_{n_k}; k = 1, ..., j\}$. Based on the *New Oxford Dictionary of English*, synergy is defined as "the interaction or cooperation of two or more organizations, substances, or other agents to produce a combined effect greater than the sum of their separate effects." "*The whole is greater than the sum of the parts*" is the fundamental source for *business value creation*. Successful business value creation depends on two things: *modules* and the *platform* to combine these modules. Consider the consumer characterized by Fig. 2 as an example. To satisfy him, what is needed are all of the modules listed in Table 1. Even though the technology has already advanced to the level S_7, knowing the use of processor F_4 to combine X_3 and S_7 can still satisfy the consumer to a higher degree, and hence create a greater business value.

A modular preference may appear many times in a preference. For example, $S_{2,4}$ in Table 1 appears twice in Fig. 2. In this case, it can simultaneously be the largest subtree of more than one modular preference. For example, $S_{2,4}$ is the largest subtree of both $S_{3,2}$ and $S_{4,1}$. Let S_k be the largest subtree of $S_{h_1}, S_{h_2}, ...,$ and S_{h_j}. Denote this relation as

$$S_k = \sqcap_1^j S_{h_i} \rhd \{S_{h_1}, S_{h_2}, ...S_{h_j}\}. \tag{16}$$

Given a preference $[U^l]$, the associated utility function is said to satisfy the *consistent condition* iff

$$U(Y_{n_i} \mid S_k \rhd S_{h_1}) = ... = U(Y_{n_i} \mid S_k \rhd S_{h_j}), \tag{17}$$

where $Y_{n_i} \mid S_k \rhd S_{h_1}$ is the commodity which matches the corresponding modular preference S_k in the designated position, $S_k \rhd S_{h_i}$. The consistency condition reiterates the synergy effect. No matter how intensively the commodity Y_{n_i} may significantly contribute to the value creation of a synergy commodity, its value will remain identical and lower when it is served *alone*.

Given a preference $[U^l]$, the associated utility function U is said to be *well-behaved* iff *it satisfies the monotone, synergy and consistency condition.* It generates a sequence of numbers $\{U(Y_{n_i})\}_{i=1}^h$ where Y_{n_i} matches the respective modular preference $S_{d,j}$. $S_{d,j}$ is the jth modular preference with depth d. The utility assigned in Table 1 is an illustration of a well-behaved utility function derived from the preference shown in Fig. 2. In fact, this specific utility function is generated by the following exponential function with base 2.

$$U(S_{d,j}) = 2^{d-1} \qquad (18)$$

Utility function (18) sheds great light on the synergy effect. Thus, primitive materials or rudimentary commodities may only satisfy the consumer to a rather limited extent. However, once suitable processing or integration takes place, their value can become increasingly large to the consumer. The exponential function with base 2 simply shows how fast the utility may be scaled up, and hence may provide a great potential incentive for producers to innovate. Of course, to be a well-behaved utility function, U can have many different functional forms.

7 Firms and Production

On the production side, the economy is composed of n_f producers, each of which is initially assigned an equal operating capital, K_0.

$$K_{1,0} = K_{2,0} = \dots = K_{n_f,0} = K_0. \qquad (19)$$

With this initial capital, the producers are able to buy materials and processors from the input markets up to the amount that they can afford. There are two types of input markets at the initial stage, namely, the *raw-material market* and the *rudimentary processor market*. For simplicity, we assume that the supply curves of the two markets are infinitely elastic with a fixed unit cost (c) for each raw material and for each rudimentary processor:

$$C_{X_1} = C_{X_2} = \dots = C_{X_\kappa} = C_{F_1} = C_{F_2} = \dots = C_{F_k} = c. \qquad (20)$$

With the materials and the rudimentary processors purchased from the input market, the producer can produce a variety of commodities, defined by the associated sequence of processors. The cost of each commodity is then simply its total amount of materials and the number of processors, or, in terms of GP, the *node complexity* of the parse tree. However, to allow for the *scale effect*, each additional unit of the same commodity produced by the producer should be less costly. This can be done by introducing a monotonically decreasing function $\tau(q)$ ($0 \le \tau(q) \le 1$), where q is the qth unit of the same commodity produced. The cost of each additional unit produced is simply the cost of the previous unit pre-multiplied by $\tau(q)$. With this description,

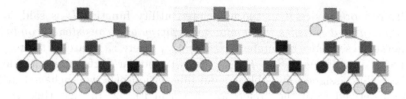

Fig. 4. EvolTech: Preferences Initialization

the *capacity constraint* for a *fully-specialized producer* i ($i \in [1, ...n_f]$), i.e. the producer who supplies only one commodity, should be

$$K_0 \geq \sum_{q=1}^{\bar{q}} C_q, \tag{21}$$

where $C_q = \tau(q)C_1$ is the unit cost of the qth unit and $\tau(1) = 1$. For a *fully-diversified producer*, i.e. the producer who produces a variety of commodities and one for each, the capacity constraint is

$$K_0 \geq \sum_{m=1}^{\bar{m}} C_{m,1}, \tag{22}$$

where $C_{m,1}$ is the cost of the first unit of commodity m. In general, the capacity constraint for the producer i is

$$K_0 \geq \sum_{m=1}^{\bar{m}} \sum_{q=1}^{\bar{q}_m} C_{m,q}, \tag{23}$$

where $C_{m,q} = \tau_m(q)C_{m,1}$.

In Equation (23), the strategic parameters are $\bar{m}$, $\bar{q}$ and C_m. To survive well, producers have to learn how to optimize them. $\bar{m}$ can be be taken as a measure of the degree of *diversification*, whereas $\bar{q}$ can be taken as the degree of *specialization*. C_m, i.e. the node complexity of the commodity m, is also a behavioral variable. Given the capacity constraint, the producer can choose to supply a large amount of primitive commodities (a quantity-oriented strategy), or a limited amount of highly delicate commodities (a quality-oriented strategy). Therefore, the choice of C_m can be regarded as a choice of the level of *quality*.

8 Demonstration

The idea presented above has been written into a computer program called **EvolTech**, which stands for "Evolution of Technology." In this section, we

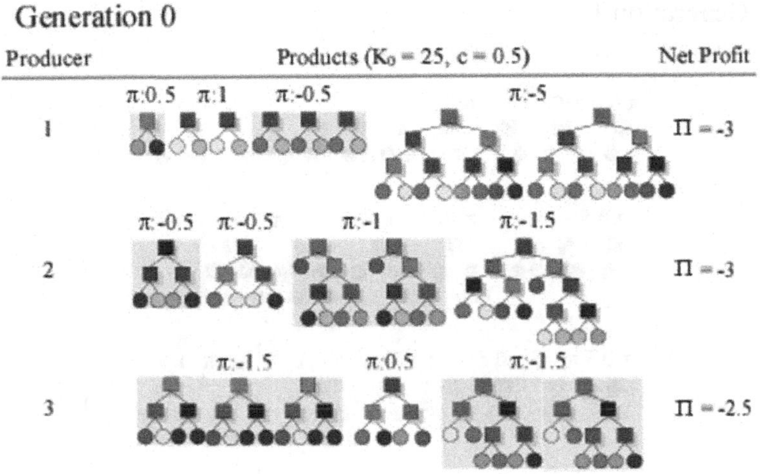

Generation 0

Producer	Products (K₀ = 25, c = 0.5)	Net Profit

Fig. 5. EvolTech: Generation 0

demonstrate a *vanilla* version of EvolTech. What we mean by *vanilla* will become clear as we give the demo.

First, as in all computational economic models, we start with a simple description of *initialization*. The initialization in EvolTech includes the generation of preferences. Based on the formulation given in Sections 5 and 6, the preferences of three consumers are randomly generated, as shown in Figure 4. The complexity of the preference has been severely restricted to a depth of 5 and is fixed throughout the entire evolution.[3] Notice that these preferences are characterized by colorful nodes. Each different color is refereed to a different primitive, sampled from a given primitive set.

In addition to the three consumers, there are three producers in the economy. Based on the same given primitive set, commodities are also randomly generated by these producers, as shown in Figure 5. Notice that the three dimensions of production behavior, i.e. quantity, quality, and diversity, are all randomly determined as long as they together satisfy the capacity constraint (23). The initial capital capacity K_0 is set to 25, and the unit cost c is set to 0.5. The scale effect is ignored. Each of the commodities is then served to the consumers whose preferences are displayed in Figure 4.

Without the details about how trade actually proceeds, it is not easy to describe how the final price and hence the profit is determined. Therefore, in this vanilla version of EvolTech we simply take the highest reservation price as the market price.[4]

---

[3] In other words, we do not consider the general preferences as discussed in Equation (8), neither the evolution of preferences in this vanilla version.

[4] We have proposed an algorithm to compute Equation (9) for a well-behaved utility function U, as defined in Section 6. This algorithm, called the *module-matching*

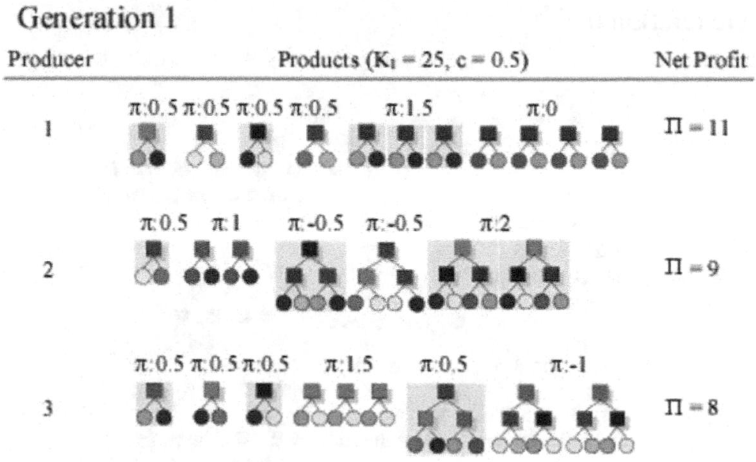

Generation 1

Producer	Products ($K_1 = 25$, $c = 0.5$)	Net Profit

Fig. 6. EvolTech: Generation 1

This simplification will facilitate our calculation of profits, π. In Figure 5, π is shown on the top of each commodity. We can see that most commodities have negative profits. This is not surprising because at this initial stage all commodities are randomly designed, and the chance of meeting consumers' needs to any significant degree is naturally low with the given combinatoric complexity.

Sophisticated commodities may be even worse than those simple designs because they induce higher production costs, and can only satisfy consumers to a very limited degree. So, as evidenced in Figure 5, commodities that suffer great economic losses tend to be those with sophisticated designs, i.e. trees with a large degree of node complexity. By summing the profits over all commodities, we get aggregate profits for each firm, which are shown on the right-hand side of the figure. In this specific case, all three firms make a loss. This finishes our description of the initial generation.

While moving to the next market period (next generation), firms start to learn from *experience*. Their profit profiles provide them with fundamental clues on how to re-design their products for the next generation. What is shown in Figure 6 is the result of their adaptation. It is interesting to notice that these new-generation commodities seem to become simpler compared with those of the previous generation (Figure 5). This is mainly because sophisticated designs do not contribute to profits but losses. Therefore, firms tend to replace those sophisticated designs with simpler ones. The economy as a whole can be described as a *quantity-based economy*, since all firms choose

---

algorithm, is very intuitive. It looks for the *projection* of the commodity Y_i to $[U^l]$, i.e. a measure of distance between a commodity Y_i and the preference $[U^l]$. Once the operational meaning of Equation (9) becomes clear, it is possible to infer the *reservation price* which a consumer would like to pay for commodity Y_i from $[U^l]$.

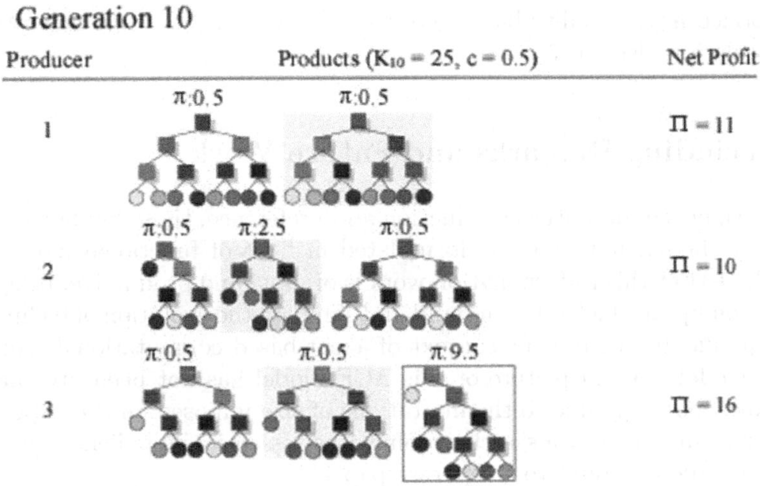

Generation 10

Producer	Products ($K_{10} = 25$, $c = 0.5$)	Net Profit

π:0.5 π:0.5

1     $\Pi = 11$

π:0.5 π:2.5 π:0.5

2     $\Pi = 10$

π:0.5 π:0.5 π:9.5

3     $\Pi = 16$

Fig. 7. EvolTech: Generation 10

to produce a large number of rudimentary commodities (i.e. they repeat doing simple standard things).

However, this strategy turns out to work well. While each simple commodity can earn a firm a tiny profit, summing them together is still quite noticeable. So, in the end, the profits of all three firms improve quite significantly. This process is then further reinforced, and in the coming generations, more resources are devoted to rudimentary commodities. Sophisticated designs are almost entirely given up. However, since there are not too many rudimentary commodities to develop in the market, when all firms concentrate on producing rudimentary commodities, the limited number of rudimentary-commodity markets become highly competitive, and the profits from producing these commodities decline as a result of the keen competition. At this point, the economy actually moves toward *an era of zero profit*.

Once producing primitive commodities is no longer profitable, the selection bias towards it also becomes weaker. Some sophisticated designs occasionally coming out of the crossover and mutation operators may find it easier to survive. That improves the chances of satisfying consumers to a higher degree. When that indeed happens, not only do firms make a breakthrough by successfully having a sophisticated (delicate) design, but the lucrative profits also attract more resources that can be devoted to quality products. While this does not always happen and the process is not always smooth, the process may be reinforcing. So, commodities with more delicate designs and higher profits may come one after the other. In the end, the economy is gradually transformed into a quality-based economy, as shown in the 10th generation of our simulation (Figure 7).[5]

9 Concluding Remarks and Future Work

In this paper, commodities, production and preference, those fundamentals of economic theory, have been re-formulated in light of functional modularity. We believe that this re-formulation work is original and productive. It lays the foundation upon which one can build and simulate the evolution of technology, more specifically, within the context of agent-based computational economic (ACE) models. A full picture of this ACE model has not been presented in this paper, partially due to the limitations of size imposed on the paper. We, therefore, can only give a sketch of some other essential ingredients, and leave a more detailed account to a separate paper.

References

1. Allen, F. and R. Karjalainen, (1999). "Using Genetic Algorithms to Find Technical Trading Rules," *Journal of Financial Economics*, **51(2)** 245–271.
2. Aoki, M., (2002a). "Open Models of Share Markets with Two Dominant Types of Participants," *Journal of Economic Behavior and Organization*, **49** 199–216.
3. Aoki, M., (2002b). "Applications of Ewens–Pitman–Zabell Inductive Methods in New Economic Dynamics," *Proceedings of the Sixth International Conference on Complex Systems*, 29–35.
4. Banville, J., (1993). *Kepler: A Novel (Vintage International)*. Vintage Books.
5. Basalla, G., (1988). *The Evolution of Technology*. Cambridge University Press.
6. Chen, S.-H. and B.-T. Chie, (2004) "Functional Modularity in the Test Bed of Economic Theory– Using Genetic Programming," in R. Poli, et al. (eds.), *GECCO-2004: Proceedings of the Genetic and Evolutionary Computation Conference*, June 26-30, 2004, Seattle, Washington, USA.
7. Ferreira, C., (2001). "Gene Expression Programming: A New Adaptive Algorithm for Solving Problems," *Complex Systems*, **13 (2)**, 87-129.
8. Levy, S., (1992). *Artificial Life: A Report from the Frontier Where Computers Meet Biology*, Vintage, New York.
9. Neely, C., P. Weller, and R. Dittmar, (1997). "Is Technical Analysis in the Foreign Exchange Market Profitable? A Genetic Programming Approach," *Journal of Financial and Quantitative Analysis*, **32(4)** 405–426.
10. Willihnganz, A., (1999). "Software that Writes Software," *salon.com*, WWW Article.
11. Zabell, S., (1992). "Predicting the Unpredictable," *Synthese*, **90** 205–232.

---

[5] The progress may not be smooth. Severe fluctuations can happen. The progress may not be sustained long enough either. The economy may stagnate after a short but fast take-off, and consumers are only supplied with some "basic needs." For a more detailed demonstration of the complex variety, see Chen and Chie (2004).

Herding Without Following the Herd: The Dynamics of Case-Based Decisions with Local Interactions

Andreas Krause

School of Management, University of Bath, Bath BA2 7AY, Great Britain
mnsak@bath.ac.uk

Summary. We use case-based decision theory to evaluate the repeated choices of individuals who are using the experience of a selected set of other individuals over a given time horizon as the basis for their own decisions. It is observed that for certain parameter constellations a large fraction of individuals make identical decisions over a long period of time, which is not implied by the information. The result of this behavior is herding, which, however, has its origin in the way individuals process information rather than a desire to be part of the herd and imitating the behavior of others.

Key words: Case-based decision theory, local interactions, networks, decision-making, herding, fads

The most common approach to model decision-making under uncertainty in economic theory is to choose the alternative that provides the highest expected utility of all alternatives available to an individual. Numerous empirical works however show that expected utility theories cannot satisfactorily explain actual behavior in many cases, see e.g. [4, 5, 21] for an overview. A particular challenge is the modelling of decisions where information on possible outcomes is limited, e.g. by not knowing all possible outcomes and/or their probabilities of occurrence. In a series of papers [8, 9, 10, 11, 12, 13, 14, 15] an alternative to the expected utility framework has been proposed, called case-based decision theory (CBDT), a special case of the theories on bounded rationality. Although some applications of CBDT have been proposed in the literature, see section 4 for a brief overview, dynamical aspects of the so determined decisions have thus far not received much attention. Only [3] considers the learning process in CBDT, but does not focus on the aggregate dynamics of decisions.

Another common observation, particularly in financial markets, is that individuals show a clear tendency towards making similar decisions, although an objective evaluation of the information available does not support the rationality of such a behavior. This so called herding behavior has received more attention in recent years and several models have been developed to explain this observation, see e.g. [23, 24] for models of imitating others or [1, 2, 6] for models of informational cascades. A common feature of many models is that they are essentially based on psychological factors and use the idea that individuals copy the behavior of other individuals out of a desire to conform with the majority or specific individuals in their peer group. The obvious result of these models is that individuals neglect information on which to base their decisions, but rather "follow the herd".

In this paper we use CBDT to model the decisions of individuals and show how they can give rise to a large fraction of individuals making similar decisions, which is not consistent with an objective evaluation of all available information. Another property is that under fairly general conditions these decisions are repeated over time, giving rise to prolonged periods of large fractions of individuals making identical decisions. The main feature of this model is that the origins of this observation is not the desire to reproduce the behavior of other individuals but the way information is processed by individuals. This establishes another reason for herding, which in our model is fully rational. It is well beyond the scope of this paper to provide a full investigation into the detailed properties of our model, we rather restrict ourselves to highlight some key properties which merit further investigation.

In the coming section we continue with a very brief introduction to CBDT and the specific model used. Section 2 then presents the results of the computer simulations conducted to evaluate this model while section 3 provides explanations for these observations. We discuss some potential applications in section 4 and finally conclude our findings in section 5.

1 The Decision-Making Process

CBDT is founded on David Hume's reasoning that similar causes are likely to produce similar effects. Hence an individual can use the experience from past decisions, which he or other individuals made to evaluate the current decision problem. A key to this decision-making problem is the similarity between the current problem and the past problems he uses. If we let $\varXi$ denote the set of all possible problems, we define the similarity of two problems as a function

$$s' : \varXi \times \varXi \to [0; 1], \tag{1}$$

where for any $i, j \in \varXi$ we interpret $s'(i, j) = 0$ as the absence of any similarity and $s'(i, j) = 1$ as an identical problem; obviously it is $s'(i, i) = 1$. We can now normalize this similarity to the "average" similarity in order to avoid any problems from the scaling of this function later on:

$$s_\tau(i, j) = \begin{cases} \frac{s'(i,j)}{\sum_{j' \in \varXi} s'(i,j')\omega_\tau(i,j')} & \text{if } \sum_{j' \in \varXi} s'(i, j')\omega_\tau(i, j') > 0 \\ 0 & \text{if } \sum_{j' \in \varXi} s'(i, j')\omega_\tau(i, j') = 0 \end{cases}, \tag{2}$$

where $\omega_\tau(i,j)$ denotes an indicator function, which is one if the decision made when facing problem j at time τ was the equivalent to the one considered currently, and zero otherwise. This indicator function thus ensures that only those outcomes are considered that result from equivalent decisions, e.g. the decision to buy a product. The outcome of any decision, $r_{\tau,i} \in \Theta$, at time τ for problem $i \in \Xi$ can be evaluated in the common way by using a utility function:

$$u : \Theta \to \Re. \tag{3}$$

Let us further assume that when considering past decisions the similarity of problems reduces by a factor of $\lambda \in [0;1]$ per time period. This reduction in similarity can, for example, be the result of a changing environment. With a total memory of M time periods, we obtain the following function which can be used to evaluate the decisions of an individual:

$$V(r_{t,i}) = \sum_{\tau=0}^{M} \left(\lambda^\tau \sum_{j' \in \Xi} s_\tau(i,j') u(r_{\tau,j'}) \omega_\tau(i,j') \right). \tag{4}$$

Here again the indicator function $\omega_\tau(i,j)$ ensures that we only consider decisions with equivalent decisions to address the present problem. It is now possible to compare the values of this function for all $r_{t,i} \in \Theta$ and choose the decisions which provides the highest score. In cases where two decisions have an identical score, the choice is made randomly with all such choices having equal probability of being selected. This decision-making rule as presented here is a marginal but in essence unimportant variation of the CBDT developed in [8, 12, 14], which also provide an axiomatic justification for using this decision rule.

We will in this paper make a small re-interpretation of this model to suit our needs. Let us assume that a set of N individuals, called "agents", have to make a repeated decision which can take one of two forms, "buy" or "sell", hence $\Theta = \{\text{buy}; \text{sell}\}$. The outcome of an individual decision is either "success", corresponding to an outcome of $+1$, or "failure", corresponding to an outcome of -1, such that $\Re = \{-1; +1\}$. Rather than interpreting Ξ as a set of decision problems, we assume it to be a set of agents, who all have to make this decision. The similarity function s' then merely indicates how similar the circumstances of these agents making their decisions are.

Let us propose that an agent evaluates the past M decisions of himself and selected other agents; these agents are selected randomly and do not change over time. Once an agent i uses the experience of an agent j they are assigned a similarity of $s'(i,j) \in [0;1]$ and if he is not using his experience we set $s'(i,j) = 0$. We can arrange the $s'(i,j)$ in a matrix and interpret it as the adjacency matrix of a graph which represents a network of agents interacting with each other in their decision-making. It has to be noticed that we use a directed graph as in general $s'(i,j) \neq s'(j,i)$. We here assume that connections between any two nodes of this graph exist with a probability of p and assign a random similarity taken from a uniform distribution on $[0; 1]$ to each edge. We now determine the outcome of any decision randomly in each time period. Hence u is a random variable and we assume that for any given decision the ex-ante probabilities of success and failure are identical to one half. The outcomes are serially independent, but we allow for a common correlation of ρ across agents.

We use this model to evaluate the decision-making process in a simple Monte-Carlo simulation using a wide range of parameters. The focus of our attention will be on the properties of the aggregate decision of agents to buy or sell, in particular the fraction of agents buying.

2 Model Evaluation

From the structure of the above model it is obvious that any agent knowing its set-up would make random decisions. Thus a situation in which all N agents make random choices, buying or selling with equal probability, serves as our benchmark to which we will compare the results of our model.

In order to evaluate the model presented above, we conducted a Monte-Carlo simulation with all parameters taken from independently and identically distributed uniform distributions across the following ranges: $M \in [0; 1000]$, $\lambda \in [0; 1]$, $\rho \in [0; 1]$, and $p \in [0; 1]$. The number of agents remains fixed at $N = 100$; as it is a common feature in most network models that its properties do not depend on the number of agents but only on the term Np, we can reasonably infer that the number of agents has no significant influence on our results. The number of simulations conducted is $1,000$. Each simulation consists of $11,000$ simulated time periods of which the first $1,000$ are discarded from the further analysis in order to eliminate any effects from initial conditions. These initial conditions consist only of the memory which we assume to be empty, i.e. it only consists of zeros.

In addition to the parameters as used in the simulation we also found it useful to determine the "average memory", which we define similarly to the duration as follows:

$$D = 1 + \frac{1-\lambda}{\lambda} \sum_{\tau=1}^{M} \lambda^{\tau} \tau. \tag{5}$$

We investigate the properties of the fraction of agents buying using the simulated time series in dependence of the parameters above. The following variables of the time series and how they are affected by these parameters are investigated:

1. Volatility (VOLA): average absolute change in the fraction of agents buying,

2. Volatility upon switch of majority (SWITCH): average absolute change in the fraction of buyers, given that buyers gain or loose the majority,
3. Deviation from benchmark (BIAS): average absolute deviation of the fraction of buyers from 0.5,
4. Length of deviation from benchmark (LENGTH): average number of time periods between a switch from the fraction of buyers exceeding (below) to below (exceeding) 0.5,
5. Autocorrelation (AUTO): first-order autocorrelation of the fraction of agents buying.

We also determined the appropriate values for our benchmark case of all agents making random choices as a comparison. It became apparent that the results from our model are significantly different from that of a random choice. Secondly, the often strong correlations between the investigated variables suggest a common origin of these relationships, which obviously have to be found in the parameters.

Some obvious relationships between the variables are easily explained. A higher serial correlation of the fraction of buyers (AUTO) causes the volatility (VOLA) to reduce, as subsequent changes are becoming smaller. The same argument can be used to show that the length of a majority of buyers or sellers being present (LENGTH) increases with the autocorrelations as the lower volatility makes it less common for the majority to change once it has been established. The larger the deviation from the benchmark of 50% buyers on average is (BIAS), the larger the change in this fraction has to become in order to cause the majority to switch (SWITCH), as this large deviation has first to be overcome. If this were to be achieved in many small steps the average deviation would be reduced, accordingly the volatility of the fraction of buyers has to be higher. As with many large changes in the fraction of buyers (VOLA) a majority can easily change, we observe an inverse relationship with the length of any majority (LENGTH). Finally, as changes in the fraction of buyers causing the majority to switch (SWITCH) forms part of the volatility (VOLA), it is only natural to observe a positive relationship between these two variables.

Of more importance than the relationship between these variables, however, is how they relate to the parameters. This information could then be used to determine which constellations are necessary to explain certain effects in decision-making. We have plotted the five variables against the two most important parameters for this variable in figure 1.

The correlation of outcomes can be shown in a more detailed analysis to have a statistically significant negative influence on the volatility (VOLA) and a positive influence on the length of a deviation from the benchmark (LENGTH). These influences are however of much less importance than of those parameters presented in figure 1. The length of the memory itself, M, does not influence the variables significantly, but of course shows its effect via the average memory D, which is an important parameter to consider.

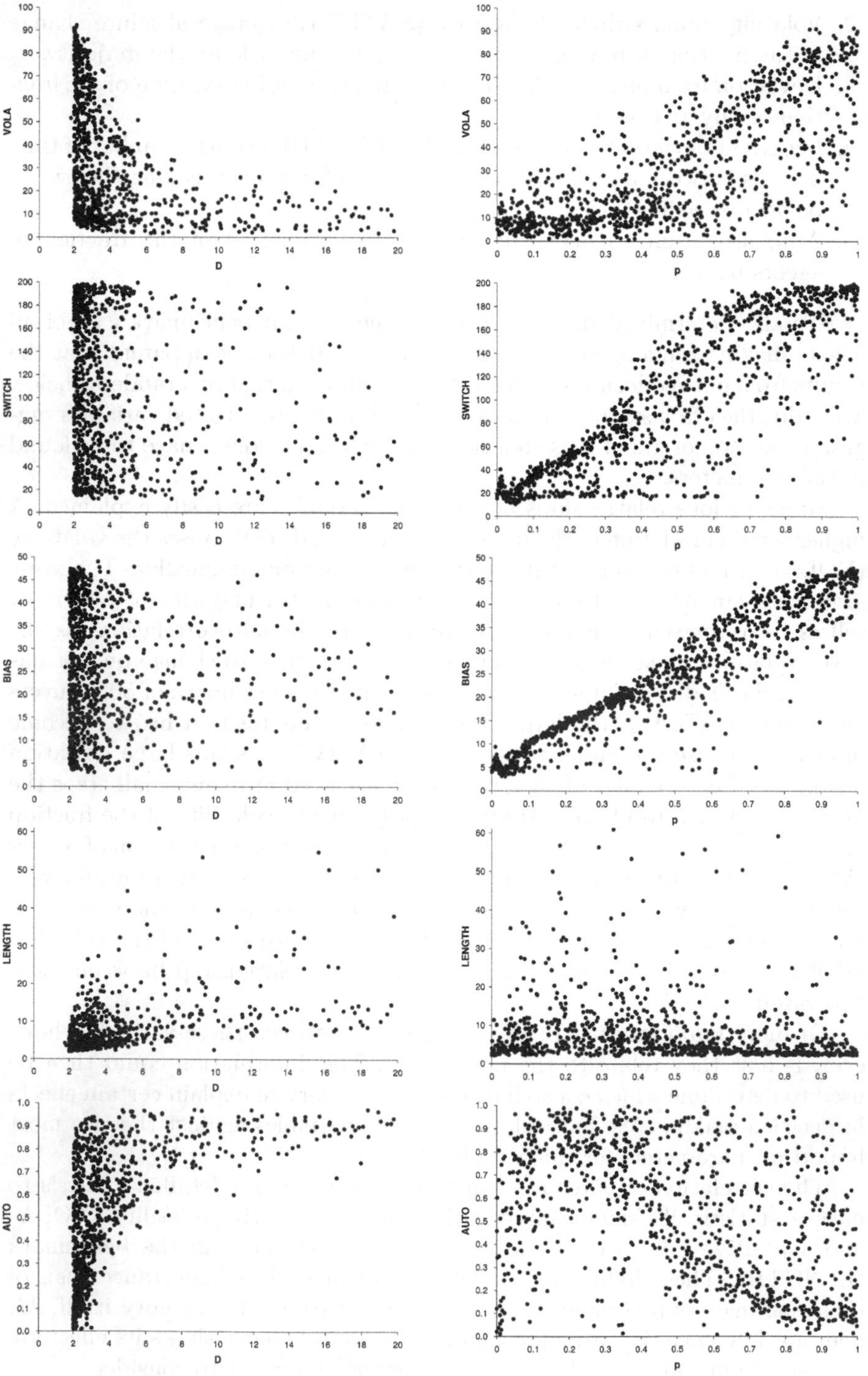

Fig. 1. Relationships between variables and parameters

We observe that the volatility of the fraction of buyers (VOLA) is increasing in the probability of interaction between two agents (p) and decreasing in the average memory (D). It is not possible to determine a single parameter as being dominant. In contrast to this, the volatility upon a switch of the majority (SWITCH) is dominated by the positive relationship with the probability of interaction (p) with the other parameters not showing an important influence. A similar effect can be observed on the deviation from the benchmark (BIAS), which is effectively determined only by this probability.

A very different picture emerges when investigating the length of a deviation from the benchmark (LENGTH). The probability of interaction (p) has only a minimal influence, but it is instead dominated by the average memory (D). The same situation is reflected in the autocorrelation of the fraction of buyers (AUTO), although the influence of the probability of interaction is larger in this case. We should have expected such a similarity given the very strong relationship between the two variables as described above.

When conducting an OLS regression of all parameters on the variables, our above findings are confirmed, all coefficients are highly significant and the goodness of fit is reasonably high. A more detailed analysis of time series properties shows the best parameter constellation to generate the typical herding behavior (a large and prolonged deviation from the benchmark) is achieved by a moderate interaction of agents ($p \approx 0.5$) and a long memory ($D > 10$). A higher interaction causes the deviation to become too short and volatile while a shorter memory reduces the length of the deviation too much to be reasonably be described as herding.

Having established these findings, we now have to turn our attention to their explanation, which we address in the coming section.

3 Explaining the Findings

We can explain the above findings quite easily by investigating the effects the individual parameters have on the variables. We firstly can easily deduct that with a higher probability of interaction (p) we will on average observe more actual interactions, hence the decision-making is influenced by the experience of more agents, which is aggregated. As all agents use more and more similar experiences given the higher interactions, their decisions are likely to become more alike; hence any changes of choice are to be conducted by a larger number of agents simultaneously, causing the volatility (VOLA) to increase and naturally also the volatility upon a switch of majority (SWITCH). A direct consequence of the more homogeneous decisions being made is that the deviation from the benchmark (BIAS) increases as more agents behave alike. The increased volatility, on the other hand, makes such changes more likely and the autocorrelation of the fraction of buyers (AUTO) decreases. For the same reason also the length of any deviation from the benchmark (LENGTH) reduces with the increased volatility making such changes more frequent.

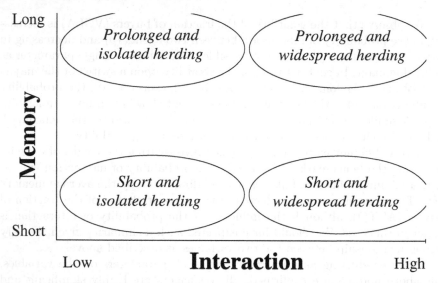

Fig. 2. The effect of memory and interaction on herding

A longer average memory (D) has the obvious consequence that the decision is based on a longer history and thus the relative importance of recent events is reduced. As the history does not change over time, decisions are based on very similar considerations and should therefore not vary that much. With subsequent decisions of agents being very similar, we see immediately that the volatility (VOLA) reduces, the length of deviations (LENGTH) increase as any deviation is unlikely to be reversed quickly and the autocorrelations (AUTO) are increasing. On the other hand the volatility upon a switch of the majority (SWITCH) is hardly affected because whether the change in the choice of the agents is determined predominantly by recent experiences or experiences in the distant past is of no relevance. Similarly the size of the deviation from the benchmark (BIAS) is not affected.

We can summarize our findings in figure 2, where we use the term "herding" to indicate the deviation from the benchmark (BIAS), "widespread" herding is to be interpreted as having a large BIAS and "isolated" herding as having a low BIAS; "short" herding has a low LENGTH and "prolonged" herding has a large LENGTH.

We would normally expect a higher probability of interaction (p) and longer memory (D) to improve the decisions made by agents as they can base their choice on a broader range of experiences. This should result in a reduced volatility (VOLA) and deviation from the benchmark (BIAS) and we should particularly see an improvement over time due to the learning effect in this stationary environment. Furthermore, the learning should be facilitated by the simplicity of the choices and outcomes. But in the data simulated we were not able to detect any learning at all, later choices showed exactly the same properties as earlier choices.

The reason for this counterintuitive observation can be found in the differences between what agents should learn to achieve the benchmark outcome of random choices and what they actually learn from case-based decision-making. The Bayesian view is that they should use the experiences to update their beliefs on the distribution of outcomes for any given choice, thereby learning the true distribution. In contrast to this, CBDT does not propose agents learning the true distribution of outcomes, but rather the best choice directly without reference to the underlying distribution which is determined by the structure of the economy.

With this discrepancy it becomes apparent why we do not observe any learning in our model – there does not exist a single best choice. Both alternatives are of equal value, an aspect which the CBDT cannot detect given its decision-making rules. Although Bayesian learning is clearly superior to the learning implied by CBDT we see that the deviation from the benchmark of all agents making random choices does neither reduce the welfare of any individual agent nor the social welfare; but it also does not increase the welfare. This is because whatever decision is taken, the ex-ante outcome is always a success and failure with equal probability, even in cases where all agents make the same choice and outcomes are perfectly correlated.

It remains an open question how the behavior would change with the existence of a superior choice. [3] provides first insights into the learning in such a situation, but does not address the evolution of aggregate decisions over time. At least in our model without the existence of a best choice, Bayesian learning is not welfare improving and thus not conducted by individuals. This absence of learning enabled the properties we investigated above to emerge and to be maintained over time.

4 Potential Applications

Although CBDT has been developed about a decade ago, with predecessors being proposed well before that, applications to real economic problems are quite limited. Most of the literature on CBDT focuses on axiomatic and other theoretical aspects. Among those isolated applications in the literature are the works on financial markets [18, 19, 20], voting [17], production theory [22], and consumer theory [7, 25]. [3] considers explicitly the learning effects in CBDT using a similar model to ours with specific interaction structures and a single superior choice. He shows the convergence towards equilibrium under certain conditions.

The results we obtained in this paper have many potential applications to explain patterns of behavior that are generally not consistent with the expected utility hypothesis. As we have seen that with a sufficiently long memory and high interactions a significant majority of agents make a particular choice over a longer period of time, we can use our model to explain various types of herding, fads and hypes.

Consumers preferring one brand over another for no obvious reason, hence there is no objectively superior choice, is a situation which is very similar to the outcome of our model in this paper. We could therefore easily explain fashion trends by consumers making decisions in line with CBDT as outlined above. As we furthermore in reality also observe that such trends are reversed as quickly as they emerged, we have a very good fit of our model with reality. Of course our model can then be applied to all kind of short-lived fads and hypes, like the sudden demand for certain computer games, fashion accessories or Christmas toys, who are usually very quickly reversing.

These short-lived fashions lasting often less than a single season are consistent with a relatively short memory of our model, which causes quick changes. Proposing a short memory for such fashions seems reasonable given the significant changes that occur over time, resulting in a low similarity over time (λ). Similarly, these fashions are often not that widespread in a society, which is consistent with a low overall interaction of agents. With fashions usually spreading in smaller peer groups, e.g. a particular age group, this seems to fit with the requirements of our model to produce such properties of these fashions.

We can easily extend the type of choices not only to cover consumer decisions, but also corporate decisions. Examples of fads in the past include the decisions to merge into conglomerates and the more recent trend for spin-offs in order to focus on the core business. We may in the same way look at investments into specific regions, like South America or South-East Asia, or into certain industrial sectors during particular periods in history, or trends in production technologies like Just-in Time production, to name but a few.

Given the importance of such decisions for a company, we would expect those decisions to be based on as much experience as possible, thus corresponding to a high interaction of agents. Furthermore, by adjusting carefully for any details that may reduce the ability to learn from more distant experiences, we would require agents to have a long memory. Our model would thus suggest a widespread and long-lasting herding of agents. This proposal is consistent with observations from such trends in companies, which can last for years and are widespread.

Another application is primarily relevant in financial markets, but to a lesser degree also in other areas: the explanation of herding with the emergence and burst of speculative bubbles. It is common to explain herding behavior of individuals with psychological factors and a desire by individuals to conform with the majority or a specific peer group. In essence these models of herding are in many cases models of imitating other agents, see e.g. [23, 24]. Another group of models uses informational cascades to obtain similar results, see [1] for one of the first of such models or [2, 6] for an overview. In our model we also observe herding as individuals make identical decisions over prolonged periods of time, which are not based on objective information, but the source of this behavior is very different. Agents are not trying to imitate the behavior of other agents but rather seek to learn from their experience by evaluating their decisions. We thus observe herding not as the result of imitation, but as the result of information evaluation. In this sense we could argue that agents are not following the herd, although their behavior resembles that of herding.

The properties of this herding can vary substantially. We can observe herding for many time periods as e.g. in the recent internet boom, but it can also be a much shorter event, e.g. in bear market rallies or corrections in a rising market. In some cases herding is very widespread affecting most investors, while at other time it is restricted to a small group of investors, e.g. specialist fund managers. We can replicate these observations by varying the memory and interactions between agents in our model accordingly; usually the parameter constellations required are consistent with our understanding of the underlying reason in financial markets.

Using these considerations we can summarize by stating that our model would be applicable to situations that involve repeated choices by a large number of individuals who are in interaction with each other.

There are strong epistemological reasons for supporting CBDT over expected utility theory as a positive theory of decision-making. If we accept that agents are applying CBDT in the form presented in this paper, we have found an alternative explanation of herding, which is not based on imitation but the way decisions are made by agents and further research in the decision-making process may be able to decide how well CBDT actually describes this process and thus our model explains herding behavior.

5 Conclusions

We have presented a model in which agents base their decisions on CBDT, which uses the past experience of a selected number of other agents over a given memory length. We used a simple stationary environment in which repeated binary choices had to be made by all agents and investigated the aggregate choices of all individuals. We found that agents are not learning the true structure of the economy and their aggregate choice can deviate substantially from the benchmark of random choices. We found that these deviations were primarily influenced by the degree of interaction between agents and their memory. It was pointed out that the aggregate choices resembled herding behavior under certain parameter constellations and that this herding behavior was not the result of agents imitating each other's behavior but arise from the way decisions are made in CBDT, thus agents are not blindly following the herd. The parameter constellations causing a behavior resembling herding well used usually a long memory and medium interactions between agents. We also pointed out that this model might be suitable to explain the emergence of a multitude of fads like fashions, besides herding.

The scope of this paper is very limited and the results can only be a first step into fully understanding its implications. As we allowed for a very general network structure, it would be very interesting to investigate in as far the network structure affects the results, indications from [3] show that they may have an impact on the outcomes. It is also of importance to fully understand the reasons behind the usually relatively quick change from a majority of buyers to a majority of sellers. Another obvious strand of future research would be a much more detailed investigation into the properties this model, which were outside the scope of this paper. Apart from these more theoretical considerations we face the need to apply it to specific economic problems in order to evaluate its capability to explain actual events.

References

1. Abhijit V. Banerjee. A simple model of herd behavior. *Quarterly Journal of Economics*, 107:797–818, 1992.
2. Sushil Bikhchandani, David Hirshleifer, and Ivo Welch. Learning from the behavior of others: Conformity, fads, and informational cascades. *Journal of Economic Perspectives*, 12:151–170, 1998.
3. Mattias Blonski. Social learning with case-based decisions. *Journal of Economic Behavior & Organization*, 38:59–77, 1999.
4. Colin F. Camerer and M. Weber. Recent developments in modelling preferences under risk. *OR Spektrum*, 9:129–151, 1987.
5. Colin F. Camerer and M. Weber. Recent developments in modelling preferences: Uncertainty and ambiguity. *Journal of Risk and Uncertainty*, 5:325–370, 1992.
6. Andrea Devenow and Ivo Welch. Rational herding in financial economics. *European Economic Review*, 40:603–615, 1996.
7. Itzhak Gilboa and Amit Pazgal. Cumulative discrete choice. *Marketing Letters*, 12:119–130, 2001.
8. Itzhak Gilboa and David Schmeidler. Case-based decision theory. *Quarterly Journal of Economics*, 110:605–639, 1995.
9. Itzhak Gilboa and David Schmeidler. Case-based optimization. *Games and Economic Behavior*, 15:1–26, 1996.
10. Itzhak Gilboa and David Schmeidler. Act similarity in case-based decision theory. *Economic Theory*, 9:46–61, 1997.
11. Itzhak Gilboa and David Schmeidler. Cumulative utility consumer theory. *International Economic Review*, 38:737–761, 1997.
12. Itzhak Gilboa and David Schmeidler. An overview of case-based decision theory. In Liuigi Luini, editor, *Uncertain Decisions: Bridging Theory and Experiments*, pages 215–235. Kluwer, 1999.
13. Itzhak Gilboa and David Schmeidler. Case-based knowledge and induction. *IEEE Transactions on Systems, Man, and Cybernetics - Part A: Systems and Humans*, 30:85–95, 2000.
14. Itzhak Gilboa and David Schmeidler. *A Theory of Case-Based Decisions*. Cambridge University Press, Cambridge, Great Britain, 2001.
15. Itzhak Gilboa and David Schmeidler. Inductive inference: An axiomatic approach. *Econometrica*, 71:1–26, 2003.

16. Itzhak Gilboa, David Schmeidler, and Peter P. Wakker. Utility in case-based decision theory. *Journal of Economic Theory*, 105:483–502, 2002.
17. Itzhak Gilboa and Nicolas Vieille. Majority vote following a debate. *Social Choice and Welfare*, 23:115–125, 2004.
18. Ani Guerdjikova. Case-based decision theory - an application for the financial markets. Mimeo, University of Heidelberg, 2002.
19. Ani Guerdjikova. Asset pricing in an overlapping generations model with case-based decision markers. Mimeo, University of Heidelberg, 2003.
20. Ani Guerdjikova. On the definition and existence of an equilibrium in a financial market with case-based decision makers. Mimeo, University of Heidelberg, 2003.
21. David Harless and Colin F. Camerer. The predictive utility of generalized expected utility theories. *Econometrica*, 62:1251–1290, 1994.
22. Hermann Jahnke, Anne Chwolka, and Dirk Simons. Coordinating demand and capacity by adaptive decision making - an application of the case-based decision theory. Mimeo, University of Bielefeld, 2003.
23. Andreas Krause. Rational choice, social structure and 'animal spirits'. Mimeo, University of Bath, 2003.
24. Andreas Krause. Herding behavior of financial analysts: A model of self-organized criticality. In Mauro Galegatti, Alan Kirman, and Matteo Marsili, editors, *The Complex Dynamics of Economic Interaction: Essays in Economics and Econophysics*, pages 257–268. Springer Verlag, Berlin, 2004.
25. Robert Meyer, Tulin Erdem, Itzhak Gilboa, Wesley Hutchinson, Steven Lippman, Carl Mela, Amit Pazgal, Drazen Prelec, and Joelm Steckel. Dynamic influences on individual choice behavior. *Marketing Letters*, 8:349–360, 1997.

Cultural Evolution in a Population of Heterogeneous Agents

Gábor Fáth[1,2] and Miklos Sarvary[2]

[1] Research Institute for Solid State Physics and Optics - P. O. Box 49, H-1525 Budapest, Hungary fath@szfki.hu
[2] INSEAD - Boulevard de Constance, 77305, Fontainebleau, France miklos.sarvary@insead.edu

Summary. A general theory of cultural evolution is formulated using a cognitive dimension reduction scheme. Rational but cognitively limited agents iteratively invent and redefine abstract concepts in order to best represent their natural and social environment. These concepts are used for decision making and determine the agents' overall behavior. The collection of concepts an agent uses constitutes his/her cultural profile. As the importance of social interactions increase and/or agents become more intelligent we find a series of dynamical phase transitions by which the coherence of concepts advances in the society. Our model explains the so-called "cultural explosion" in human evolution 50,000 years ago as a spontaneous ordering phenomenon of the individual mental representations.

1 Introduction

Culture is the sum of knowledge, beliefs, values and behavioral patterns built up by a group of human beings and transmitted from one generation to the next. Its evolution is investigated in many disciplines using different paradigms. Economics uses game theory methods to understand the development of rules, social norms and other cultural institutions assuming rational behavior of the constituting individuals (agents) [1]. Cognitive and behavioral sciences study how culture is represented in the mind [2], and how these mental representations (concepts, memes, heuristic decision rules, etc.) change under social interactions [3]. Physics views culture and its evolution as a complex, dynamical system, and aims to identify its basic universal properties using simple minimal models [4, 5, 6, 7].

The emergence of culture seems to involve a chicken-and-egg problem: culture becomes what the constituting agents make it to be, and agents, based on their capacity of learning, constantly adapt to the culture they happen to live in. In this sense culture is the dynamical attractor of the complex social dynamics, determined by the agent's actual constraints and social interactions.

The aim of this paper is to investigate the properties of such cultural attractors in an adequately formulated social dynamics model, which integrates the relevant concepts from the social sciences with those from the physics perspective.

One of the popular minimal models of culture is that of Axelrod [4, 5], which takes a stochastic imitation process the basic motif of social interactions, and investigates a population of biologically and economically uniform agents whose cultural profile is initially random. The interesting question is under what circumstances global cultural diversity gets preserved despite the homogenizing interaction [6, 7]. However, the question can also be posed from the opposite direction: in a population with innate heterogeneity and under social interactions which are not necessarily imitational but reflect selfish individual interests how can a coherent culture emerge?

In order to study this question, our model differs from that of Axelrod. Instead of simply assuming imitative interactions, we focus on individual decision making mechanisms and identify *dimension reduction* as the governing heuristics for cognitively limited but otherwise rational agents. Obviously, an agent's cultural profile manifests itself in decision making situations where he has to choose from a set of alternatives. Different biological preferences and cultural background give rise to different conceivable alternatives and different evaluation of these. People with unequal views of the world and diverse convictions about what is right or wrong, true or false, likely or unlikely, important or negligible are biased to make different choices.

In this paper we consider agents who are rational but cognitively limited (bounded rationality [1, 8]). They are constrained to describe and understand their world along a finite number of *abstract concepts* (ACs). These ACs constitute the agent's cultural profile and establish an interface to the objective world: agents use ACs to build an adequate mental representation of the world, which they use to evaluate decision alternatives. Obviously, decision making is successful if evaluation is precise, thus rational agents face the problem of optimizing their AC sets given the actual state of the world. We think about the ACs as flexible mental constructs which are continuously adapting in a learning process for maximizing the agents' predictive power. As such, our model treats the evolution of cultural profiles (the set of ACs) as the result of a rational behavior governed by economic benefit. The ACs are continuous variables, and agents are assumed heterogeneous in their (biological) preferences by default.

Our theory is very similar in spirit to the *renormalization group* schemes of statistical physics as it is based on the following two key elements: (i) it combines the original state variables (attributes of decision alternatives) into important and less important degrees of freedom; agents keep the former (defined to be the ACs), and discard the latter; (ii) this is done in an iterative way aiming to identify fixed points, which are then associated with culture. The criterion for truncating the degrees of freedom is local: each agent tries to maximize his evaluation accuracy, i.e., to minimize the representation/evaluation error within his cognitive limits.

2 A Cognitive Model

We consider I agents, each restricted to use a number K of ACs only. Agents divide the world into a number X of *contexts* in which they evaluate decision alternatives according to their personal preferences. We assume that alternatives are characterized by their objective (physical) attributes $a = \{a_1, \ldots, a_D\}$. Agent i's *theoretical payoff* from choosing alternative a in context x is posited to be a linear function

$$\pi_i^{(x)}(a) = \omega_i^{(x)} \cdot a, \tag{1}$$

where $\omega_i^{(x)}$ is the agent's *context vector*, reflecting actual circumstances of the context and the agent's personal (biological) preferences. For each agent there are X context vectors each of dimension D, which are assumed fixed in the model. An agent does not know his context vectors explicitly (as this would require a detailed understanding of the effect of all attributes on the payoff), but by collecting experience on choices he has made previously, he learns to approximate his payoffs using an appropriate *mental representation* (this learning process is not modeled explicitly). The mental representation is built around the world's K *most important degrees of freedom*, constituting the agent's ACs. Assuming again a linear relationship, the *approximate payoff* that the agent can compute directly is

$$\tilde{\pi}_i^{(x)}(a) = \tilde{\omega}_i^{(x)} \cdot a = \sum_{\mu=1}^{K} v_{i\mu}^{(x)} \gamma_{i\mu} \cdot a \tag{2}$$

where $\tilde{\omega}_i^{(x)}$ is the agent's *approximate context vector*. This is decomposed using *mental weights* $v_{i\mu}^{(x)}$ in a reduced subspace of dimension K. The weight $v_{i\mu}^{(x)}$ reflects the significance of AC μ in context x. Equation (2) implies that agent i possesses a number K of AC vectors, $\{\gamma_{i\mu}\}_{\mu=1}^{K}$, assumed normalized $|\gamma_{i\mu}| = 1$, which the agent uses to evaluate alternatives. We emphasize that the ACs are defined in a context-independent way, and are used in *all* occurring contexts. Note that by Eq. (2) the number of parameters defining the agent (the v's and γ's) is reduced to $(X + D)K$, which can be much less than XD, the number of parameters in $\omega_i^{(x)}$, if $K << X, D$.

The structure of our model is depicted in Fig. 1. We have defined three layers: the layer of Physical Attributes of the decision alternatives, the layer of Abstract Concepts (AC vectors) which constitute the renormalized degrees of freedom, and the layer of context- and agent-specific Payoffs. At the lowest level agents are homogeneous and have identical information about alternatives. In the highest layer agents are heterogeneous and have individual payoff

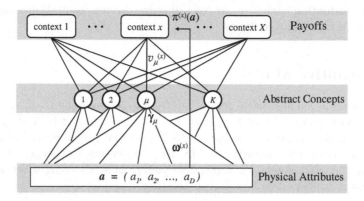

Fig. 1. Decision making heuristics using Abstract Concepts (ACs) as renormalized degrees of freedom.

functions based on their innate preferences. The middle layer shows partial coherence, whose measure, as we will see, is determined by the strength of social interactions. This is the layer we intend to monitor across society for observing the emergence of a (common) culture.

An example may be enlightening at this point. Think about the world of chess where decision alternatives are the possible moves a player can choose in a given situation. These moves are characterized by the positions of the pieces on the board (physical attributes). Calculating the payoff of a move (the probability of winning the game with that move) is basically a two step procedure for human players and chess programs alike. First the move is evaluated along some abstract Concepts like "material advantage", "positional advantage", "pinned piece", etc. Indeed, research in psychology has demonstrated that a human grand master possesses tens of thousands of such chess-related concepts [9, 10]. Second, these concept scores are weighted (mental weights) in an appropriate mental model. The weighting scheme depends on the player's personal skills/style (agent heterogeneity) and on the actual adversary (context dependence). There is a theoretically best move (preference vector) for that player with that adversary in that situation, but that is not known explicitly to the player. Mastering chess amounts to finding an optimal mental representation, i.e., optimal Concepts and optimal weights (approximate preference vectors) which allow a decision close to the theoretical best. In chess there are different schools (cultures) where the applied Concepts have somewhat different meaning. Of course, decision in chess is very non-linear, and our simplified linear model can only grasp the big picture.

The hierarchical structure depicted in Fig. 1 formally resembles a linear two-layer neural network. Note, however, that it is not a microscopic neural network model of a cognitive function, but a phenomenological (macroscopic) model based on a psychologically/economically supported, realistic human decision making strategy.

Due to the reduction of dimensionality, $K < D$, the approximate payoff $\tilde{\pi}_i^{(x)}$ deviates from the theoretical payoff $\pi_i^{(x)}$. The agents' goal is to find the best possible set of ACs and mental weights which minimizes the error of this mental representation under the constraint that only K ACs can be used. The natural measure of agent i's *representation error* is the variance

$$E_i = \sum_{x=1}^{X} \left\langle (\pi_i^{(x)} - \tilde{\pi}_i^{(x)})^2 \right\rangle_x , \tag{3}$$

where $\langle . \rangle_x$ is the average over alternatives in context x. We restrict our attention to the case, where the attributes are delta correlated $\langle a_d \rangle_x = 0$, $\langle a_d a_{d'} \rangle_x = \delta_{dd'}$, and nontrivial structure is assumed in the agent-specific context vectors only. This simplifies the forthcoming analysis without losing essential features. With this proviso, the representation error turns out to be

$$E_i = \sum_{x=1}^{X} \left(\omega_i^{(x)} - \sum_{\nu=1}^{K} v_{i\nu}^{(x)} \gamma_{i\nu} \right)^2 . \tag{4}$$

The agents learn to minimize the error E_i by adapting their AC vectors $\gamma_{i\nu}$ and mental weights $v_{i\nu}^{(x)}$ to the $\omega_i^{(x)}$ structure.

3 Optimal Concepts as a PCA Problem

Theoretically the minimizing parameters can be easily determined. Differentiating w.r.t. $v_{i\mu}^{(x)}$ and assuming that the $K \times K$ metric tensor $G_{i\nu\mu} = \gamma_{i\nu} \cdot \gamma_{i\mu}$ is invertible (note that the ACs are not necessarily orthogonal), the optimal mental weights read

$$v_{i\mu}^{(x)} = \omega_i^{(x)} \cdot \gamma_i^{\mu}, \tag{5}$$

where we have introduced the *dual* AC vectors $\gamma_i^{\mu} = \sum_{\nu=1}^{K} [G_i^{-1}]_{\mu\nu} \gamma_{i\nu}$. Writing this back to Eq. (4) we can write the error as

$$E_i = \sum_{x=1}^{X} |\omega_i^{(x)}|^2 - \sum_{x=1}^{X} \sum_{\mu=1}^{K} (\omega_i^{(x)} \cdot \gamma_{i\mu})(\omega_i^{(x)} \cdot \gamma_i^{\mu}). \tag{6}$$

The first term is a constant, the second term should be maximized for $\{\gamma_{i\nu}\}_{\nu=1}^{K}$. The problems becomes more transparent, if we write the second term in the form $\sum_{\mu=1}^{K} \gamma_{i\mu} \cdot W_i \gamma_i^{\mu}$, by introducing the positive semi-definite, $D \times D$ dimensional *world matrix*

$$W_i = \sum_{x=1}^{X} \omega_i^{(x)} \circ \omega_i^{(x)}.$$ (7)

This matrix encompasses all information about agent i's personal relation to the World. This maximization problem is well-known in many domains of science under the name *Principal Component Analysis* (PCA). A particular solution for the optimal AC vectors is given by the K most significant (largest eigenvalue) eigenvectors of W_i. Note, however, that any non-singular basis transformation yields a different solution with the same error. We can freely define linear transformations in the subspace of concepts; the mental weights adapt, and the overall representation error remains unchanged. It is obvious that if W_i is a non-singular, rank-D matrix, then for $K < D$ the error is strictly positive even with an optimally chosen set of ACs.

The identification of the important degrees of freedom is closely analog to what occurs in White's Density Matrix Renormalization Group Method (DMRG) [11]. In the DMRG the renormalized degrees of freedom are the K most significant eigenvectors of the reduced density matrix. A formal mapping between the DMRG and our method can be established by identifying the DMRG's density matrix for block i with the world matrix W_i, and the renormalized (kept) degrees of freedom with the ACs. The DMRG's *truncation error* is analog to our representation error.

4 Social Interactions

As discussed above, by minimizing the representation error each agent learns to approximate his world matrix in a lower dimensional space. If agents i and j have different preferences, and thus different world matrices, they would necessarily invent different optimal ACs. Cultural evolution boils down to the interaction of these different ACs in a social network. In order to introduce social interactions we cast contexts into two basic categories: those whose context vector only depends on a single agent (*individual contexts*), and those where the context vector for agent i depends on the (simultaneous) decision of at least another agent, say j (*social contexts*). For simplicity only pair interactions will be considered. In social contexts the understanding of other agents' valuation and decision making processes is an asset. Agent i has strategic advantage from being able to estimate j's payoff for her alternatives. Since j, as any other agent, bases her evaluation on her mental representation using her actual ACs, the task appears for i as the approximation problem of the context vector $\tilde{\omega}_i^{(j)}$ (the context now being agent j). Notice that $\tilde{\omega}_i^{(j)}$ is necessarily within j's AC subspace. The incentive to approximate agent j's context vector accurately introduces a "force" which tries to deform i's AC set towards that of j in order to improve i's predictive power in j-related social contexts. The interplay between the two competing goals - optimizing the mental representation rather for individual or for social contexts - is the ultimate factor which determines the society's cultural coherence.

In terms of the above classification, agent i's overall world matrix is a sum of individual and social contributions. As seen above, individual contexts contribute $\boldsymbol{W}_i^0 = \sum_{x=1}^{X} \boldsymbol{\omega}_i^x \circ \boldsymbol{\omega}_i^x$ (x = individual). The contribution of a j-related social context is $\tilde{\boldsymbol{\omega}}_i^{(j)} \circ \tilde{\boldsymbol{\omega}}_i^{(j)}$, which can be expressed explicitly in terms of j's AC vectors $\{\boldsymbol{\gamma}_{j\mu}\}_{\mu=1}^{K}$ using Eq. (2). We realize that this contribution is a rank-K operator, which can also be written as $\tilde{\boldsymbol{\omega}}_i^{(j)} \circ \tilde{\boldsymbol{\omega}}_i^{(j)} = \boldsymbol{B} \boldsymbol{S}_j$, the product of $\boldsymbol{S}_j = \sum_{\mu=1}^{K} \boldsymbol{\gamma}_{j\mu} \circ \boldsymbol{\gamma}_j^{\mu}$ projecting onto agent j's AC subspace, and another operator $\boldsymbol{B}$ acting within this subspace. This latter depends on the actual $v_{j\mu}$ weights, but for simplicity in the sequel we assume that it is an identity operator. This assumption simply states that all directions in j's AC subspace have *equal importance* for i, and is reasonable if i has to predict j's behavior in many different j-related contexts which average out the weight dependence.

Eventually, agent i's overall world matrix takes the form

$$\boldsymbol{W}_i = \boldsymbol{W}_i^0 + \sum_{j \in \mathcal{N}_i} h_{ij} \boldsymbol{S}_j, \tag{8}$$

the sum being over his social network $\mathcal{N}_i$. The parameters h_{ij} measure the relative strength (importance) of social interaction between agents i and j. The above "equal importance" assumption assures that our $\boldsymbol{W}_i$ remains invariant w.r.t. nonsingular local linear transformations of the individual AC vectors. This means that in our model the actual choice of Concepts does not count, only the subspace they span does.

We view the society as a dynamical system in which agents continuously act as best-response optimizers, trying to minimize their representation error given the actual state of nature and other agents. A steady state of this *best-response dynamics* [12] is necessarily a *Nash equilibrium*, where agents have no incentive to unilaterally change their mental representations any further because these are already mutually optimal.

In order to monitor cultural coherence we introduce a *coherence order parameter* (COP) as the population average of AC subspace projector operators

$$\boldsymbol{O} = \langle \boldsymbol{S}_j \rangle_I = \left\langle \sum_{\mu=1}^{K} \boldsymbol{\gamma}_{j\mu} \circ \boldsymbol{\gamma}_j^{\mu} \right\rangle_I. \tag{9}$$

Our COP is a tensor order parameter which measures the collective overlap of the individual AC subspaces. In fact it is a high-dimensional generalization of de Gennes's tensor order parameter introduced in the theory of liquid crystals [13]. The eigenvalue structure of $\boldsymbol{O}$ is useful to characterize the level of coherence in the society and to distinguish different phases. Obviously, if there is perfect order and the individual AC subspaces are all lined up, the COP has eigenvalues $o_1 = o_2 = \cdots = o_K = 1$, and $o_{K+1} = o_{K+2} = \cdots = o_D = 0$. In the

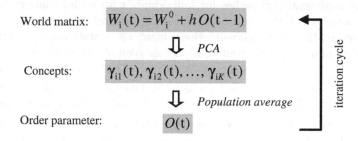

World matrix: $W_i(t) = W_i^0 + h\,O(t-1)$

$\Downarrow$ PCA

Concepts: $\gamma_{i1}(t), \gamma_{i2}(t), \ldots, \gamma_{iK}(t)$

$\Downarrow$ *Population average*

Order parameter: $O(t)$

iteration cycle

Fig. 2. Iteration cycle of the best response dynamics in a mean-field social network. Fixed points of the dynamics are Nash equilibria.

opposite extreme of complete disorder all directions are equivalent and the eigenvalues are $o_1 = o_2 = \cdots = o_D = K/D$. The trace of the COP is always K.

5 Cultural Ordering as a Phase Transition

The equilibrium and dynamical properties of the system depend crucially on the parameters D, X, K, the statistical properties of W_i^0, and the social interaction matrix h_{ij} (connectivity structure). In the following we analyze the case when the agents' individual context vectors $\omega_i^{(x)}$ are Gaussian random vectors with zero mean and unit variance, i.e., without interaction agents choose random AC subspaces (complete disorder). This is a benchmark case which can demonstrate in its purest form how cultural coherence can emerge spontaneously in an interacting social system.

Assuming this, the random matrix W_i^0, which is quadratic in the context vectors, has Wishart distribution [14]. We restrict our attention to a *mean-field* network (all agents are connected) and set $h_{ij} = h/I$. With this the world matrix in Eq. (8) simplifies to

$$W_i = W_i^0 + h\,O. \tag{10}$$

We let the system evolve according to the best-response dynamics: we start from a certain initial condition of individual AC sets. In each step we first calculate the actual COP, then update each agent's AC set according to the K largest principal components of W_i. This gives agent i's best response for the actual environment (individual plus social). This procedure is repeated until a fixed point is reached. A fixed point of the dynamics is necessarily a Nash equilibrium, which we associate with the dynamically emerging *culture* of the population. The scheme of the iteration is illustrated in Fig. 2.

An example for the COP spectrum as determined by a numerical simulation is depicted in Fig. 3. For small h there is complete disorder, while for

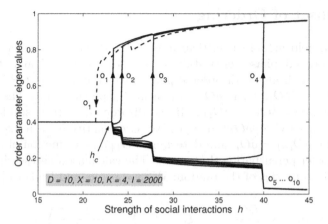

Fig. 3. Fixed point spectrum of the order parameter matrix O for $D = X = 10$, $K = 4$, $I = 2000$. Solid lines indicate increasing h, dashed line (only shown for o_1) decreasing h. The phase transitions are subcritical bifurcations.

$h \to \infty$ the fixed point is completely ordered. In between, however, we observe a series of dynamical phase transitions with hysteresis. Each transition can be associated with a first order jump in the eigenvalues. A qualitatively similar behavior, with a series of first order transitions was found numerically for other parameter values, too.

Our results suggest that the social dynamics has a unique equilibrium only for small couplings. Above the critical value h_c spontaneous ordering occurs in the system, and the number of equilibria increases. On the one hand, there is a discrete number of possible fixed point "families" as represented by the eigenvalue structure of the COP: an adiabatically slow change in h leads to bifurcations at some critical values, where the system jumps from one equilibrium to another. These transitions are irreversible and give rise to hysteresis, which assures the coexistence of at least two different equilibrium families within the hysteresis loop. On the other hand, each family of equilibria is infinitely degenerate in itself, since the associated *eigenvectors* of the COP can point in essentially any direction (respecting orthogonality). Since we assumed that preferences are random, there are no preferred directions on the social level by default. This symmetry of the D-space is broken spontaneously above h_c. As $h \to \infty$ there is only one possible structure for the eigenvalues of the COP (there is only one family), meaning that the society is fully coherent. However, this coherent K-dimensional subspace, representing the most important dimensions of the society is not fixed a priori. The ultimate state of culture gets selected as a result of the idiosyncratic fluctuations present at the time of the symmetry breaking bifurcations.

6 The Onset of Ordering

In the mean field model the critical interaction strength h_c, where the completely disordered phase loses stability can be calculated analytically for $I \to \infty$. Starting from the disordered phase we can write the COP in step l as $O = (K/D)1 + \varepsilon \delta O_l$, where δO_l is an arbitrary perturbation. The dynamics defines a mapping $\delta O_l \to \delta O_{l+1}$. If the disordered state is stable we have $||\delta O_l|| \to 0$, otherwise $||\delta O_l||$ diverges. We can identify h_c as a fixed point of this mapping $\delta O_{l+1} = \delta O_l$, and determine δO_{l+1} as a function of δO_l and h using first order perturbation theory in ε. The calculation necessarily involves the spectral properties of the random W_i^0 ensemble, and leads to [15]

$$h_c = \frac{D^2}{\sum_{\nu=1}^{K} \xi_\nu}, \qquad \xi_\nu = 2 \left\langle \sum_{m \neq \nu}^{D} \frac{1}{\lambda_{i\nu} - \lambda_{im}} \right\rangle_I, \tag{11}$$

where λ_{im} are eigenvalues (in descending order) of the Wishart matrix W_i^0. The joint probability distribution for real Wishart matrices in the Coulomb gas representation reads [14]

$$P(\{\lambda\}) \sim \exp\left(-\sum_{n=1}^{\min(D,X)} V(\lambda_n) + \prod_{n>m}^{\min(D,X)} \ln|\lambda_n - \lambda_m| \right)$$

with $V(\lambda) = (\lambda - (|X - D| - 1)\ln\lambda)/2$. For large D and X a saddle-point approximation can be made to determine the average position of particles (eigenvalues), giving the alternative expression $\xi_\nu = 1 + (D - X - 1)\langle 1/\lambda_\nu \rangle$. When $\nu << D, X$ we can write $\langle 1/\lambda_\nu \rangle \approx 1/\lambda_1 = 1/(D + X + 2\sqrt{DX})$ [14]. Finally, in the limit $D, X \to \infty$ with $X/D = r > 0$ and $K << D$ this leads [15] to the critical coupling

$$h_c \approx \frac{D^2}{K} \frac{1 + \sqrt{r}}{2}. \tag{12}$$

Keeping r fixed, h_c increases in D (quadratically) and decreases in K. When the World is complex (large D and X) and the agents are primitive (small K) the critical coupling is high and the society is likely to stay culturally disordered.

7 Conclusions

In summary, in this paper we have proposed a rather general theory of cultural evolution built upon a dimension reduction scheme. We assumed that bounded rationality forces agents to choose a limited set of abstract concepts in describing their world. These concepts and the way they are used are continuously

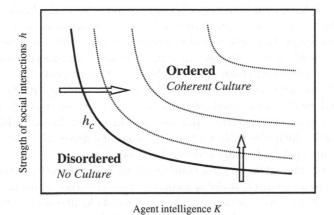

Fig. 4. Schematic phase diagram on the h vs. K plain ($D, X =$ fixed) for the mean-field model with random preferences. Solid line indicates the principal transition calculated in the text, dashed lines the additional transitions.

updated and refined in order to find an optimal representation. Individual interest in making the best possible decisions in natural (individual) and social (collective) contexts act as a driving force for cultural evolution and, at least in the particular cases studied, drive the system towards a Nash equilibrium fixed point with measurable level of social (cultural) coherence. In our model with random preferences this coherence manifested itself as a spontaneous ordering of the individual concept (sub)spaces, and was quantified by an adequate order parameter. We demonstrated that as the fundamental cognitive and social parameters change, cultural evolution goes through a number of phase transitions, which we found first order in the mean-field model. The critical value of the most relevant transition was calculated analytically.

Transition to a phase with substantially higher cultural coherence can be triggered in two different ways: social contexts can increase in significance, or alternatively, agents may develop better cognitive skills (see Fig. 4). A slow increase in the parameters leads to a bifurcation, beyond which the subspaces spanned by the agents' concepts rearrange into a state with higher average overlap, and we can speak about a boost in cultural coherence. A similar transition may explain the sudden cultural explosion for humanity approximately 50,000 years ago [16, 17]. Since in our model the ordering involves a spontaneous symmetry breaking, the emerging fixed point strongly depends on the initial perturbation. Such *path dependence* implies that cultural evolution is not completely deterministic but depends on a number of idiosyncratic factors.

Our theory differs conceptually from some traditional thinking on the possible causes underlying human cultural explosion. Most of these traditional explanations presume a sudden and substantial *biological* change (e.g., genetic mutations [17, 18], the integration of cognitive modules in the brain [16], etc.) as the initial trigger. In contrast to these, the mechanism we have proposed in this paper suggests a *spontaneous ordering* of the mental representations, which emerges when *slow changes* in some relevant variables (cognitive abilities, social interactions) reach a critical level. In this respect the "cultureless state" of the society, originally rooted in the heterogeneity of its members in individual preferences, needs, experiences, and the extreme simplicity of the mental representations they possess, becomes unstable at a bifurcation point, beyond which the collective social dynamics drives the system into a drastically different, culturally ordered fixed point.

It would be interesting to go beyond mean field, and investigate the model's dynamics in more realistic social networks, where the fixed point may possess a more complex internal structure. A non-mean-field-like social structure is expected to give rise to subcultures or cultural domains, where individuals are highly coherent within a domain, but the average cultural subspace the domain represents may be rather different from that of another domain, as occurs usually in real societies. The type of the occurring dynamical transitions may also change character. An evaluation of how a potential structure in the attribute space or non-linearity in the world/mental representations influence the results could also be interesting. Other intriguing problems involve the social dynamics itself, namely how the model should be extended to cope with intrinsically dynamical aspects of culture such as fads and fashions, where the social attractor is much rather a limit cycle or a chaotic attractor.

8 Acknowledgments

The authors acknowledge support from Insead Foundation and the Hungarian Scientific Research Found (OTKA) under grant Nos. T043330 and T047003.

References

1. H. A. Simon. *Models of bounded rationality, Vol. 3: Empirically grounded economic reason.* MIT Press, 1997.
2. A. K. Romney, J. P. Boyd, C. C. Moore, W. H. Batchelder, and T. J. Brazill. Culture as shared cognitive representations. *Proc. Natl. Acad. Sci. USA*, 93:4699, 1996.
3. R. Aunger, editor. *Darwinizing culture: The status of memetics as a science.* Oxford University Press, 2001.
4. R. Axelrod. The dissemination of culture - a model with local convergence and global polarization. *J. Conflict Res.*, 41:203–226, 1997.
5. R. Axelrod. *The complexity of cooperation.* Princeton University Press, 1997.
6. C. Castellano, M. Marsili, and A. Vespignani. Nonequilibrium phase transition in a model for social influence. *Phys. Rev. Lett.*, 85:3536, 2000.

7. G. Weisbuch. Bounded confidence and social networks. *Eur. Phys. J. B*, 38(2):339–343, 2004.
8. J. Conlisk. Why bounded rationality? *Journal of Economic Literature*, 34:669, 1996.
9. L. Mérő. *Ways of Thinking: The Limits of Rational Thought and Artificial Intelligence*. World Scientific, 1990.
10. H. A. Simon and W. G. Chase. Perception in chess. *Cognitive Psycology*, 4:55–81, 1973.
11. S. R. White. Density matrix formulation for quantum renormalization groups. *Phys. Rev. Lett.*, 69:2863, 1992.
12. J. Hofbauer and K. Sigmund. *Evolutionary Games and Population Dynamics*. Cambridge University Press, 1998.
13. P. G. de Gennes. *The Physics of Liquid Crystals*. Oxford University Press, 1974.
14. A. Edelman. *Eigenvalues and condition numbers of random matrices*. PhD thesis, MIT, 1989. downloadable from http://www-math.mit.edu/~edelman.
15. G. Fáth and M. Sarvary. A renormalization group theory of cultural evolution. *Physica A*, 348:611–629, 2005.
16. S. Mithen. *The Prehistory of the Mind*. Thames and Hudson, London, 1996.
17. R. G. Klein and B. Edgar. *The dawn of human culture*. John Wiley and Sons, 2002.
18. Y.-C. Ding, H.-C. Chi, D. L. Grady, A. Morishima, J. K. Kidd, K. K. Kidd, P. Flodman, M. A. Spence, S. Schuck, J. M. Swanson, Y.P. Zhang, and R. K. Moyzis. Evidence of positive selection acting at the human dopamine receptor d4 gene locus. *Proc. Natl. Acad. Sci. USA*, 99:309–14, 2002.

Part V

Auction and Two-sided Matching

Simulating Auctions

Konrad Richter

University of Kiel and McKinsey&Co
konrad_richter@mckinsey.com respectively fidibus1972@yahoo.com

Summary. Current auction theory relies crucially on the assumption that all bidders bid homogeneously according to their Nash Equilibrium bidding strategies. However, it remains silent on whether and how a priori heterogeneous bidders arrive at the NE. This paper investigates computationally whether evolutionary learning in repeated auctions could justify the assumption of NE bidding. Simulations show that Best Response learning of a priori heterogeneous bidders in first-price auctions does not converge to the NE. Instead, bidders involve in permanent mutual adaptation that shows non trivial characteristics.

1 Background

The Revenue Equivalence Theorem (see [9, 6, 8]) is one of the most fundamental Theorems in auction theory. It basically states that expected seller revenue and the final allocation of goods in pure-bidder independent private-value auction formats depends only on the allocation rule but not on the payment mechanism.[1] The Theorem relies crucially on the assumption that bidders play according to their Nash Equilibrium (NE) strategies. Also most other auction literature since then has focused on NE bidding. However, the literature is sparse when it comes to explaining whether and how a priori heterogeneous bidders arrive at NE play.

The usual justification for NE play is that players in games are perfectly rational, they know that each of their opponents is perfectly rational, they know that the opponents know that all the other players are perfectly rational and so on. Clearly, in many games, this is not a very realistic assumption. A common alternative justification for the focus on NE play in repeated games is given by the theory of evolutionary learning. In particular, Best Response

---

[1] Common payment mechanisms are for instance:
First-price auctions: The winning bidder pays her own bid
Second-price auctions: The winning bidder pays the second highest bid
All-pay auctions: All bidders – also the loosing ones – pay their bid

(BR) Dynamics – where each bidder plays the strategy that would have maximized her payoff in the past – converges to the NE in many games (see, e.g., [4, 11, 2]). BR can be interpreted as describing perfectly rational players who just don't know whether their opponents are perfectly rational as well. In that sense the only difference to standard game theory is the disposal of the assumption of a priori player homogeneity.

In [7] I develop a mathematical formalism to quantify the behavior of BR bidders in the particular case of first- and second-price private-value auctions (1PAs resp. 2PAs) with two bidders and values drawn anew in each round from a uniform distribution. By restricting players to linear (value-independent) bidding strategies I am able to derive closed-form expressions for the strategies if one bidder always updates to a BR against a fixed strategy of the opponent. The restriction to linear bidding functions is motivated by the fact that in 1PAs and 2PAs with uniform value distributions and risk neutral bidders, the NE is given by linear bidding strategies $b_i(v_i) = \beta_i v_i$. In the specific case of two bidders with uniformly distributed private values, the NE in a 1PA is given by $\beta_i = 0.5$. In 2PAs, the NE is bidding the true value ($\beta_i = 1$), irrespective of the number of participants in the auction or the underlying value distribution.

In addition to different auction formats, I also investigate the influence of memory capabilities on bidders' behavior. At the one extreme I consider myopic Best Response (mBR) bidders who in each round play the strategy that would have maximized the payoff in the last auction. At the other extreme I consider perfect memory Best Response (pBR) bidders who play the strategy that would have maximized the cumulated payoff of the whole game up to the present round. Perfect memory Best Response is in the literature also often denoted as Fictitious Play.

Both, mBR and pBR, are extreme cases which are not a plausible model for the behavior of human players. For the interpolation between these two extremes, I consider truncated Fictitious Play (tFP): Players update to the strategy that would have maximized their cumulated payoff in the last nRds rounds. Note that nRds= 1 corresponds to mBR while nRds= L corresponds to pBR if L denotes the total number of rounds played in the game so far.

The mathematical results show that the Revenue Equivalence Theorem breaks down under Best Response Dynamics: BR in repeated 2PAs leads bidders to pursue their NE bidding strategies of bidding their true value as long as they are not completely myopic. The underlying reason is that bidding the true value is a (weakly) dominant strategy and therefore is included in every Best Response set - irrespective of the opponent's bid.

In contrast, bidders in 1PAs fail in general to find the NE under BR: If bidders in 1PAs pursue pBR, they converge to the NE from below. However, it takes them infinitely long to arrive there since the opponent's below-NE strategies from early stages of the game push the respondent's strategy below the NE forever thereafter. If, however, bidders are modeled with limited memory capabilities via tFP, they can aggregate too little information to accurately estimate the opponent's strategy from her bids. This uncertainty about their opponent's true strategy pushes play constantly below the NE resulting in permanent fluctuations around a mean below the NE. As a consequence the players' strategies are not Best Responses to each other any more. Therefore, under the realistic assumption of finite memory capabilities, bidders in 1PAs lock in permanent mutual adaptation of heterogeneous bidding strategies which in turn leads to a suboptimal allocation of goods.

The positioning of this research is at the boundary of auction theory and the theory of evolutionary learning. The research differs in two technical aspects from the usual focus on matrix games in evolutionary learning theory: First, bidders can not observe the opponent's bidding strategy directly but only her bid $b_i(v_i) = \beta_i v_i$ that depends on the strategy and the (unobservable) private value of the opponent. Second, each bidder has a continuum of feasible strategies as opposed to the finite strategy set that is usually considered in the literature. On the other hand, the research differs also from most other papers on auction theory in that it doesn't investigate how the auction setup influences NE-bidding[2]. Instead, it investigates if and under which assumptions initially heterogeneous bidders arrive at the NE and what the dynamics of repeated auctions looks like.

Since this is a rather unexplored research area, I used two complementary methodologies for the investigation. On the one hand, I derived in [7] a mathematical formalism to analyze the behavior of BR bidders in 1PAs. On the other hand, I developed a computational model, the Auction Simulator (AS) that is – among other setups – capable of simulating Best Response strategy updating with linear bidding strategies. The simulation results are the focus of this paper.

The correct replication of the mathematical results by the simulation serves as a double-check for both, the mathematical formalism and the simulation. Therefore, the AS can be used to reliably investigate auctions at a more detailed level than would be possible by purely mathematical methods.

In this paper I present the results of simulation runs with the AS which investigate the internal dynamics of 1PAs. In section 2 I present the basic structure of the AS. Section 3 compares the simulation results with the mathematical predictions of [7]. Section 4 presents results of simulations of truncated Fictitious Play and shows how the AS can be used to assess Nash Equilibria of auctions. Section 5 concludes and gives an outlook for future research.

---

[2] A notable exception in the literature is [3]. This paper focuses on learning in repeated first-price auctions where bidders' values are the same in each auction. Under this assumption the paper shows that evolutionary learning converges to the NE of the one-shot auction. My research instead investigates auctions where values are drawn anew in each auction.

2 The Auction Simulator

The AS is programmed in SWARM. SWARM is basically a library of Objective C and Java routines that was designed at the Santa Fe Institute for facilitating agent based modeling. It is distributed freely under the GNU General Public License and is downloadable from http://wiki.swarm.org. It runs under Unix, Linux, Windows and Apple operating systems. The source code of the AS can be obtained from me on request.

2.1 Simulation Flow

The AS simulates an arbitrary number of bidders. Each bidder i has access to a private set of nStrat strategies. Strategies are real numbers β_i. Each bidder i chooses an active strategy β_i^* that determines the bid $b_i(v_i) = \beta_i^* v_i$ that she is actually placing in a specific auction. All other strategies are evaluated as well to see how they would have performed if they would have been the active ones. Changes of the active strategy are possible only every nRds rounds. For the strategy updating there are two alternative mechanisms available. With Fixed Strategies (FS), a new strategy is chosen as active while the strategy set remains fixed. With Genetic Algorithms (GAs), the strategy population is additionally replaced by a new generation.

A simulation run with nGen generations consists of the following steps:

1. in the beginning, participants initialize their active and nStrat-1 non-active strategies
2. for nGen generations
   a) bidders reset their current payoffs for all strategies
   b) for nRds auctions
      i. bidders' values are randomly chosen from the underlying value distribution
      ii. bidders submit their bids $b_i(v_i) = \beta_i^* v_i$ according to their active strategy β_i^*
      iii. the winner is determined
      iv. bidders update the current payoffs of their active and passive strategies
   c) bidders evaluate the payoff, their strategies (would) have generated during the nRds auctions and choose an active one for the next nRds auctions (see below)
   d) if the GA is used, the strategy population is updated by the genetic operators crossover and mutation
   e) proceed with (a)

2.2 Strategy Updating

A straight-forward implementation of the mathematical model of Best Response is implemented by a fixed strategy (FS) grid for players. Under FS

I partition the strategy space $(0, 1)$ into a grid of equidistant strategies, e.g. nStrat=1000 ranging from 0 to 1.

In each auction the AS calculates for each bidder the payoff that each of the 1000 strategies would have generated if it would have been active in the auction. Focusing on bidder 0 in a two-bidder auction with bidder 0 and bidder 1[3], the current payoff $PO_0^{\text{curr};(i)}$ in each auction round i for each of bidder 0s strategies β_0 is given by

$$PO_0^{\text{curr};(i)}(\beta_0) = \begin{cases} (v_0 - \beta_0 v_0)\Theta(\beta_0 v_0 - \beta_1^* v_1) & \text{in 1PAs} \\ (v_0 - \beta_1^* v_1)\Theta(\beta_0 v_0 - \beta_1^* v_1) & \text{in 2PAs .} \end{cases}$$

The Unitstep-function

$$\Theta(\beta_0 v_0 - \beta_1^* v_1) = \begin{cases} 1 & \text{if} \quad \beta_0 v_0 \geq \beta_1^* v_1 \\ 0 & \text{if} \quad \beta_0 v_0 < \beta_1^* v_1 \end{cases}$$

guarantees that a strategy only gets nonzero payoff if it would have won the auction.

Each nRds auctions, each strategy's payoffs of the last nRds auctions are added up and yield the cumulated payoff of that strategy:

$$PO_0^{\text{cum}}(\beta_0) = \sum_{i=1}^{\text{nRds}} PO_0^{\text{curr};(i)}(\beta_0) .$$

After, the strategy that generated the highest payoff is selected as the active one. If no strategy has positive payoff[4], the currently active strategy remains active, $\beta_0^{\text{new}} = \beta_0^{\text{old}}$. If more than one strategy would have maximized the cumulated payoff, the AS stays with the previous strategy if it is included in the Best Response set.

This simulation setup allows the simulation of all kinds of memory capabilities: Setting nRds= 1 simulates myopic Best Response, nRds> 1 allows the simulation of truncated Fictitious Play. Additionally, the AS allows for the aggregation of the payoff over all auctions to simulate perfect memory Best Response.

As an alternative tool for the simulation of learning I employ Genetic Algorithms (GAs). However, simulations with GAs run the danger that the results are governed by the internal GA dynamics and therefore do not correspond to the theoretical concept of Best Response anymore. This is especially true if we consider myopic Best Response because then the fitness landscape fluctuates too frequently to allow the GA to converge to the optimum. Still, by setting nRds sufficiently high, GAs are a fast tool to assess the approximate position of the NE. I will say more on this in section 4.2.

---

[3] Setups with more bidders are straight-forward generalizations

[4] The case that no strategy has positive payoff happens for instance in 1PAs under mBR if $v_0 < \beta_1 v_1$ because in this case, no feasible strategy $\beta_i \in [0, 1]$ could have won the auction

3 Reproduction of Analytic Results

In [7], I derive for a variety of different learning models closed form solutions if two bidders bid in 1PAs with uniform value distributions. In particular, I obtain formulas that predict how one bidder adapts under mBR or pBR if the opponent plays a fixed strategy. For all responses against fixed strategies, the AS replicates the analytic predictions very accurately with FSs. The simulation results are within a 1%-range of the analytic predictions.

For mutual adaptation, the mathematical results predict that strategies under pBR converge slowly from below to the NE whereas under mBR they converge from above to 0.

Table 1 reports the mean strategy for mBR in dependence of the number of strategies nStrat that are used throughout the simulation.

Table 1. Myopic Best Response

nStrat	9	99	999	9999	49999
β_0^{FS}	0.3009	0.1194	0.0608	0.0387	0.0282

The results show that strategies – in contrast to the mathematical prediction – in general do not converge to 0 under mutual adaptation. However, the simulations for increasing numbers of strategies show that this is a finite size effect. As the number of strategies goes to infinity, the strategies converge to 0 under mutual adaptation and are therefore in line with the theoretical prediction. How can we understand this?

The mathematical analysis predicts that strategies converge to 0 for two mutually adapting bidders who update according to myopic Best Response. In the simulation, however, there exists in each bidder's strategy set a lowest strategy $\beta^{\min} > 0$. In the case of 1.000 strategies this would be $\beta^{\min} = \frac{1}{1.000}$. A bidder who would be driven by mBR to play a strategy below this lowest strategy is forced to overbid because the lowest strategy guarantees her positive expected payoff whereas bidding 0 gives 0 payoff for sure. This effect constitutes a barrier against convergence to 0. Now, imagine for instance that player 1 plays her minimum strategy $\beta_1^{(t)} = \frac{1}{1.000}$ and that the random values in a particular round are given as $v_1^{(t)} = 0.5$ and $v_0^{(t)} = 0.0007$. Then, the best response of bidder 0 is given by $\beta_0^{(t+1)} = \frac{\beta_1^{(t)} v_1^{(t)}}{v_0^{(t)}} \sim 0.715$. So, even if the opponent plays her minimum strategy, there always is a positive probability that each bidders' strategy breaks away from 0. This probability doesn't decrease over time. As soon as one strategy breaks away from 0, the other strategy subsequently also breaks away and the strategies involve in mutual oscillations. These oscillations eventually die out, leading the system near 0 again.

Therefore, the simulation of mutually adapting mBR bidders shows clustered volatility in the time series of strategies and asset prices[5].

Note that the effects of decimalization quickly vanish as the number of bidders or nRds increases.

To investigate perfect memory Best Response, I set nRds= 1.000 and cumulate the payoffs over the whole game. I simulate 24.000 generations, that is a total of 24.000.000 auctions. The following table reports the averages of the first to the fourth quartile of 6.000 generations each.

Table 2. Perfect Memory Best Response

generation	0-6.000	6.000-12.000	12.000-18.000	18.000-24.000
$\bar{\beta}_0{}^{FS}$	0.4574	0.4771	0.4818	0.4842

In agreement with the mathematical prediction, strategies converge to the NE of 0.5 but it takes them infinitely long to arrive there.

In summary, the comparison between mathematical predictions and simulation results shows that the AS is free of bugs. However, in the case of mutually adapting mBR bidders, the simulation results differ from the mathematical predictions due to the quantization of the strategy space. This shows that bid decimalization can be an important factor in determining the outcome of repeated auctions.[6] The decimalization effects vanish quickly if either bidders are equipped with nonzero memory capabilities or if more than two bidders participate in the auctions.

4 Simulation Results

Since the AS is free of bugs, I can use it to investigate also setups where the mathematical analysis would be too tedious. In the following section I follow this route by looking at more general auction setups.

4.1 Truncated Fictitious Play

In the following I consider a simulation of 10 bidders who mutually adapt by truncated fictitious play with nRds=100: Bidders calculate in each auction for

---

[5] The price of the asset, i.e., the seller revenue, is given by the higher of the two bids.

[6] Actually, I model here the effects of strategy quantization, not those of bid quantization. However, the effect that bidders do not converge to 0 is the same in both cases.

every strategy the payoff it would have generated if it had been the active one in the previous 100 rounds. Every 100 rounds they update their strategy to the strategy that would have maximized their payoff if it had been the active one. Figure 1 shows typical time series over 200 time steps - i.e. over 20.000 auction rounds - between 18.500 and 18.700.

The upper left plot shows the strategy of bidder 0. Of course, under mutual

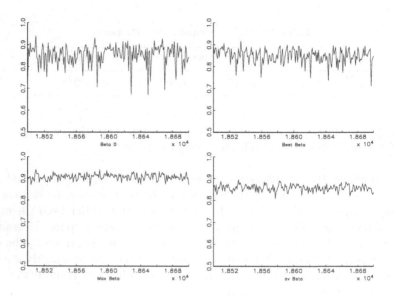

Fig. 1. Truncated Fictitious Play with Uniform Value Distribution

adaptation, the strategies of the other bidders look qualitatively the same. As predicted by the mathematical analysis, strategies fluctuate permanently. Moreover we see from the plot that the fluctuations are asymmetric: There are occasional breakouts to extremely low strategies whereas corresponding breakouts to high strategies are not observed. In consequence, the histogram of bidder 0's strategies (not shown) has an asymmetric shape, it is skewed to the left. Table 3 quantifies the skewness: the mean of β_0 is below the median of β_0. This shows that there are less strategies below the mean than above the mean. Therefore, the low strategies must show more significant deviations from the mean than those above the mean - the distribution is skewed to the left.

The upper right plot shows for each time step the strategy of the bidder who has obtained the highest payoff in the previous **nRds** rounds. The plot supports the theoretical result that the best response against an environment of mutually adapting bidders is in general not the NE. Again we see visually from the plot or from Table 3 that the distribution is skewed to the left. Note that the fluctuations of the best strategies to low strategies are much more modest than those of the individual strategies. This is intuitively clear: A player with high values throughout the **nRds** auctions of a certain time step, will play a low strategy in the next round. Such a low strategy occurs then in the time series of the bidder's strategies. However, the chance that she will be the most successful among all the other players with this low strategy in these next **nRds** auctions – which would put this strategy into the time series of best strategies – is very low, since in these next auctions new value realizations are drawn.

The lower two plots in Figure 1 shows the maximum strategy and the average of the strategies used by all bidders in the corresponding time steps. Visually and by comparison with Table 3 we see that these strategies are distributed symmetrically around their mean.[7]

The maximum strategy is on average *above* the NE. So, though mathematically the best response function doesn't get higher than the NE, the finiteness of **nRds**= 100 lets strategies also increase beyond the NE because then the aggregation of the values over the 100 rounds is subject to stochastic shocks. The average of all strategies played by the bidders, however, is usually below the NE. A priori it is not clear, how important the maximum strategy is in determining the seller revenue when compared to the average strategy: The bidder with the highest strategy is on average the most likely to win each auction, therefore the maximum strategy is the most important single strategy for determining seller revenue. On the other hand, due to the random value realizations, the highest strategy will not win all of the **nRds** auctions but all players will win some of the auctions. Therefore, the average strategy is also an important factor in determining the expected seller revenue. The question is, which of the two is more important in determining the seller revenue[8].

To get a feeling for which of the factors is more important for the determina-

	β_0	β^{best}	β^{max}	β^{av}
mean	0.858	0.849	0.907	0.858
median	0.864	0.854	0.907	0.859

Table 3. Strategy Means and Medians under Truncated Fictitious Play

---

[7] β_{av} might be slightly skewed to the left. However, this remains to be assessed in longer simulation runs.

[8] If, for instance, the average strategy was 5% below the NE, the maximum strategy was 1% above the NE and seller revenue was one 1% below the NE level, then the maximum strategy would be more important in determining the seller revenue.

tion of seller revenue, I computed the effective strategy. This is the strategy that would give the same seller revenue as truncated fictitious play if all bidders would use it constantly as their active strategy. To calculate it I use the seller revenue: Average seller revenue in a first-price auction with 10 bidders where bidders bid according to the NE of $\beta_i = 0.9$ is in the simulation obtained as 0.81838. Average seller revenue in the 1PA with truncated fictitious play is 0.78660. Therefore, the effective strategy of the bidders is given as

$$\beta_i^{\text{eff}} = 0.9 \cdot \frac{0.78660}{0.81838} = 0.86505 \ . \tag{1}$$

Comparison with the means in Table 3 show that this is much closer to the mean of the average strategy than to the mean maximum strategy. Therefore, the average strategy is the more important factor for seller revenue than the maximum strategy. The intuition behind this is the following: Bidders' strategies in a given time step typically are between 0.6 and 0.9 whereas values come from $U(0, 1)$. Therefore, the more important determinant of the single bid $b_i(v_i) = \beta_i v_i$ is the bidder's value. So, the bidder with the maximum strategy will not make that many more successful bids than an average bidder and therefore she is not the main driver of seller revenue. An open question is, what would happen if values come from a sufficiently narrow normal distribution. Values would be closer together but strategies as well. It remains a topic for further research to assess wether the effective strategy moves closer to the maximum strategy or wether its position between average and maximum strategy is a constant that is independent of the underlying value distribution. I want to assess this in the future when I have extended the AS for the simulation of nonlinear strategies.

The simulation result for a 2PA with 10 NE bidders is an average seller revenue – cumulated over 100 rounds – of 81.835. This is virtually the same as the 81.838 in the first-price auction in the NE. The equality highlights the validity of the Revenue Equivalence Theorem in this special setting if bidders play according to their NE bidding strategies.

The standard deviation of the seller revenue – cumulated over 100 rounds – in the 1PA where bidders use their NE strategies is 0.74 whereas in the second-price auction it is 1.112. This is simply a consequence of the fact that the variance of the second highest of 10 values is higher than the variance of the highest of 10 values. In the literature this observation has been used to claim that risk averse sellers should prefer first-price auctions over second-price auctions.

Under truncated fictitious play, this result doesn't hold anymore. In 1PAs, in addition to the random value realizations, the seller revenue is influenced by the fluctuating strategies. The standard deviation of cumulated seller revenue in the 1PA under truncated fictitious play with numBidders= 10 and nRds= 100 is 1.417 which is about 25% above the standard deviation in the second-price auction. In consequence, I would argue that only a sufficiently risk-loving seller should employ first-price auctions to allocate goods. Risk neutral and risk averse sellers would prefer second price auctions since they give more expected revenue and less risk for that revenue.

To assess the dynamics of truncated fictitious play in more detail, I estimate an $AR(p)$-model[9] for the average strategy β_{av}. The first step in estimating such a model is to decide on the number of lags of the observable. Allowing for too few parameters might neglect useful information, allowing for too many parameters dilutes the focus on the core drivers of the time evolution. For the model specification I chose p that minimizes the Bayesian Information Criterion (BIC). The BIC is defined as

$$BIC_p = \ln \sigma_p^2 + \frac{\ln T}{T} p \qquad (2)$$

Here, T is the number of time steps in the sample, σ_p the standard deviation of the residual error and p the number of considered lags. The first part in BIC_p gives lower values, the better the goodness of fit is. The second part gets the bigger, the more parameters the estimation uses.

Estimation with different settings for p in GAUSS shows that the BIC is minimized for $p = 2$ with a value of -136420. Therefore I choose the AR-Model

$$\beta_{av}^{(t)} = \alpha_0 + \alpha_1(\beta_{av}^{(t-1)} - \alpha_0) + \alpha_2(\beta_{av}^{(t-2)} - \alpha_0) + \epsilon^{(t)} \qquad (3)$$

The parameter estimates are given by GAUSS as

$$\alpha_0 = 0.858 \pm 0$$
$$\alpha_1 = 0.344 \pm 0.007 \ (0.330, 0.357)$$
$$\alpha_2 = -0.030 \pm 0.009 \ (-0.047, -.0.014) \ ,$$

where the numbers in the bracket give the upper and lower confidence interval at the 2σ-level. All coefficients are significantly different from 0 at the 2σ- level and therefore the $AR(2)$-equation for the average strategy is given as

$$\beta_{av}^{(t)} = 0.858 + 0.344(\beta_{av}^{(t-1)} - 0.858) - 0.03(\beta_{av}^{(t-2)} - 0.858) + \epsilon^{(t)} \qquad (4)$$

That high average strategies promote high average strategies in the next round is intuitively clear: If each bidder faces at time t an environment of high bidding competitors, she will in the next round also use a high strategy. Conversely, by low bidding competitors she is led to use a low strategy in the next round. Compared to a random walk, this $AR(2)$ process leads to longer consecutive periods of the average strategy above the mean respectively below it. So, in repeated auctions we might expect periods of fiercer competition among bidders that interchange with periods of lower competition. This process is purely internally generated and stems only from the mutual adaptation of bidding strategies.

Analogously I estimated an $AR(2)$-model of the cumulated seller revenue over the time steps of 100 rounds and obtained the parameter estimates

---

[9] I do not consider general ARIMA-models since the time series is obviously stationary. Estimation of an $ARMA(1,1)$-model gives the same BIC-criterion as an $AR(2)$-model, so alternatively we could also investigate this model.

$$\alpha_0 = 78.6 \pm 0$$
$$\alpha_1 = 0.285 \pm 0.006 \quad (0.273;\ 0.298)$$
$$\alpha_2 = -0.009 \pm 0.004 \quad (-0.017; -0.001)\ .$$

The process is qualitatively the same as for the average strategy: Periods of relatively high seller revenue change with periods of relatively low seller revenue. However, the parameter α_1 is lower than in the $AR(2)$ model for β_{av}. The reason is that the random value realizations dilute the effects of the time evolution of the strategies. I did not estimate AR-models for the individual strategy β_0 and the minimum strategy β_{min} since they are not symmetrically distributed around their mean and therefore an AR model with a normal noise specification is not apt for their investigation.

By setting nRds= 1.000, I obtain the following $AR(2)$-models for β_{av}

$$\beta_{av}^{(t)} = 0.858 + 0.374(\beta_{av}^{(t-1)} - 0.858) - 0.042(\beta_{av}^{(t-2)} - 0.858) + \epsilon^{(t)} \quad (5)$$

and for the cumulated seller revenue

$$PO_{sell}^{t} = 807 + 0.296(PO_{sell}^{(t-1)} - 807) - 0.005(PO_{sell}^{(t-2)} - 807) + \epsilon^{(t)}\ . \quad (6)$$

Comparing the coefficients with the coefficient estimates of the simulation with nRds= 100 shows that the parameter nRds has only little effect on the time evolution of $\beta_{av}^{(t)}$ and $PO_{sell}^{(t)}$: Increasing nRds leads to a slight increase of the parameter values since the influence of the random value realizations decreases.

The coefficients depend only on the number of bidders in the auction. Table 4 shows the coefficients of the $AR(2)$ estimation for different numbers of bidders. The first-order autocorrelation decreases slowly when we increase the number

numBidders	2	5	10	20
$\alpha_0(\beta_{av})$	0.423	0.395	0.374	0.34
$\alpha_1(\beta_{av})$	-0.009	-0.032	-0.042	-0.041
$\alpha_0(PO_{sell})$	0.396	0.328	0.295	0.274
$\alpha_1(PO_{sell})$	0.003	-0.006	-0.005	-0.012

Table 4. Coefficient Estimates for Different Number of Bidders

of bidders. It remains a topic for further research to investigate wether the autocorrelation vanishes for numBidders$\to \infty$. I want to investigate this when I have access to faster computer facilities.

4.2 Computational Assessment of Nash Equilibria

[10] proves that the NE in any k^{th} price auction with n bidders and underlying uniform value distributions is given by the linear bidding strategy

$$b(v) = (1 + \frac{k-2}{n-k+1})v$$

Since the AS can simulate linear bidding strategies and we have seen that pBR converges (slowly) to the NE in a 2 bidder 1PA, it is natural to ask if we can use the AS in general to assess these NEs in general.

As an example I investigate a 4th price auction with 8 participants. The NE strategy of this auction is $b(v) = 1.4v$.
Since initially I don't know anything about the position of the NE, I use the GA to find the approximate position.[10] I put nRds=49.999 to limit the effects of fictitious play truncation and calculate the average of the last 50 of 150 generations. The result indicates that the NE is at 1.3998 ± 0.0133 which is in perfect agreement with the theoretical prediction.

In general, for all k^{th} price auctions with n bidders and uniform value distributions, I was able to get the analytically predicted result for the NE.
For general auction setups like, e.g., asymmetric or normal value distributions it is often difficult if not impossible to calculate the NEs. Therefore the AS could provide significant help in identifying NEs of more complex auction setups. However in these setups, the NE is usually not given by linear bidding strategies. Therefore, I want to extend the AS in the future to allow for nonlinear bidding functions.

5 Conclusion

In this paper I have used the Auction Simulator to investigate Best Response learning in repeated first-price auctions. Bidders bid in repeated auctions for assets. They have different asset valuations which change for every auction. Over time, each bidder tries to learn her payoff-maximizing bidding strategy. Bidders can be modeled with different memory capabilities (myopic, perfect memory or in between with truncated Fictitious Play).

The simulation results confirm the mathematical predictions that were derived in [7]. Bidders in 1PAs under Best Response dynamics show excess volatility of strategies and therefore also of seller returns if bidders don't have perfect memory. Each bidder's Best Response depends on all the other bidders' strategies and the learning process doesn't converge to the NE because too much stochasticity is introduced by the random value realizations. The permanent mutual adaptation of bidding strategies generates a rich internal dynamics. 2 bidders under myopic play lock in mutual strategy oscillations that are characterized by clustered volatility. As the memory strength of bidders increases, extended periods of fiercer competition alternate with periods of lower bidding. In general, bidders in 1PAs pursue heterogeneous bidding

---

[10] For FS I would have to determine a strategy range in the beginning of the simulation.

strategies. This heterogeneity leads to a suboptimal allocation of goods because the bidder with the highest value does not necessarily obtain the auctioned good.

In contrast, bidders in repeated 2PAs find the NE easily by following Best Response strategy updating with any positive memory strength. Since bidding strategies are homogeneous in the NE, the 2PA is in the long run maximally efficient. The underlying reason for the convergence is that the 2PA has a dominant strategy of bidding the true value. Therefore the Best Response doesn't depend on the opponents' strategies.

In general, auction formats with dominant strategies are maximal efficient under Best Response. For multi-unit auctions, a dominant strategy auction format would be the Ausubel auction (see [1]) or the generalized Vickrey-auction. For single-unit double auctions, McAfee has proposed a dominant strategy mechanism (see [5]). The development of corresponding formats for more general auction setups is a topic for further research.

Particularly interesting in this context are double-sided multi-unit affiliated-value auctions since they correspond to order books in financial markets. By switching to dominant-strategy order books, the strategic mutual adaptation of traders' bidding strategies could be minimized. The resulting reduction of the volatility of asset prices could benefit the world economy since it leads to a more efficient capital allocation and to less investment risk.

Further promising fields for auction redesign include order books for electricity markets, the restructuring of supply chains, IPOs, treasury bill auctions, procurement auctions and many more.

Simulation can be a powerful tool in the search for more efficient auction formats. In the future I want to extend therefore the Auction Simulator to allow the simulation of a wider range of auction formats, for non-linear bidding strategies and for interdependent values.

Table 5. Abbreviations

Abbreviation	Meaning
1PA	First-price Auction
2PA	Second-price Auction
AS	Auction Simulator
BR	Best Response
mBR	Myopic Best Response
pBR	Perfect Memory Best Response
tFP	truncated Fictitious Play

References

1. L. M. Ausubel. An efficient ascending-bid auction for multiple objects. *forthcoming in American Economic Review, Working Paper Version at www.ausubel.com/auction-papers/efficient-ascending-auction-r.pdf*, 2004.
2. L. E. Blume. The statistical mechanics of best-response strategy revision. *Games and Economic Behavior*, 11:111–145, 1995.
3. S. Hon-Snir and A. Sela. A learning approach to auctions. *Journal of Economic Theory*, 82:65–88, 1998.
4. M. Kandori, G. J. Mailath, and R. Rob. Learning, mutation and long run equilibria in games. *Econometrica*, 61(1):29–56, 1993.
5. P. McAfee. A dominant strategy double auction. *Journal of Economic Theory*, 6:434–450, 1992.
6. R. B. Myerson. Optimal auction design. *Mathematics of Operations Research*, 6:58–73, 1981.
7. K. Richter. Revenue equivalence revisited. *Workingpaper; mail to* konrad_richter@mckinsey.com *respectively* fidibus1972@yahoo.de, 2004.
8. J. G. Riley and W. F. Samuelson. Optimal auctions. *American Economic Review*, 71:381–392, 1981.
9. W. Vickrey. Counterspeculation, auctions and competitive selaed tenders. *Journal of Finance*, 16(1):8–37, 1961.
10. E. Wolfstetter. Third- and higher-price auctions. *Working Paper; www.wiwi.hu-berlin.de/wt1/papers/1996/third_and_higher_price_auctions.pdf*, 1996.
11. H. P. Young. The evolution of conventions. *Econometrica*, 61(1):57–84, 1993.

Counterclockwise Behavior Around the Beveridge Curve

Koji Yokota

Otaru University of Commerce yokota@res.otaru-uc.ac.jp

It is an established fact that the time-series plot on the uv plane moves around the Beveridge curve. We show that such a behavior is pertinent to the matching process under imperfect information.

1 Introduction

It is natural to consider that stable matching will arise as an equilibrium when agents in the matching market have perfect information about other agents. The original setting of this problem by [6], called stable marriage problem, is that there are two groups of agents, men and women, each having preference over the agents in the other group and being single. The objective of this problem is to make matches between men and women, one by one, in a sustainable way. Namely, the *matching*, i.e. a pattern of matches, should satisfy such a property that, in equilibrium, there should be no agents who are not matched as a pair but both of whom have incentive to form a new pair changing from the current partner. We call such a matching *stable matching*. In the current literature, it is assumed that not only agents in the matching market but also their preferences are common knowledge. [3] introduced a notion of equilibrium, called *bargaining equilibrium*, when utility is transferable between agents in a matched pair. This transferable utility case applies to the matching in the labour market because part of utility of an agent in a matched pair can be transferred to the other in the form of wage payments. In the bargaining equilibrium, a pair equally divides the surplus of total payoff of this pair over the sum of their threats and distribute it on each threat. The threat is the payoff the agent receives in a new matching once the original pair becomes unavailable. Since we need to know the threat in the new matching game, we need to calculate threat recursively until we reach to a trivial matching game in which only one pair is available. The above operation determines the pattern of utility transfer within all possible pairs, which transforms the utility profile of each agent when utility is not transferable into another utility profile. The bargaining equilibrium is a stable matching equilibrium upon this transformed utility.

Then, the question arises: can the bargaining equilibrium explain the data observed in the labour market? The answer is unfavourable. [4] and [12] showed that, under certain probability distributions that generate evaluation of the counterpart, such a matching process leads to *complete matching* when the market becomes large in size. It implies that there should be no observed prolonged unemployment. [4] assumed that the utility that an agent feels from a potential partner is drawn from a special case of extreme value distribution. The utility to be single is also drawn from the same distribution but multiplied by a scaling factor. The great contribution of [4] is that he obtained an asymptotic but analytical formula for the number of matches. As far as only number of matches concerns, the assumption on the probability distribution is harmless because what matters is the preference order of each agent and the level of utility does not affect the pattern of matching. However, the assumption that utility of being single is random may be arguable. In search context, the factors which affect the decision to stay single are the expected utility to match in the next period, unemployment benefits and intertemporal discount rate, which are all not random. However, [12] confirmed that the same result holds for the cases where the utility of being single is constant. He also found that the convergence to complete matching occurs fairly quickly and under empirically conceivable size of matching market, all agents in the smaller group should be matched.

These stochastic models assume independence of probability distributions from which evaluation of an agent by another is drawn. Since utility is transferred within a pair in the form of wages, this is a fictitious story under the job matching situation. [13] showed that the bilateral utilities within any pair after the transfer of utility will exhibit positive correlation. It has the effect to raise the matching probability and it only strengthens the tendency to reject the hypothesis that job matching is described by a stable matching process.

All the above results suggest that some kind of disruption of perfect information plays a key role in job matching. Therefore, we want to see what happens if searching agents cannot find others costlessly. We will find that the matching process which arises from there brings prolonged unemployment and unfilled vacancies, constant returns to scale of the aggregate matching function and counterclockwise trajectory on the uv plane. Traditionally, it is argued that deterioration of matching efficiency are the cause of the shift of the Beveridge curve after 1970's. Our result places precautious warning to interpret the position of estimated Beveridge curve directly as an indicator of efficiency. The same mechanism which makes the trajectory in the uv plane go around the Beveridge curve causes apparent "shift" of the Beveridge curve without any deterioration of matching efficiency. There are empirical studies which cast doubt on the validity of the Beveridge curve to represent structural or stable relationship in the labour market. Using the panel data of nine regions in West Germany, [1] showed that the observed shift of the Beveridge curve oc-

curred too quickly to be explained by structural changes and that the change
in the composition of unemployment pool, i.e. structural factors on the side
of workers, have little explanatory power of the shifts. He raised a question
on the stability of the Beveridge curve under the presence of business cycles.
He pointed out the possibility that, after a shock, the trajectory on the uv
plane may not converge back to the original Beveridge curve in general, since
parameters change. In the same spirit, we will find that, given the existence
of fluctuation in inflows to the labour market, the apparent shift of the Beveridge curve is inevitable. In our case, what makes the trajectory converge
to a different point is the fact that the market is in a different phase of the
inflow cycle when the market crosses the same vu ratio from above and below.
It makes the trajectory on the uv plane form a cycle, especially in the counterclockwise direction. Counterclockwise behaviour of the trajectory has been
found in a number of countries. [7] pointed out the cyclical behaviour from
the data of United Kingdom from 1953 to 1964. He explained the emergence
of the cyclical behaviour using a model, but failed to show why the direction
of the cycle is counterclockwise, not clockwise. To explain the direction of the
cycle, [2] assumed that vacancies responds more quickly to shocks than unemployment. [11] explained it from the asymmetry between matching process
and breakup process. He assumed that breakup of a job-worker pair can be
undertaken immediately, whereas matching is a time-consuming process. We
will show in this paper that the counterclockwise behaviour can arise solely
from the matching process regardless the breakup process. It comes from the
time lag until change in flows (inflow of jobs and workers) is reflected in stocks
(u and v).

2 The Model

Now, we consider such a matching process that firms advertise vacant jobs
and workers apply to a job responding to the advertisement. A worker applies
to his best choice among advertising firms at the beginning of a period. At
the end of the period, he receives a notification of appointment or rejection. If
he is appointed, he leaves the labour market and engages in production from
the next period. If he is rejected, he repeats the same application process in
the next period. For simplicity, we put a memoryless assumption. Namely, a
worker who has rejected by a firm can apply to the same job next period if
the vacancy remains unfilled. This simplification assumption is harmless when
the market size is sufficiently large. Or, it can be thought as a case that there
is a possibility that the change of environment by passage of time can change
the evaluation of the worker. A firm will assign one of the applicants to a
job if and only if the evaluation of the best applicant exceeds the threshold
value. Job-seekers will apply if and only if his evaluation of the best job posted
exceeds his threshold value. We express both evaluation in terms of utility,
denoting a firm's evaluation of a worker by x and a worker's evaluation of a

job by y. The threshold value of a firm is denoted by x^* and that of a worker is denoted by y^*. If the evaluation of the best applicant by a firm is $x \geq x^*$, it will employ him. If $x \leq x^*$, it will employ nobody and leave the position vacant for another period. The same holds for a worker. We assume that each agent is risk neutral. x and y are random variables that have i.i.d. probability distributions $F(\cdot)$ and $G(\cdot)$ respectively. We denote the number of unfilled job vacancies in the labour market by v and that of job-seekers by u.

Proposition 1. *There are v vacancies and u job-seekers in the labour market. On each encounter, the evaluation of a worker by a firm x is drawn from an i.i.d. distribution F and the evaluation of a firm by a worker y is drawn from an i.i.d. distribution G. The threshold evaluation of a firm and a worker are x^* and y^* respectively. Then, the probability for a vacancy to match is given by*[1]

$$1 - \left[1 - \frac{(1 - F(x^*))(1 - G(y^*)^v)}{v} \right]^u. \tag{1}$$

Proof. The probability that the evaluation of a job by a worker is above the threshold to apply is $1 - G(y^*)$. When there are v job vacancies, a worker will apply to one of them with probability $1 - G(y^*)^v$. Therefore, when there are u job-seekers, the probability that exactly m job-seekers apply to one of the vacancies is $\binom{u}{m} [1 - G(y^*)^v]^m G(y^*)^{v(u-m)} =: \Pr(A)$. On the other hand, each vacancy is homogeneous ex ante. Once having m effective job-seeking applicants in the labour market ($u - m$ workers do not apply), the probability that n workers apply to a particular vacancy is $\binom{m}{n} (1/v)^n (1 - 1/v)^{m-n} =: \Pr(B)$. The firm assigns one of those n applicants who is ranked top, if his evaluation exceeds the threshold. The firm approves a worker with probability $1 - F(x^*)^n =: \Pr(C)$.

The event that m job-seekers actually apply and n among them apply to a particular firm is exclusive each other for different (m, n) pairs. Thus, the probability that a firm assigns a job to a worker is given by

$$\sum_{m=1}^{u} \sum_{n=1}^{m} \Pr(A) \Pr(B) \Pr(C) = 1 - \left[1 - \frac{(1 - F(x^*))(1 - G(y^*)^v)}{v} \right]^u \tag{2}$$

The second line comes from the fact that $1 - F(x^*)^n = 0$ when $n = 0$ and $\binom{m}{n} = 1$ when $m = 0$.

Corollary 1. *Suppose that vu ratio θ is given. The probability that a firm receives applications and rejects all of them is $e^{-(1-F(x^*))/\theta} - e^{-1/\theta} > 0$ when market size is large.*

Proof. The probability is calculated by $\sum_{m=1}^{u} \sum_{n=1}^{m} \Pr(A) \Pr(B) \Pr(\bar{C})$. Taking $v/u = \theta$ into account, its limit as $v \to \infty$ is $e^{-(1-F(x^*))/\theta} - e^{-1/\theta}$.

---

[1] The probability when $F(x^*) = 0$ is obtained by $F(x^*) \to +0$.

Proposition 2. *Suppose $u \neq 0$ and $v \neq 0$. If $G(y^*) \neq 1$, then for any given vu ratio θ, the probability that there exist workers who voluntarily do not apply to any jobs converges to zero as u and v go infinity.*

Proof. The probability that all workers apply to a vacancy is $[1 - G(y^*)^v]^u$. We define a notation $r := G(y^*)^\theta$. As u and v go infinity keeping θ constant, the probability that all workers apply to a vacancy becomes

$$\lim_{u,v \to +\infty} [1 - G(y^*)^v]^u \bigg|_{v/u=\theta} = \lim_{u \to +\infty} \left[1 - \left\{G(y^*)^\theta\right\}^u\right]^u = 1$$

As the number of vacancies increases, workers have more and more choices, making it easier to find a satisfactory candidate. However, as the number of workers increases, there are more and more chances that at least one worker cannot find an appropriate job. This proposition says that the former effect is larger. When the labour market is sufficiently large so that it allows a worker to have sufficiently many choices, a job-seeker will not give up application voluntarily. This makes sharp contrast to Corollary 1. There always exist firms which voluntarily reject all applicants.

Theorem 1. *When vu ratio θ is given, matching probability of vacancies P_v is asymptotically given by*

$$P_v = 1 - \exp\left(-\frac{1 - F(x^*)}{\theta}\right) \tag{3}$$

as u and v go infinity. Using the asymptotic matching probability of vacancy, the aggregate matching function μ is given by

$$\mu(u, v; F(x^*)) = P_v v = \left[1 - \exp\left\{-\frac{u}{v}[1 - F(x^*)]\right\}\right] v \tag{4}$$

Proof. The result is obtained by taking a limit $u, v \to +\infty$ keeping θ constant in Proposition 2.

In Proposition 1, note that P_u and P_v depend on θ, but not on the value of u or v themselves. The following proposition shows that most of ordinary presumption about an aggregate matching function are supported but not all. We can easily show the following result.

Proposition 3. *The asymptotic aggregate matching function $\mu(u, v)$ has the following properties:*

1. μ is homogeneous of degree one,
2. $\mu(u, 0) = \mu(0, v) = 0$
3. $m < \min(u, v)$

4. $\dfrac{\partial \mu}{\partial u} > 0, \ \dfrac{\partial \mu}{\partial v} > 0, \ \dfrac{\partial^2 \mu}{\partial u^2} < 0$

5. $\dfrac{\partial^2 \mu}{\partial v^2} < 0$ if and only if $\dfrac{u}{v} > \dfrac{2}{1 - F(x^*)}$

Since most of aggregate matching models are based on the assumption of homogeneity of degree one of the matching function, property (1) of Proposition 3 may be of special interest. It gives a justification of such a presumption. It excludes the possibility of multiple equilibria caused by increasing returns to scale of the matching function.

3 Inflow Cycles

Empirical data show that inflows of new job-seekers and vacancies to the labour market fluctuate persistently and fairly regularly along with business cycles. I said "*fairly* regularly" but in general, it is hopeless for the inflows to have *strictly* regular cycles. Nevertheless, it is still useful to focus on a case in which inflows have regular cycles, since what is important is the recurrence nature of cycles. We assume inflows η and ξ to fluctuate and to have same averages over a cycle. They are balanced in long run although the length of a cycle may not coincide. Then, a trajectory on the uv plane beginning from any initial point converges to a unique cycle, the period of which is the least common multiple of the period of vacancy inflow and the period of job-seeker inflow. We show the following lemma without a proof.

Lemma 1. *Let* $\{\mathsf{T}^{(0)}, \mathsf{T}^{(1)}, \dots, \mathsf{T}^{(k-1)}\}$ *be a sequence of contraction mappings with modulus* $\beta_\kappa \in (0, 1)$ *(*$\kappa = 0, 1, \dots, k-1$*) from a Banach space* X *to itself. For any initial value* $x_0 \in X$, *we define a dynamics by*

$$x_{t+1} = \mathsf{T}^{(t \bmod k)} x_t. \qquad (t = 1, 2, \dots)$$

Then, the sequence $\left\{\mathsf{T}^{(\kappa)}\right\}_{\kappa=1}^{k-1}$ *has a unique and globally stable cycle* $\{\bar{x}^{(0)}, \bar{x}^{(1)}, \dots, \bar{x}^{(k-1)}\} \in X^k$ *where* $\bar{x}^{(\kappa)}$ *(*$\kappa = 0, 1, \dots, k-1$*) is a fixed point of the mapping* $\mathsf{T}^{(\kappa-1)} \circ \mathsf{T}^{(\kappa-2)} \circ \dots \mathsf{T}^{(0)} \circ \mathsf{T}^{(k)} \circ \mathsf{T}^{(k-1)} \circ \dots \circ \mathsf{T}^{(\kappa)}$.

We denote the state of the labour market at the beginning of each period $t = 0, 1, 2, \dots$ by $(u_t, v_t) \in \mathbb{R}^2_{++}$ where u_t is the number of job-seekers and v_t is the number of unfilled vacancies in the labour market. The initial value (u_0, v_0) is arbitrarily given. At the beginning of each period, firms advertise their vacancies and job-seekers apply to a vacancy which they think best. Successfully matched m_t pairs of a job-seeker and a vacancy exit the market and the rest remains in the market. At the end of the period, new inflow of job-seekers ξ_t and new inflow of vacancies η_t are added to the remaining stocks, forming the stock of job-seekers and vacancies (u_{t+1}, v_{t+1}) in the next period. We denote by $\mathrm{LCM}(x, y)$ the least common multiple of integers x and y.

Theorem 2. *Suppose that the inflow of new vacancies into the labour market has a cycle of period $S \in \mathbb{N}$. The inflows with in a cycle are given by $\left\{\eta^0, \eta^1, \ldots, \eta^{S-1}\right\} \in \mathbb{R}_+^S$. The inflow of new job-seekers into the labour market has a cycle of period $T \in \mathbb{N}$. The inflows within a cycle are given by $\left\{\xi^0, \xi^1, \ldots, \xi^{T-1}\right\} \in \mathbb{R}_+^T$. Then, if the condition*

$$\frac{\sum_{i=0}^{S-1} \eta^{(i)}}{S} = \frac{\sum_{i=0}^{T-1} \xi^{(i)}}{T} \tag{5}$$

is satisfied, a trajectory on the uv plane beginning from any initial value $(u_0, v_0) \in \mathbb{R}_+^2$ converges to a cycle of period $\mathrm{LCM}(S, T)$.

Proof. The transition of (u, v) plot is given by

$$v_{t+1} = v_t + \eta^{(t \bmod S)} - m_t \tag{6}$$
$$u_{t+1} = u_t + \xi^{(t \bmod T)} - m_t \tag{7}$$

for any $t = 0, 1, \ldots$. Using the condition (5), we obtain $u_{t+M} = u_t - v_t + v_{t+M}$. Recursive operation gives

$$u_{\kappa+Mn} = u_\kappa - v_\kappa + v_{\kappa+Mn}. \tag{8}$$

for any $\kappa = 0, 1, \ldots, M - 1$ and for any $n \in \mathbb{N}$. Namely, the data picked up from time series with interval M stay on the same line. As obvious from derivation, it is one of isoinflux lines.

From equation (4), for any t,

$$v_{t+1} = \eta^{(t \bmod S)} + \exp\left\{-\left(1 - \frac{v_{t \bmod M} - u_{t \bmod M}}{v_t}\right)\left(1 - F(x^*)\right)\right\} v_t \tag{9}$$

where $v_{t \bmod M} - u_{t \bmod M}$ is a constant obtained by $v_{t \bmod M} - u_{t \bmod M} = v_0 - u_0 + \sum_{i=0}^{t \bmod M-1}\left(\eta^{(i)} - \xi^{(i)}\right)$.

Equations (8) and (9) describe the dynamics. Since (8) is continuous, u converges as v does. Defining the right-hand side of (9) by $T^{(t \bmod M)} v_t$, we have $dT^{(t \bmod M)} v_t / dv_t \subset (0, \exp\{-[1-F(x^*)]\} \subset (0, 1)$. For any $a, b \in \mathbb{R}_{++}$, there exists $c \in (a, b)$ such that $T^{(t \bmod M)} a - T^{(t \bmod M)} b = (dT^{(t \bmod M)} c / dc)(a - b)$ by the mean-value theorem. However, $0 < c < \exp\{-(1 - F(x^*))\} < 1$. Thus, for any a and b, $|T^{(t \bmod M)} a - T^{(t \bmod M)} b| < \beta |a - b|$ where $0 < \beta < 1$, i.e. $T^{(t \bmod M)}$ is a contraction mapping. From Lemma 1, the sequence of mappings $\left\{T^{(\kappa)}\right\}_{\kappa=0}^{M-1}$ has a unique cycle $\left\{\bar{v}^{(0)}, \bar{v}^{(1)}, \ldots, \bar{v}^{(M-1)}\right\} \in \mathbb{R}_+^M$ and the paths from any initial value (u_0, v_0) converge to it. Furthermore, $\bar{v}^{(\kappa)}$ $(\kappa = 0, 1, \ldots, M - 1)$ is a fixed point of $T^{(\kappa)}$. This proves that the asymptotic cycle has a period no greater than M. However, the period of the cycle should be exactly M as shown below.

Suppose that the period of the cycle is $M' < M$. Then, M' must be a divisor of M. Otherwise, it contradicts the above fact that any $v \pmod M$ share the same fixed point. For such M' to exist,

$$\left\{\bar{v}^{(0)}, \bar{v}^{(1)}, \ldots, \bar{v}^{(M'-1)}\right\} = \left\{\bar{v}^{(M')}, \bar{v}^{(M'+1)}, \ldots, \bar{v}^{(2M'-1)}\right\}$$

and

$$\left\{\bar{u}^{(0)}, \bar{u}^{(1)}, \ldots, \bar{u}^{(M'-1)}\right\} = \left\{\bar{u}^{(M')}, \bar{u}^{(M'+1)}, \ldots, \bar{u}^{(2M'-1)}\right\}$$

where $\bar{u}^{(n)} := u_{(n)} - v_{(n)} + \bar{v}^{(n)}$ for $n = 0, 1, \ldots, M-1$ must hold. It implies that $u_{(n)} - v_{(n)} = u_{(M'+n)} - v_{(M'+n)}$ for given n. From equation (9), for $\left(\bar{u}^{(n+1)}, \bar{v}^{(n+1)}\right) = \left(\bar{u}^{(M'+n+1)}, \bar{v}^{(M'+n+1)}\right)$ to hold, the condition $\eta^{(n)} = \eta^{(M'+n)}$ must be satisfied. On the other hand, $u_{(n)} - v_{(n)} = u_{(M'+n)} - v_{(M'+n)}$ requires $\sum_{i=0}^{M'-1} \eta^{(i)} = \sum_{i=0}^{M'-1} \xi^{(i)}$ for all n, which implies $\xi^{(n)} = \xi^{(M'+n)}$. The result $\left(\eta^{(n)}, \xi^{(n)}\right) = \left(\eta^{(M'+n)}, \xi^{(M'+n)}\right)$ contradicts the fact that M is the *least* common multiple of S and T.

It is an established fact that the trajectory on the uv plane empirically moves around the Beveridge curve. During a recession, u gradually increases whereas v decreases. Thus, the trajectory moves to the lower right of the uv plane. On recovery, opposite pressure carries the trajectory to the upper right. However, when recovery takes place, u does not recover quickly whereas v begins to increase. Thus, the trajectory during a recovery phase goes above the one during a recession phase. Often, this asymmetric response of the labour market to business cycles is attributed to loss of skills of unemployed workers. However, Theorem 2 shows that cycle in the uv plane can emerge even without the existence of loss of skills. Only cyclical fluctuations of new vacancies and new job-seekers causes the emergence of a cycle in the uv plane.

Theorem 2 does not tell anything about the shape of the cycle. Especially, it does not tell to which direction the cycle circulates. We will study conditions that bring counterclockwise trajectory on the uv plane now. The dynamics when inflows periodical fluctuates are given by equation (9). Note that $\mathsf{T}^{(t \bmod M)}$ depends only on $v_{(t \bmod M)} - u_{(t \bmod M)}$ and $\eta^{(t \bmod M)}$. $v_{(t \bmod M)} - u_{(t \bmod M)}$ determines the position of the isoinflux line to which sequence $\left\{\left(u_{(t \bmod M)}, v_{(t \bmod M)}\right), \left(u_{2(t \bmod M)}, v_{2(t \bmod M)}\right), \left(u_{3(t \bmod M)}, v_{3(t \bmod M)}\right), \ldots\right\}$ belongs to. The initial point in the sequence $v_{(t \bmod M)} - u_{(t \bmod M)}$ is calculated by the equation

$$v_{(t \bmod M)} - u_{(t \bmod M)} = v_0 - u_0 + \sum_{i=0}^{t \bmod M} \left(\eta^{(i)} - \xi^{(i)}\right). \qquad (10)$$

Once the isoinflux line is determined above, $\mathsf{T}^{(t \bmod M)}$ solely depends on $\eta^{(t \bmod M)}$.

Consider a simple example that the inflow of vacancies is fixed at $\eta = 1$ and the inflow of job-seekers follows $\xi = 1 + \sin t$. This is a good example to begin with since constant η makes mapping $\mathsf{T}^{(t \bmod M)}$ fixed for all periods. Note that it has balanced inflows. Suppose that there are four matching sessions within a cycle of the inflows as in the figure below.

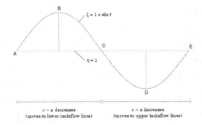

The matching market opens at points A, B, C, D and E. We assume that initial point is balanced: $v = u$. We plot the status of the matching market that corresponds to points A-E in Figure 1 that shows recursive application of functions . We label corresponding plots in the figure as points A-E as well. Point A in the uv plane (it represents all periods 0 mod 4) is on the 45 degree line $v = u$. At point B (period 1 (mod 4)), job-seeker inflow $\xi^{(1)}$ is larger than vacancy inflow $\eta^{(1)}$ by one. Thus, the labour market is on isoinflux line $v = u - 1$. At point C (period 2 (mod 4)), inflows are balanced. It does not change the isoinflux line the point belong to, so point C stays on the same line as point B. At point D (period 3 (mod 4)), vacancy inflow $\xi^{(3)}$ is less than job-seeker inflow $\eta^{(3)}$ by one, thus the isoinflux line shifts up by one returning back to the 45 degree line. In sum, point A and D and point B and C are on the same isoinflux line respectively. We use the following small lemma.

Lemma 2. *Let $X \subset \mathbb{R}$. If $\mathsf{T} : X \to X$ is a contraction mapping, then $\mathsf{T}x - x$ is x a strictly decreasing function.*

Proof. If $\mathsf{T}x - x$ is not a decreasing function of x, then there exist $a, b \in X$ such that $a > b$ and $\mathsf{T}a - a > \mathsf{T}b - b$. However, it implies $\mathsf{T}a - \mathsf{T}b \geq a - b$. Since $a > b$, it means $\|\mathsf{T}a - \mathsf{T}b\| \geq \|a - b\|$. This contradicts the fact that T is a contraction mapping. Therefore, the lemma holds.

We characterise the limit cycle by the following proposition.

Proposition 4. *Let $\mathfrak{T} = \left\{ \mathsf{T}^{(0)}, \mathsf{T}^{(1)}, \dots, \mathsf{T}^{(k-1)} \right\}$ be a sequence of contraction mappings from $X \subset \mathbb{R}$ to itself with modulus $\beta_\kappa \in (0, 1)$ $(\kappa = 0, 1, \dots, k-1)$. Suppose each mapping is an increasing function. Consider a dynamics defined by*

$$x_{t+1} = \mathsf{T}^{(t \bmod k)} x_t, \qquad (t = 1, 2, \dots)$$

for any initial value $x_0 \in X$. Let $\bar{y}^{(\kappa)}$ $(\kappa = 0, 1, \dots, k-1)$ be a fixed point of $\mathsf{T}^{(\kappa)}$ and at least for one pair of $i, j = 0, 1, \dots, k-1$ $(i \neq j)$, $\bar{y}^{(i)} \neq \bar{y}^{(j)}$. Then the unique stable limit cycle $\left\{ \bar{x}^{(0)}, \bar{x}^{(1)}, \dots, \bar{x}^{(k-1)} \right\} \in X^k$ of $\left\{ \mathsf{T}^{(\kappa)} \right\}_{\kappa=1}^{k-1}$ satisfies the condition

$$\min_{i=0,1,\dots,k-1} \bar{y}^{(i)} < \bar{x}^{(\kappa)} < \max_{i=0,1,\dots,k-1} \bar{y}^{(i)}$$

where $\kappa = 0, 1, \dots, k-1$.

Proof. If we show $\min_{i=0,1,\ldots,k-1} \bar{y}^{(i)} < \bar{x}^{(\kappa)}$ ($\kappa = 0,1,\ldots,k-1$), then $\bar{x}^{(\kappa)} < \max_{i=0,1,\ldots,k-1} \bar{y}^{(i)}$ can be shown by a similar argument.

We define $\bar{y}^{(p)} := \min_{i=0,1,\ldots,k-1} \bar{y}^{(i)}$ and denote by $\mathsf{T}^{(p)}$ the operator that has $\bar{y}^{(p)}$ as a fixed point. Also denote by $\mathfrak{T}\left(\mathsf{T}^{(p)}\right)$ a composite operator $\mathsf{T}^{(p-1)} \circ \mathsf{T}^{(p-2)} \circ \cdots \circ \mathsf{T}^{(0)} \circ \mathsf{T}^{(k-1)} \circ \mathsf{T}^{(k-2)} \circ \cdots \circ \mathsf{T}^{(p)}$. Suppose that $\hat{x} \in \left(-\infty, \bar{y}^{(p)}\right] \cap X$ is a fixed point of $\mathfrak{T}\left(\mathsf{T}^{(p)}\right)$. Then, for arbitrary $n \in \mathbb{Z}_+$, $\left(\mathsf{T}^{(p)}\right)^n \hat{x} \in \left[\hat{x}, \bar{y}^{(p)}\right]$ holds. If the fixed point $\bar{y}^{(p+n)} \pmod{M}$ of $\mathsf{T}^{(p+n)} \pmod{M}$ satisfy $\bar{y}^{(p+n)} > \bar{y}^{(p)}$, $\mathsf{T}^{(p+n)} \circ \left(\mathsf{T}^{(p)}\right)^n \hat{x} > \hat{x}$ holds by Lemma 2. Since any operator $\mathsf{T}^{(i)}$ where $i \neq q$ has a fixed point $\bar{y}^{(i)} \geq \bar{y}^{(p)}$, $\tilde{x} := \mathsf{T}^{(p-2)} \circ \cdots \circ \mathsf{T}^{(0)} \circ \mathsf{T}^{(k-1)} \circ \mathsf{T}^{(k-2)} \circ \cdots \circ \mathsf{T}^{(p)} \hat{x} \geq \mathsf{T}^{(p+n)} \circ \left(\mathsf{T}^{(p)}\right)^n \hat{x} > \hat{x}$. For $\bar{y}^{(p)}$ to be a fixed point of $\mathfrak{T}\left(\mathsf{T}^{(p)}\right)$, there must exist $\tilde{\mathsf{T}} \in \mathfrak{T}$ such that $\hat{x} = \tilde{\mathsf{T}}\tilde{x}$ holds. It means $\tilde{\mathsf{T}}\hat{x} - \hat{x} = \tilde{\mathsf{T}}\hat{x} - \tilde{\mathsf{T}}\tilde{x} < 0$. Then, from Lemma 2, the fixed point $\tilde{y}$ of $\tilde{\mathsf{T}}$ satisfies $\tilde{y} < \hat{x} \leq \bar{y}^{(p)}$. This contradicts to the definition of $\bar{y}^{(p)}$. Therefore, in the region $\left(-\infty, \bar{y}^{(p)}\right] \cap X$, the fixed point of $\mathfrak{T}\left(\mathsf{T}^{(p)}\right)$ cannot exist. By a similar argument, it is shown that it does not exist in the region $\left[\max_{i=0,1,\ldots,k-1} \bar{y}^{(i)}, +\infty\right) \cap X$.

In our example, there are two contraction mappings. We denote them by $\overline{\mathsf{T}}$ and $\underline{\mathsf{T}}$. $\underline{\mathsf{T}}$ operates on points B and C and $\overline{\mathsf{T}}$ operates on points A and D. In general, $\mathsf{T}^{(t \bmod M)}$ is a function of $\eta^{(t \bmod M)}$ and $v_{(t \bmod M)} - u_{(t \bmod M)}$. Since $\eta^{(t \bmod M)}$ is fixed for the moment, let us write the operator $\mathsf{T}^{(t \bmod M)}$ as $\mathsf{T}^{(t \bmod M)}\left(v_{(t \bmod M)} - u_{(t \bmod M)}\right)$ to be explicit on the parameter. Consider two cases $\mathsf{T}^{(t \bmod M)}(K_1)$ and $\mathsf{T}^{(t \bmod M)}(K_2)$ where $K_1 > K_2$. Suppose $v_t \neq 0$. From (9), we can show

$$\frac{\mathsf{T}^{(t \bmod M)}(K_1) v_t}{\mathsf{T}^{(t \bmod M)}(K_2) v_t} = \exp\left\{\left(1 - F(x^*)\right)\frac{K_1 - K_2}{v_t}\right\} > 1.$$

Thus, $\mathsf{T}^{(t \bmod M)}(K_1) v_t > \mathsf{T}^{(t \bmod M)}(K_2) v_t$ for fixed η. This fact implies $\overline{\mathsf{T}} > \underline{\mathsf{T}}$. It is shown in Figure 1(a).

By Proposition 4, all points on the limit cycle is between $\underline{y}$ and $\bar{y}$. Thus, points A and D are on the left-hand side of the fixed point of $\overline{\mathsf{T}}$. Point A is obtained by operating $\overline{\mathsf{T}}$ twice on v_4 in the graph whereas point D is obtained by operating it only once. Thus, point A is closer to fixed point $\bar{y}$. By the same reason, point C is closer to fixed point $\underline{y}$ than point B. Therefore, $v_1 > v_4$ and $v_2 > v_3$. The dynamics of v is translated to the dynamics of u by isoinflux lines. The trajectory in the uv plane is shown in Figure 1(b). Point A' corresponds to point A in the previous figure and so on. As easily observed, the direction of the cycle is opposite to the empirical cycle. This is obvious from equation 9. If $v - u$ is greater, the slope of equation 9 is larger. Since η is fixed, it implies that the corresponding mapping T is located above of other equations that have smaller $v - u$. If $v - u$ is higher in period t than period s, it means that the point of the trajectory at period t is on an isoinflux line which is upper-left of the isoinflux line at period s. When the position of

Fig. 1. Dynamics of limit cycles

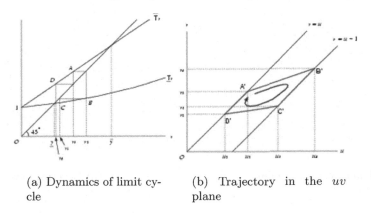

(a) Dynamics of limit cy-
cle

(b) Trajectory in the uv
plane

an isoinflux line is *relatively* above of other isoinflux lines that the trajectory visits, then v in next period will be larger than the current period. Thus, in the uv plane, as the trajectory goes to upper left (i.e. visits an isoinflux line upper-left) the path tends to shift outward. Note that only *relative* position among related isoinflux lines is crucial in determining the direction of the shift. The *absolute* position of the isoinflux line, namely the value of $v - u$, is not related. This observation suggests that the fluctuation of vacancies η is necessary to generate an counterclockwise behaviour of the trajectory in the actual data. Therefore, we take another simple example in which η fluctuates.

Now, suppose that vacancy inflow fluctuates following $\eta = 1 + \sin(t + \pi)$ instead of $\eta = 1$ in the previous example. This pattern of ξ and η is similar to the data observed in the actual labour market. Figure 2 shows the limit cycle of v and the same movement in the uv plane. The cycle in Figure 2 (b) shows a counterclockwise movement. This result is more general. We can provide a sufficient condition for the emergence of counterclockwise trajectory in the uv plane. We use the following lemma, which we show without a proof.

Lemma 3. *Let $X \subset \mathbb{R}$. Consider a family of contraction mappings $\mathsf{T}(\eta)$: $X \to X$ which has $\eta \in \mathbb{R}$ as a parameter. Suppose that $\mathsf{T}(\eta)$ is continuous on the parameter and if $\eta < \eta'$ then $\mathsf{T}(\eta) < \mathsf{T}(\eta')$. Suppose η is a continuous function of $t \in \mathbb{R}$ and there exists $T \in \mathbb{R}$ such that for all $n \in \mathbb{Z}$, $\eta(t + nT) = \eta(t)$ Pick up s points $t_i \in [0, T)$ where $i = 0, 1, \ldots, s - 1$ and consider a dynamics defined by $x_{k+1} = \mathsf{T}(\eta(t_{k \bmod s})) x_k$. Then, it has a unique stable limit cycle $\{\bar{x}_s^{(0)}, \bar{x}_s^{(1)}, \ldots, \bar{x}_s^{(s-1)}\}$ and if s increases such that $\max |t_{i+1} - t_i| \to 0$ holds, then $\bar{x}_s^{(i)} \to \bar{y}_s^{(i)}$ holds where $\bar{y}_s^{(i)} = \mathsf{T}(\eta(t_i)) \bar{y}_s^{(i)}$.*

Theorem 3. *Suppose that η and ξ are time continuous and $d\eta/dt \gtreqless 0 \Leftrightarrow d\xi/dt \lesseqgtr 0$. Then, if matching sessions are held sufficiently frequently within a*

Fig. 2. More realistic case

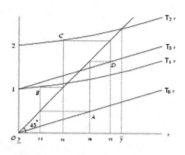

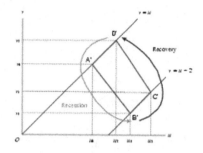

(a) Limit cycle in dynam-   (b) Beveridge cycle
ics of v

cycle of η and ξ, then the point that the trajectory on the uv plane crosses a particular isoinflux line from the right comes upper right of the point that the path crosses it from the left in the uv plane.

Proof. Since η is time continuous, for any $\varepsilon > 0$ and t, if the interval of the matching session is made sufficiently short, then it is possible to make $\left|\eta^{(t \bmod M+1)} - \eta^{(t \bmod M)}\right| < \varepsilon$. That is,

$$\eta - \varepsilon + \exp\left\{-\left(1 - \frac{v_{(t \bmod M)} - u_{(t \bmod M)}}{v_t}\right)\left(1 - F\left(x^*\right)\right)\right\} v_t$$
$$< v_{t+1} < \eta + \varepsilon + \exp\left\{-\left(1 - \frac{v_{(t \bmod M)} - u_{(t \bmod M)}}{v_t}\right)\left(1 - F\left(x^*\right)\right)\right\} v_t.$$

From $v_{(t \bmod M)} - u_{(t \bmod M)} = v_0 - u_0 + \sum_{i=1}^{t \bmod M}\left(\eta^{(i)} - \xi^{(i)}\right)$ and $d\eta/dt \gtrless 0 \Leftrightarrow d\xi/dt \lessgtr 0$, the differentiation by $\eta^{(t \bmod M)}$ of the left-most or the right-most hand side of the above equation gives

$$1 + (1 - F\left(x^*\right)) K_t \exp\left\{-\left(1 - \frac{v_{(t \bmod M)} - u_{(t \bmod M)}}{v_t}\right)\left(1 - F\left(x^*\right)\right)\right\} > 0$$

where K_t is a positive constant. Therefore, taking ε sufficiently small, T can be ordered by $\eta^{(t \bmod M)}$. Namely, if $\eta_1 > \eta_2$, then $\mathsf{T}(\eta_1) > \mathsf{T}(\eta_2)$.

Now, suppose that the uv path passes over the isoinflux lines $v - u = k_1$ and $v - u = k_2$ where $k_1 < k_2$, and it is currently on the line $v - u = k$ where $k_1 < k < k_2$ after passing over $v - u = k_2$ for the first time. Then $\eta^{(t \bmod M)} - \xi^{(t \bmod M)} < 0$. Since η and ξ cyclically fluctuates, the path must return to a point on $v - u = k_2$ from Theorem 2. It means that the path crosses $v - u = k$ with $\eta - \xi > 0$. From $d\eta/dt \gtrless 0 \Leftrightarrow d\xi/dt \lessgtr 0$, the η in the latter case has larger value. From Proposition 3, it implies that, if matching session is held frequently enough, then the fixed point of v in the latter case must be larger than the former case.

The theorem says that when the trajectory forms a simple cycle, namely when it crosses any isoinflux line at most twice within a cycle, then the direction of such a cycle is counterclockwise under the conditions provided by the theorem. Generally, when the path crosses any isoinflux line from the right, its crossing point should be upper right of any points where the path crosses it from the left, and vice versa. Note that we assumed no changes in matching technology. The cycle is brought by the asymmetry between job-seekers and vacancies embedded in the matching process. Figure 3 shows an example of

Fig. 3. Trajectory in the uv plane when inflows are symmetric

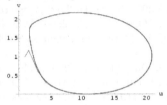

the trajectory in the uv plane when inflows to the labour market satisfy the condition of Theorem 3. The case $\eta = 1 + \sin t$ and $\xi = 1 + \sin(t - \pi)$ is drawn in the figure.

4 Conclusion

In this paper, we studied a matching process which arises when there is a cost to collect information about agents in the labour market. We found that its implication is consistent with data in many ways including the behaviour on the uv plane. The main reason that the behaviour on the uv plane becomes not trivial is that the it plots two *stock* variables. It is considered that wage rate under the existence of labour market friction becomes a function of the uv ratio. Since a shock does not reflect immediately to the uv ratio, wage rate responds slowly. It deteriorates the adjustment power of wage rate between imbalance of labour supply and demand.

References

1. Axel H. Börsch-Supan (1991) Panel data analysis of the beveridge curve: Is there a macroeconomic relation between the rate of unemployment and the vacancy rate?, Economica 58:279–97

2. Robert J. Bowden (1980) On the existence and secular stability of $u - v$ loci, Economica 47:33–50

3. Vincent P. Crawford, Elsie Marie Knoer (1981) Job matching with heterogeneous firms and workers, Econometrica 49:437–450

4. John K. Dagsvik (2000) Aggregation in matching markets, International Economic Review 41(1):27–57

5. Fitoussi, Jean-Paul, Jestaz, David and Phelps, Edmund S. and Zoega, Gylfi (2000) Roots of the Recent Recoveries: Labor Reforms or Private Sector Forces?, Brookings Papers on Economic Activity 1:237-311

6. David Gale, Lloyd S. Shapley (1962) College admissions and the stability of marriage, American Mathematics Monthly 69:9–15

7. Bent Hansen (1970) Excess demand, unemployment, vacancies, and wages, Quarterly Journal of Economics 84(1):1–23

8. Nickell, Stephan J., Layard, Richard (1999) Labor Market Institutions and Economic Performance, in Ashenfelter, O. and Card, D. (eds) Handbook of Labor Economics 3029-3084 North-Holland Amsterdam

9. Nickell, Stephan J., van Ours, Jan C. (2000) Why Has Unemployment in The Netherlands And the United Kingdom Fallen So Much?, Canadian Public Policy 26:201-220

10. Nickel Stephen, J., van Ours, Jan C. (2000) The Netherlands and the United Kingdom: A European Unemployment Miracle?, Economic Policy 30:137-175

11. Pissarides, Christopher A. (1985) Taxes, Subsidies and Equilibrium Unemployment Review of Economic Studies 52:121-133

12. Koji Yokota (2001) Competition between markets: Why can matchmakers survive in the temporary help service industry?

13. Koji Yokota (2004) Aggregate properties of stable matching

The Waiting-Time Distribution of Trading Activity in a Double Auction Artificial Financial Market

Silvano Cincotti[1*], Sergio M. Focardi[2], Linda Ponta[1], Marco Raberto[1], and Enrico Scalas[3]

[1] DIBE, Università di Genova, Via Opera Pia 11a, 16145 Genova, Italy
[2] The Intertek Group, rue de Javel 94, 75015 Paris, France
[3] DISTA, Università del Piemonte Orientale, Piazza Ambrosoli 5, 15100 Alessandria, Italy

Summary. In this paper, the statistical properties of high-frequency data are investigated by means of computational experiments performed with the Genoa Artificial Stock Market (Raberto et al. 2001, 2003, 2004). In the market model, heterogeneous agents trade one risky asset in exchange for cash. Agents have zero intelligence and issue random limit or market orders depending on their budget constraints. The price is cleared by means of a limit order book. The order generation is modelled with a renewal process where the distribution of waiting times between two consecutive orders is a Weibull distribution. This hypothesis is based on recent empirical investigation made on high-frequency financial data (Mainardi et al. 2000, Raberto et al. 2002, Scalas et al. 2003). We investigate how the statistical properties of prices and of waiting times between transactions are affected by the particular renewal process chosen for orders. Results point out that the mechanism of the limit order book is able to recover fat tails in the distribution of price returns without ad-hoc behavioral assumptions regarding agents; moreover, the kurtosis of the return distribution depends also on the renewal process chosen for orders. As regarding the renewal process underlying trades, in the case of exponentially distributed order waiting times, also trade waiting times are exponentially distributed. Conversely, if order waiting times follow a Weibull, the same does not hold for trade waiting times.

Introduction

In this paper, we present a model of trading in the Genoa Artificial Stock Market [4, 10, 12, 14, 15] characterized by a double auction clearing mechanism, i.e., the limit order book [3, 13]. The limit order book is a snapshot at a given instant of the queues of all buy limit orders and sell limit orders, with their respective price and volume. Limit orders are organized in ascending

---

* Corresponding author. E-mail: cincotti@dibe.unige.it

order according to their limit prices. All buy limit orders are below the best buy limit order, i.e., the buy limit order with the highest limit price (the bid price). The best buy limit order is situated below the best sell limit order, i.e., the sell limit order with the lowest limit price (the ask price). All other sell limit orders are above the best sell limit order. Orders are stored in the book. A transaction occurs when a trader hits the quote (the bid or the ask price) on the opposite side of the market. If a trader issues a limit order, say a sell limit order, the order either adds to the book if its limit price is above the bid price, or generates a trade at the bid if it is below or equal to the bid price. In the latter case, the limit order becomes a marketable limit order or more simply a market order. Conversely, if the order is a buy limit order it becomes a market order and is executed if its limit price is above the ask price, otherwise it is stored in the book. Limit orders with the same limit price are prioritized by time of submission, i.e., the oldest order has the highest priority. Order's execution often involves partial fills before it is completed, but partial fills do not change the time priority.

In recent years, some studies about the statistical properties of the limit order book have appeared in the literature [1, 2, 7, 8, 11]. An important empirical variable is the waiting time between two consecutive transactions. In fact, trading via the order book is asynchronous, i.e., a transaction occurs only if a trader issues a market order. For liquid stocks, waiting times vary in a range between some seconds to a few minutes. Raberto et al. [16] analyze the intra-day trades of General Electric stock prices and find that waiting times exhibit a 1-day periodicity, corresponding to the daily stock market activity, and a survival probability distribution which is well fitted by a stretched exponential. They also find a significative cross-correlation between waiting times and the absolute value of log-returns.

In this paper, the effect of a more general distribution of order waiting times is investigated. In particular, we focus our attention on the Weibull distribution that admits the exponential distribution as a limit case. Results show that in the case of exponentially distributed order waiting times, also trade waiting times are exponentially distributed. Conversely, if order waiting times follow a Weibull, the same does not hold for trade waiting times.

The paper is organized as follows: Section 1 presents the model, Section 2 shows computational experiments and Section 3 provides a discussion of results and the conclusions of the study.

1 The Model

A model of artificial trading by means of a limit order book is presented in this section. Agents trade one single stock in exchange for cash. They are modelled as liquidity traders; as a consequence, the decision making process is nearly random and depends on the finite amount of financial resources (cash

+ stocks) they own. At the beginning of the simulation, cash and stocks are uniformly distributed among agents.

1.1 The Order Generation Process

Trading is organized in M daily sections. Each trading day is subdivided in T elementary time steps, say seconds. During the trading day, at given time steps t_h, a trader i is randomly chosen for issuing an order. Order waiting times $\tau_h^O = t_h - t_{h-1}$ are determined according to a Weibull distribution, whose probability density function (PDF) $p(\tau)$ is:

$$p(\tau) = \frac{\beta}{\eta} \left(\frac{\tau}{\eta} \right)^{\beta-1} e^{-\left(\frac{\tau}{\eta} \right)^{\beta}}, \tag{1}$$

where η is the scale parameter and β is the shape parameter, also known as the slope, because the value of β is equal to the slope of the regressed line in a probability plot. The expected value $\langle \tau \rangle$ of a random variable following a Weibull PDF is given by:

$$\langle \tau \rangle = \eta \cdot \Gamma \left(1/\beta + 1 \right), \tag{2}$$

where Γ is the Gamma function. The survival probability distribution $P_>(\tau) = \int_\tau^\infty p(\tau)$ is given by:

$$P_>(\tau) = e^{-\left(\frac{\tau}{\eta} \right)^{\beta}}. \tag{3}$$

The exponential distribution is a particular case of the Weibull distribution for $\beta = 1$. In the case $\beta < 1$, the Weibull distribution assumes the form of the so-called stretched exponential and great values of τ occur with higher probability than in the case of $\beta = 1$. In our computational experiments, we considered only values of the shape parameters β less or equal than one.

The order generation process is then described as a general renewal process where the times between two consecutive orders τ_h^O are independent and identically distributed random variables following a Weibull distribution. In the case $\beta = 1$, the order generation process is assumed to be a Poisson process with an exponential waiting-time distribution. For further details on renewal processes, see Cox and Isham 1980 [5].

1.2 The Trading Decision Making Process

A trader issues a buy or a sell order with probability 50%. Let us denote with $a(t_{h-1})$ and with $d(t_{h-1})$ the values of the ask and of the bid prices stored in the book at time step t_{h-1}. Let us suppose that the order issued at time step t_h be a sell order, then we assume that the limit price $s_i(t_h)$ associated to the sell order is given by:

$$s_i(t_h) = n_i(t_h) \cdot a(t_{h-1}),\tag{4}$$

where $n_i(t_h)$ is a random draw by trader i at time step t_h from a Gaussian distribution with constant mean $\mu = 1$ and standard deviation σ. If $s_i(t_h) > d_i(t_{h-1})$, the limit order is stored in the book and no trades are recorded; else, the order becomes a market order and a transaction occurs at the price $p(t_h) = d(t_{h-1})$. In the latter case, the sell order is partially or totally fulfilled and the bid price is updated. The quantity of stocks offered for sale is a random fraction of the quantity of stocks owned by the trader.

If the order is a buy order, we assume that the associated limit price $b_i(t_h)$ is given by:

$$b_i(t_h) = n_i(t_h) \cdot d(t_{h-1});\tag{5}$$

where $n_i(t_h)$ is determined as in the previous case. If $b_i(t_h) < a(t_{h-1})$, the limit order is stored in the book and no trades are recorded; otherwise the order becomes a market order and a transaction occurs at the price $p(t_h) = a(t_{h-1})$. The quantity of stocks ordered to buy depends on cash endowment of the trader and on the value of $b_i(t_h)$.

It is worth noting that, in our framework, agents compete for the provision of liquidity. If an agent issues a buy order, its benchmark is the best limit buy order given by the bid price. Being $\mu = 1$, half times, he offers a more competitive buy order (if $b_i(t_h) > d(t_{h-1})$), which may result in a trade if $b_i(t_h) \geq a(t_{h-1})$. The same applies for sell limit orders.

2 Computational Experiments

The timing parameters of every simulation have been set as follows: $M = 50$ daily sections, each characterize by a length of $T = 25,200$ s (corresponding to 7 hours of trading activity). Each simulation is characterized by a particular value of the shape parameter β of the Weibull distribution modelling order waiting times. We have chosen 17 different values of β ranging from 0.2 to 1 with step 0.05. The average order waiting time $\langle \tau^o \rangle$ has been set to 20 s for every simulation. The scale factor η is so determined by the choice of β and $\langle \tau^o \rangle$ according to Eq. 2. The orders lifespan has been set to 600 s $\gg \langle \tau^o \rangle$.

Sell and buy limit prices are computed following Eq. 4 and Eq. 5 respectively. The random number $n_i(t_h)$ is a random draw by trader i from a Gaussian distribution with constant mean $\mu = 1$ and standard deviation $\sigma = 0.005$.

The number of agents is set to 10,000. At the beginning of the simulation, the stock price is set at 100.00 units of cash, say dollars and each trader is endowed with an equal amount of cash and of shares of the risky stocks. These amounts are 100,000 dollars and 1,000 shares, respectively.

Computational experiments produce realistic intraday price paths, see Ref. [13] for further details. Log-returns $r_{\Delta t}$ have been computed in homogeneous

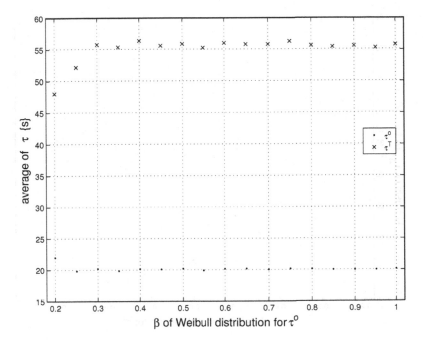

Fig. 1. Average values of order waiting times and of trade waiting times as a function of the shape parameter β of the Weibull distribution for orders.

time windows Δt, according to the previous-tick interpolation technique [6, pag. 37]. The time window Δt has been set to 100 s, which is about two times the average value of trade waiting times. Figure 1 presents the average values of orders waiting times $\langle \tau^o \rangle$ and trade waiting times $\langle \tau^T \rangle$, respectively, as a function of the shape parameter β. $\langle \tau^o \rangle$ is nearly 20 s, as expected by model construction, whereas $\langle \tau^T \rangle$, i.e., an output of the GASM model, is around 55 s. $\langle \tau^T \rangle$ appears to be independent from β when $\beta \geq 0.3$; the pattern of increasing values of $\langle \tau^T \rangle$ when $\beta < 0.3$ may be due to numerical effects. The choice of $\Delta t > \langle \tau^T \rangle$ has been made in order to avoid spurious effect in the statistical properties of returns due to long period of trading inactivity, especially in the case of small value of β.

Figure 2 presents the values of kurtosis of log-returns as a function of the shape parameter β. The figure shows that the distribution of log-returns is characterized by increasing values of kurtosis as β decreases. This is the main result of the paper. In previous works [3, 13], we already showed that the mechanism of the limit order book was able to recover fat tails in the distribution of log-returns without ad-hoc behavioral assumptions regarding agents. In that cases, β was set to 1, i.e., the order generation process was modelled as a Poisson process with exponentially distributed waiting times. Figure 2 generalizes such a result showing that the same conclusion also holds for a more general renewal process, i.e., Weibull distributed waiting times.

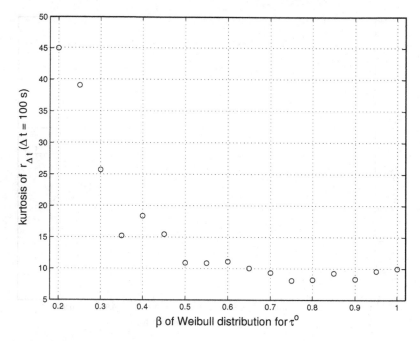

Fig. 2. Kurtosis of the distribution of log-returns as a function of the shape parameter β.

Furthermore, the tails of the distribution of returns become fatter when β decreases from 1, i.e., the distribution of order waiting times becomes a stretched exponential.

Figure 3 and Figure 4 show estimates of the survival probability distribution of order waiting times (dots) and of the survival probability distribution of trade waiting times (crosses) in the case $\beta = 0.4$ (Fig. 3) and $\beta = 1$ (Fig. 4). As expected by the assumptions of the model, survival probability distributions of order waiting times follow the corresponding Weibull distribution, represented by the continuous line. The survival probability distribution of trade waiting times is well fitted by a Weibull distribution only in the case $\beta = 1$. Figure 3 shows that, in the case $\beta = 0.4$, the survival probability distribution of trade waiting times departs from a Weibull distribution determined from data by means of the maximum likelihood principle and represented in the Figure with the dashed curve. In this case, the Kolmogorov-Smirnov test rejects the null hypothesis of Weibull distribution at the significance level of 5 %. Generally speaking, we find that trade waiting times are Weibull distributed only in the case $\beta = 1$.

In the case $\beta = 1$, the process of trading is a Poisson process as the order arrival process. An identical conclusion follows from theoretical considerations. In fact, consider that every transaction occurs when a new order that arrives in the book finds a matching order in the queue of orders of the opposite type.

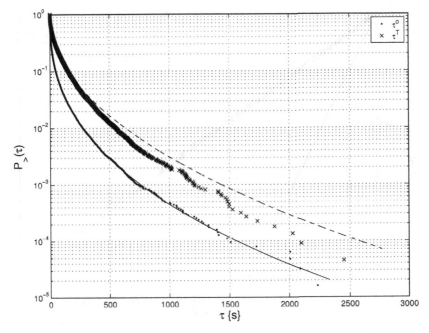

Fig. 3. Survival probability distribution of order waiting times (dots) and of trade waiting times (crosses) in the case $\beta = 0.4$. The two curves represent the corresponding Weibull fits.

Therefore, any new order will be satisfied or not in function of the state of the book in that moment. The state of the book varies for each moment and for each simulation path. Given the absence of significant feedbacks in the market, however, it is reasonable to assume that the average state of the book is time invariant. Therefore, each incoming order will be satisfied on average with a constant probability and the trading process can be regarded as a random extraction from the Poisson process of orders issuing. The procedure of random extraction from a Poisson process is called "thinning" and it is well known that a thinning from a Poisson process is a new Poisson process; see Ref. [5] for further details.

3 Discussion and Conclusion

In this paper, the results of Raberto et al. 2004 [13] are confirmed and strengthened. Indeed, the return distribution is leptokurtic not only when the order generation process is a Poisson process but also in the presence of memory, as in the case of Weibull-distributed waiting times. Moreover, with memory, return tails are fatter.

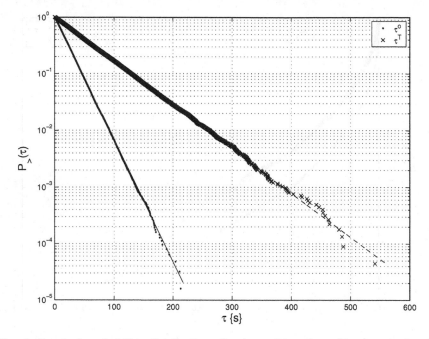

Fig. 4. Survival probability distribution of order waiting times (dots) and of trade waiting times (crosses) in the case $\beta = 1$ (exponential distribution). The two lines represent the corresponding exponential fits.

This result deserves attention because previous empirical analyses [9, 16, 17] have shown that the distribution of trade waiting times is non-exponential. Conversely, exponentially distributed trade waiting times only result from a finite thinning of a Poisson order process. Consequently, the distribution of order waiting times should be a more general distribution, e.g., a Weibull distribution, than an exponential distribution underlying a simple memoryless Poisson process.

The hypothesis of non exponentially distributed order waiting times is empirically accessible and can be directly checked if full book information were available. Moreover, one can try to solve the inverse problem: given a trade waiting time distribution, which is the originating order waiting time distribution? Our future activity will follow these lines.

Acknowledgments

This work has been supported by the University of Genova and by the Italian Ministry of Education, University and Research (MIUR) under grant FIRB 2001 and under grant FISR "Ultra-high frequency dynamics of financial markets".

References

1. B. Biais, P. Hillion, and C. Spatt, "An empirical analysis of the limit order book and the order flow in the Paris bourse," *J Financ*, vol. 50, no. 5, pp. 1655–1689, December 1995.
2. J.-P. Bouchaud, M. Mezard, and M. Potters, "Statistical properties of stock order books: empirical results and models," *Quantitative Finance*, vol. 2, no. 4, pp. 251–256, August 2002.
3. S. Cincotti, C. Dose, S. M. Focardi, M. Marchesi, and M. Raberto, "Analysis and simulation of a double auction artificial financial market," 6-10 July 2003, Istanbul, Turkey, EURO/INFORM 2003.
4. S. Cincotti, S. M. Focardi, M. Marchesi, and M. Raberto, "Who wins? study of long-run trader survival in an artificial stock market," *Physica A*, vol. 324, no. 1-2, pp. 227–233, June 2003.
5. D. R. Cox and V. Isham, *Point processes.* Chapman and Hall, 1980.
6. M. M. Dacorogna, R. Gencay, U. Muller, R. B. Olsen, and O. V. Pictet, *An Introduction to High Frequency Finance.* Academic Press, 2001.
7. G. Iori, M. G. Daniels, J. D. Farmer, L. Gillemot, S. Krishnamurthy, and E. Smith, "An analysis of price impact function in order-driven markets," *Physica A*, vol. 324, no. 1-2, pp. 146–151, June 2003.
8. A. W. Lo, A. C. MacKinlay, and J. Zhang, "Econometric models of limit-order executions," *J Financ Econ*, vol. 65, no. 1, pp. 31–71, July 2002.
9. F. Mainardi, M. Raberto, R. Gorenflo, and E. Scalas, "Fractional calculus and continuous-time finance ii: the waiting-time distribution," *Physica A*, vol. 287, no. 3-4, pp. 468–481, December 2000.
10. M. Marchesi, S. Cincotti, S. M. Focardi, and M. Raberto, *The Genoa artificial stock market: microstructure and simulation*, ser. Lecture notes in economics and mathematical systems.   Springer, February 2003, vol. 521, pp. 277–289.
11. S. Maslov and M. Mills, "Price fluctuations from the order book perspective: empirical facts and a simple model," *Physica A*, vol. 299, no. 1-2, pp. 234–246, October 2001.
12. M. Raberto, "Modelling and implementation of an artificial financial market using object oriented technology: the Genoa artificial stock market," Ph.D. dissertation, University of Genoa, Italy, 2003.
13. M. Raberto, S. Cincotti, C. Dose, S. M. Focardi, and M. Marchesi, "Price formation in an artificial market: limit order book versus matching of supply and demand," in *Proceedings WEHIA 2003.* Springer, 2004.
14. M. Raberto, S. Cincotti, S. M. Focardi, and M. Marchesi, "Agent-based simulation of a financial market," *Physica A*, vol. 219, pp. 319–327, October 2001.
15. M. Raberto, S. Cincotti, S. Focardi, and M. Marchesi, "Traders' long-run wealth in an artificial financial market," *Computational Economics*, vol. 22, no. 2-3, pp. 255–272, October-December 2003.
16. M. Raberto, E. Scalas, and F. Mainardi, "Waiting-times and returns in high-frequency financial data: an empirical study," *Physica A*, vol. 314, no. 1-4, pp. 749–755, November 2002.
17. E. Scalas, R. Gorenflo, F. Mainardi, M. Mantelli, and M. Raberto, "Anomalous waiting times in high-frequency financial data," 2003, submitted to Quantitative Finance, http://xxx.lanl.gov/abs/cond-mat/0310305.

Minority Games and Collectie Intelligence

Part VI

Minority Games and Collective Intelligence

Theoretical Analysis of Local Information Transmission in Competitive Populations

Sehyo Charley Choe[1], Sean Gourley[1], Neil F. Johnson[1], and Pak Ming Hui[2]

[1] Clarendon Laboratory, Physics Department, Oxford University, Parks Road, Oxford OX1 3PU, UK
  s.choe1@physics.ox.ac.uk
[2] Physics Department, The Chinese University of Hong Kong, Shatin, New Territories, Hong Kong

Summary. We study Complex Adaptive Systems in which information is shared locally within a heterogeneous population of adaptive nodes ('agents') competing for a limited global resource. The emerging collective behaviour is found to depend strongly on the level of available resource, the connectivity between agents and the accuracy of information transmission.

1 Introduction

The study of Complex Adaptive Systems is expanding the boundaries of Physics into non-traditional areas, such as economics, sociology and biology. Statistical physics and the validity of its associated assumptions such as ergodicity, self-averaging and history-independence, is based on particles obeying deterministic equations-of-motion, time-averaged over the behaviour of the entire collective. On the other hand, in Complex Adaptive Systems the adaptive objects (or 'agents') modify individual behavioural rules ('strategy') according to past experience. It is the local-local interactions between such agents which gives rise to emergent behaviour. The traditional assumptions of statistical physics may therefore not hold, creating a rich area of study for theorists. These adaptive systems are often rich in empirical data, offering Physics a new testing ground for theories of complex systems and non-equilibrium statistical mechanics in general [1]. In particular, Arthur's El Farol bar problem (EFBP) [2], and its subsequent binary representation, namely the Minority Game (MG) [3, 4, 5, 6, 7, 8], are recognized [9] to embody the interacting, many-body nature of many real-world complex systems, in which there is often competition between agents for some global resource [2, 3, 4]. Researchers have only just started considering the effect of networks in the MG and EFBP [6, 7], despite widespread interest among physicists in socio-economic, biological, and informational networks [10].

Most network research assumes that nodes are inert, focusing rather on the structural and statistical properties of the connections. In contrast, multi-agent research has focused on adaptive decision-taking agents in the absence of network connections. Very little has been said about the combination of the two, and hence the *functionality* of such a network, despite its practical importance [11]. Furthermore, it has so far been assumed that any information shared between agents is always perfectly accurate. However, real-world complex systems do not operate at such levels of perfection. Informational networks might be used by agents to spy and mislead rather than to benefit others. This raises the interesting question of the effect of networks within competitive population where the information transmitted is corrupted.

Here we study a class of Complex Adaptive Systems comprising a competitive multi-agent population, known as Binary Agent Resource (B–A–R) systems. We give a brief overview of the analytical Crowd-Anticrowd theory [4] which incorporates the many-agent (many-body) correlations or 'crowding' arising in the system's strategy space as a result of the dynamics in the history space. We then extend the Crowd-Anticrowd theory to encompass the micro and/or macroscopic effects of network functionality and the effects of information corruption on the system behaviour.

2 Binary–Agent–Resource (B–A–R) Systems

2.1 The Model

Our B-A-R model is a networked binary version of Arthur's El Farol problem [2, 3, 4]. The model consists of N agents who repeatedly compete for some limited resource L. At each timestep t, each agent decides whether to access resource L (i.e. action $+1$) or not to access this resource (i.e. action -1). The two global outcomes at each timestep, 'resource over-used' and 'resource not over-used', are denoted as '0' and '1'. If the number of agents $n_{+1}[t]$ choosing action $+1$ exceeds L (i.e. resource over-used and hence global outcome '0') then the $n_{-1}[t]$ abstaining agents win. By contrast if $n_{+1}[t] \leq L$ (i.e. resource not over-used and hence global outcome '1') then these $n_{+1}[t]$ agents win. Each agent decides its actions in light of *global information* which takes the form of the history of the m most recent global outcomes. Adaptation is introduced by randomly assigning S strategies to each agent. Each of the 2^{2^m} possible strategies is a bit-string of length 2^m defining an action ($+1$ or -1) for each of the 2^m possible global outcome histories. For example, $m = 2$ has $2^2 = 4$ possible global outcome histories: 00, 01, 10 and 11. Consequently there are $2^{2^{m=2}} = 16$ possible strategies. Strategies which predicted the winning (losing) action at a given timestep, are assigned (deducted) one point.

2.2 Crowd-Anticrowd Formalism

The Crowd-Anticrowd theory [4] incorporates the many-agent correlations arising in the system's strategy space as a result of the dynamics in the history space. To a very good approximation, we can replace the Full Strategy Space by a Reduced Strategy Space (RSS) [3] which provides a minimal basis set of strategies for the system [4]. The appropriate choice for the RSS depends on the relative frequency of visits to the 2^m histories. With all histories visited equally often, the RSS comprises a total of $2P = 2 \cdot 2^m$ strategies [3]. As emphasized in previous MG works, small fluctuations in $D(t)$ are signatures of an efficient self-organization within the population [3, 4, 5, 6]. The excess demand $D(t)$ at timestep t is given by:

$$D(t) = n_{+1}(t) - n_{-1}(t) = \sum_{K=1}^{K=2P} a_K n_K(t) \tag{1}$$

where $n_{+1}(t)$ $(n_{-1}(t))$ is the number of agents taking action $+1$ (-1); $a_K = \pm 1$ is the action predicted by strategy K in response to history $\mu(t)$ and $n_K(t)$ is the number of agents using strategy K at time t. K labels the K'th highest scoring strategy while $\overline{K} = 2P + 1 - K$ labels the anti-correlated strategy. In the non-networked MG at low m, $D(t)$ exhibits large, crowd-driven fluctuations while $\mu(t)$ follows a quasi-deterministic Eulerian Trail[3] in which all histories $\{\mu(t)\}$ are visited equally. Hence, the time-averages $\langle n_{-1}(t) \rangle = \langle n_{+1}(t) \rangle$ yielding $\langle D(t) \rangle = 0$ which is the optimal value for $D(t)$. We will assume that the combined effect of averaging over t for a given Ψ (where Ψ is a given realization of the initial strategy allocation matrix [4]), *and* averaging over Ψ, will have the same effect as averaging over all histories. This is true for the non-networked MG, and produces a mean $D(t)$ of zero. Hence, it can be shown that the fluctuation (i.e. standard deviation) of the excess demand, σ_D, is given by:

$$\sigma_D^2 = \left\langle \sum_{K=1}^{P} \left(n_K(t) - n_{\overline{K}}(t) \right)^2 \right\rangle_{t,\Psi} \approx \sum_{K=1}^{P} \left(n_K^{\text{mean}} - n_{\overline{K}}^{\text{mean}} \right)^2 . \tag{2}$$

In an unconnected system, $n_K^{mean} = n_K$, where n_K is the intrinsic crowd-size due to crowding in the global strategy space and can be shown to be given by [4]:

$$n_K = N \left(\left[1 - \frac{(K-1)}{2^{m+1}} \right]^S - \left[1 - \frac{K}{2^{m+1}} \right]^S \right) . \tag{3}$$

---

[3] The Eulerian Trail corresponds to a particular cycle around the global information states (de Bruijn graph), of duration $2P = 2 \cdot 2^m$ timesteps, in which every possible transition between global information states is visited once. See P. Jefferies, M.L. Hart and N.F. Johnson, Phys. Rev. E **65**, 016105 (2002).

We have used the orthogonality properties of the vectors with elements a_K where $K = 1, 2, \ldots, 2P$ [4]. In addition, since $n_K(t)$ will generally fluctuate around some mean value n_K^{avg}, we have also written $n_K(t) = n_K^{\text{avg}} + \epsilon_K(t)$ and assumed that the fluctuation terms $\{\epsilon_K(t)\}$ are uncorrelated stochastic processes.

3 Networked B–A–R Systems

Here we examine the functionality of a network within the B–A–R model. In particular, whether network connections are beneficial or detrimental depends dramatically on the level of inter-nodal competition. An accurate analytic theory can only be derived *if* one incorporates both the strong correlations within the internal strategy space of the nodes *and* the global non-ergodicity. Standard assumptions related to ergodicity, self-averaging and history-independence, *fail* in this system.

3.1 Perfect Information Transmission

The basic setup is identical to the B–A–R model. However, at each time step t each agent decides its actions in light of (i) *global information* which takes the form of the history of the m most recent global outcomes, in addition to (ii) *local information* obtained via network connections. At each timestep, each agent compares the score of its own best-scoring strategy (or strategies) with the highest-scoring strategy (or strategies) among the agents to whom it is connected. The agent adopts the action of whichever strategy is highest-scoring overall, using a coin-toss to break any ties. For simplicity we here assume a random network, where the connection between any two agents exists with a probability p. We emphasize that *any* of the model assumptions can easily be generalized: the numerical results are reasonably robust.

The mean success-rate of the system, which is inversely proportional to the wastage σ_D as defined in Eq. 2, can be expressed [4] in terms of the mean number or *Crowd* of agents using the K'th highest-scoring strategy (i.e. n_K^{mean}) and the mean number or *Anticrowd* of agents using strategy $\overline{K}$ which is anticorrelated to K (i.e. $n_{\overline{K}}^{\text{mean}}$ where $\overline{K} = 2^{m+1} + 1 - K$). Explicitly

$$n_K^{\text{mean}} = n_K + n_{\to K} - n_{K\to} \tag{4}$$

where n_K is the number of agents who would have used strategy K in the absence of the network because of the initial random strategy allocation and is given by Eq. 3.

$n_{\to K}$ is a sum over all agents who use strategy K as a direct result of a network connection:

$$n_{\to K} = \left[\sum_{J > K} n_J \right] \left[(1-p)^{\Sigma_{G<K} n_G} \right] \left[1 - (1-p)^{n_K} \right] . \tag{5}$$

Hence $n_{\to K}$ represents the *increase* in Crowd-size due to the local connectivity. By contrast, $n_{K\to}$ is a sum over all agents who would have used K in the absence of a network, but who will now use a better strategy as a result of their network connections:

$$n_{K\to} = n_K \left[1 - (1-p)^{\Sigma_{G<K} n_G}\right] . \tag{6}$$

Hence $n_{K\to}$ represents the *decrease* in Crowd-size due to the local connectivity. The resulting analytic expressions for the mean success-rate are in excellent agreement with numerical results (see Fig. 2).

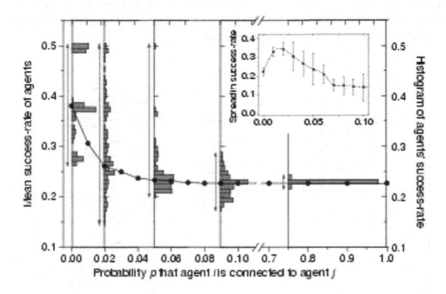

Fig. 1. Modest resources. Solid line (left axis) shows numerically-obtained mean success-rate per agent (node) as a function of the probability p that agent i is connected to agent j. Right axis: typical histograms of agents' success-rate in a typical run of 10^5 timesteps. $N = 101$, $S = 2$, $m = 1$ and $L = 50$. Inset: spread in success-rate as a function of p. Error bars obtained from 20 separate runs.

Figure 1 shows the mean success-rate per agent, averaged over many numerical realizations of the underlying connections and strategy distributions (left axis: solid curve) and the success-rate distribution within the population for a typical run (right axis: histograms) as a function of the inter-connectivity p, in a modest-resource population illustrated by $L = (N-1)/2$. We focus on the small m or 'crowded' regime [3, 4] since there are relatively few strategies compared to the number of agents and hence many agents will simultaneously be competing to win with the same strategy. The mean success-rate decreases rapidly as p increases (left axis: solid curve) before saturating at $p = p_{sat} \sim 0.1$. As m increases, p_{sat} increases. The success-rate distribution for a typical run (right axis: histogram) at $p = 0$ exhibits a definite 'class structure' in terms of success. Dramatic changes then arise as p increases. The spread in the success-rate – shown explicitly in the inset and by vertical arrows in Fig. 1 – indicates a rapid increase in the population's heterogeneity in the range $0 \leq p \leq 0.02$. The success-rate distribution becomes almost continuous, washing out the $p = 0$ class structure. Above $p \sim 0.02$, there is a rapid drop in the proportion of highly-successful agents. Further increasing p leads to a decrease of the spread in success-rates. High levels of inter-connectivity therefore provide increased fairness (i.e. small spread in success-rate) but decreased efficiency (i.e. small mean success-rate). Similar results are obtained for networks with different topological properties, e.g. the Barabási-Albert network driven by preferential attachment [10].

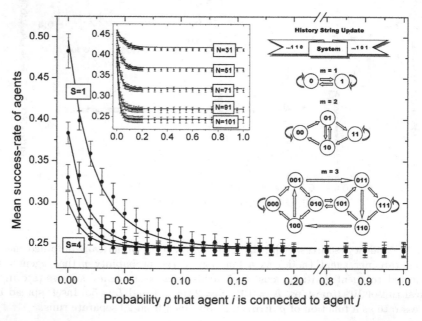

Fig. 2. Solid lines show analytic Crowd-Anticrowd theory for mean success-rate per agent (node) with modest resources, for $S = 1, 2, 3, 4$. Data-points are numerical results. $m = 1$, $N = 101$, $L = 50$. Inset: $S = 2$, various N values. Right: Eulerian Trails (i.e. high p attractor) for $m = 1, 2, 3$.

Figure 2 compares our analytical results using the Crowd-Anticrowd theory [4] to the numerical results for the mean success-rate per agent. The mean success-rate per agent can be obtained from the standard deviation of the number of agents making a given choice (e.g. +1), by averaging this quantity over the attractors of the system (i.e. averaging over the Eulerian Trail or subsets of it).

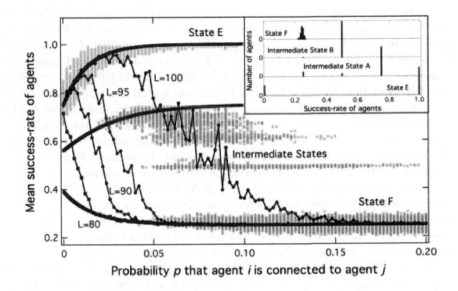

Fig. 3. Higher resources. Thin solid lines: numerically-obtained mean success-rate per agent (node) as a function of p, at resource levels $L = 80, 90, 95, 100$. Results averaged over 200 runs of 10^5 timesteps, with $N = 101$, $S = 2$, $m = 1$. Scattered circles show mean success-rate per agent for separate runs at $L = 100$ with $m = 1$, as an illustration. Thick solid lines: analytic curves (Eqs. 7 – 9) describing the three dynamical regimes. The analytic curves obscure some of the numerical results. Inset: histograms of typical success-rate distribution, for $L = 100$, $m = 1$.

Figure 3 shows the regime changes which arise as a function of the inter-connectivity p in higher resource populations. If the global resource level L exceeds a critical amount $L_{crit} = 3N/4$, the mean success-rate per agent can exhibit a *maximum* at small but finite inter-connectivity, for the following reason. When $L > L_{crit}$ and $p = 0$, the population inhabits a 'frozen' state in the sense that the global outcome is persistently 1, i.e. . . . 111111. With $S = 2$ as in Fig. 3, each agent has probability $1/4$ of being assigned two strategies which both define action -1 following a string of m '1's. Thus approximately $1/4$ of the population always lose and $3/4$ of the population always win, leading to $L_{crit} = 3N/4 \approx 75$. The global resource is therefore under-used at $p = 0$ by $\Delta L \approx (L - 3N/4)$ at each timestep. Therefore increasing p away from $p = 0$ will benefit some of the less successful agents by connecting them up to successful agents, thereby giving them access to ΔL, until a p value is reached where the connectivity is sufficient to break the outcome series of 1's. These run-averaged results can be better understood by considering run-specific results for $L = 100$ as an illustrative case (scattered circles). Specific runs appear to aggregate into groups having similar temporal dynamics and success-rate distributions – these groups correspond to different dynamical states of the system. The increase in the mean success-rate at low p, corresponds to the system following an *Efficient* State E which has an outcome series of 1's as discussed above. The success-rate distribution (inset) for State E is characterized by groups of persistent winners and losers. As p increases further, the outcome series for high L tends to move through a set of Intermediate States which combine a low spread in success-rate with a high mean (e.g. Intermediate States A and B). In Intermediate State A, the least-successful agents (which in State E had zero success) now have a success-rate which *exceeds* the average success-rate in the high connectivity limit. In the high p limit, the outcome series tends toward the period-4 Eulerian Trail given by $\ldots 00110011 \ldots$. The corresponding number of agents taking action $+1$ follows the pattern $\ldots N, N, N/2, N/2, N, N, N/2, N/2 \ldots$. The resulting *Fair* State F corresponds to all agents having access to the best performing strategies, either by being assigned that strategy or by being connected to another agent with that strategy. The resulting system is fair but inefficient, having a small spread but also a small mean success-rate ≈ 0.25.

Figure 4 shows how these transitions between regimes in high-resource populations (e.g. $L = 95$), vary with history bit-string length m. The Intermediate States, characterized by a reasonably high mean success-rate and a reasonably small spread (see Fig. 3 inset), have an increasingly dominant effect on the system's behaviour as m increases. For a particular value of m, increasing p moves the system toward cycles of increasing period and hence the fractions of 1's and 0's in the output series become more equal. The resulting State F, which represents an increasingly noisy version of the Eulerian Trail as m increases, is fair but inefficient (see Fig. 3 inset).

We have derived analytic expressions, within the Crowd-Anticrowd framework including non-ergodicity, which describe well the three regimes (thick solid lines in Figs. 3 and 4). The upper analytic branch at low p, describing the efficient State E, is given by:

$$\frac{3}{4} + \frac{1}{4}\left[1 - (1-p)^{\frac{3N}{4}}\right] . \tag{7}$$

The middle analytic branch at intermediate p, describing the Intermediate States, is given by:

$$\frac{3}{4} - \frac{9}{32}(1-p)^{\frac{7N}{16}} - \frac{1}{32}(1-p)^{\frac{15N}{16}} + \frac{1}{8}(1-p)^{\frac{3N}{4}} . \tag{8}$$

The lower analytic branch at high p, describing the fair State F, is given by:

$$\frac{1}{4} + \frac{9}{128}(1-p)^{\frac{7N}{16}} + \frac{1}{128}(1-p)^{\frac{15N}{16}} + \frac{1}{16}(1-p)^{\frac{3N}{4}} . \tag{9}$$

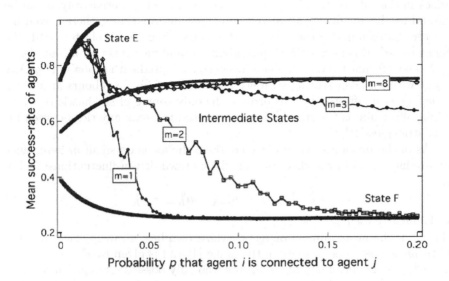

Fig. 4. High resources. Thin solid lines: numerically-obtained mean success-rate per agent (node) as a function of p, at various m. Results averaged over 200 runs of 10^5 timesteps, with $N = 101$, $S = 2$, $L = 95$. Scattered circles show mean success-rate per agent for separate runs at $m = 3$. Thick solid lines: analytic curves (Eqs. 7 – 9) describing the three dynamical regimes. The analytic curves obscure some of the numerical results.

The outcome series of 1's at low p which yield State E, can persist up to $p_{crit} \sim 1 - (1 - 4\Delta L/N)^{4/3N}$. For $L - 80, 90, 95$ and 100, this yields $p_{crit} \sim 0.002, 0.011, 0.019$ and 0.042 respectively, which are also all in excellent agreement with the numerical results.

3.2 Imperfect Information Transmission

We now investigate the effects of *microscopic* connection-driven data error on the system's *macroscopic* dynamics. We examine how the interplay between p, the connectivity between agents, and q, the probability of an error occurring, affects the fluctuations in the excess demand $D(t)$. We show that erroneous data transmission generates an abrupt global transition in the B–A–R model [8]. This phase-like transition is driven by a 'temporal symmetry breaking' in the global outcome series. The Crowd-Anticrowd theory provides a quantitative yet physically intuitive explanation of this phase transition.

As in the B–A–R model with perfect information transmission, agents use the connections they have, if any, to gather information from other agents. As before, each agent takes the action predicted by the highest-scoring strategy among his own *and* those of the agents to which he is connected. However, there now exists a parameter q which represents the probability that an *error* arises in the information he gathers from his cluster. Alternatively, q can be viewed as the weight an agent places on the information gathered from his cluster. For example, if the action of the best strategy in his cluster is $+1$, the agent records this as a -1 with probability q (and vice versa for a best action -1). The information transmission has been corrupted with probability q. Any agent with a higher-scoring strategy than those of his neighbours at a given timestep, is unaffected by this error – the only source of stochasticity which might affect him is the standard coin-toss used to break any ties between his own strategies [4][4].

As in the unconnected case, we continue the focus on small m here, since we are interested in the effect of q on these crowd-driven fluctuations. In Eq. 2,

$$n_K^{\text{mean}} = n_K + n_{\to K}^q - n_{K\to}^q - n_K^q \tag{10}$$

and similarly for $n_{\overline{K}}^{\text{mean}}$, where:

(1) n_K is the mean number of agents whose own best strategy is actually the K'th highest scoring strategy in the game [4] and is given by Eq. 3.

(2) $n_{\to K}^q$ is the mean number of agents who only possess strategies worse (i.e. lower scoring) than K, but who will use strategy K due to connections they have to one or more agents who each possess strategy K but no better:

$$n_{\to K}^q = \left(1 - q\right) \cdot n_{\to K} + q \cdot n_{\to \overline{K}} \tag{11}$$

where $n_{\to K}$ is given by Eq. 5, with $n_{\to \overline{K}}$ being obtained from Eq. (5) by setting $K \to \overline{K} = 2P + 1 - K$.

(3) $n_{K\to}^q$ is the mean number of agents who possess strategy K, but who will nevertheless use a strategy better than K due to connections:

$$n_{K\to}^q = \left(1 - q\right) \cdot n_{K\to} + q \cdot n_{\overline{K}\to} \tag{12}$$

where $n_{K\to}$ is given by Eq. 6 and similarly for $n_{\overline{K}\to}$.

(4) n_K^q accounts for the situation in which an agent is connected to other agents with the same highest scoring strategy K as him. q therefore gives the probability that this agent will take the opposite action to strategy K:

$$n_K^q = q \cdot n_K \left[1 - \left(1 - p\right)^{n_K}\right] - q \cdot n_{\overline{K}} \left[1 - \left(1 - p\right)^{n_{\overline{K}}}\right] . \tag{13}$$

Figure 5 compares the numerical and analytical results for σ_D, which is the standard deviation in excess demand. The agreement is remarkable given the complexity of σ_D as a function of p and q. As the 'noise' level q increases, the system undergoes a change in regime at a critical connectivity p defined by the critical boundary $C_{\text{crit}}(q, p)$. Moving across $C_{\text{crit}}(q, p)$, the symmetry

---

[4] We note that, in contrast to the agents' 'on-site' stochastic strategy selection arising in the Thermal Minority Game (TMG) [5, 8], the stochasticity associated with q in our game depends on the agents' connectivity.

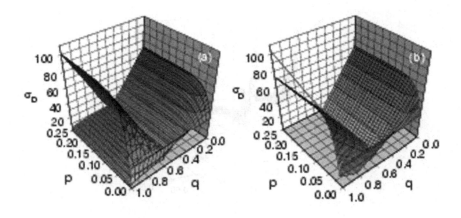

Fig. 5. Fluctuation in excess demand, σ_D, as a function of the error-probability q and the network connectivity p. (a) Numerical results averaged over 300 runs, each with 10^5 iterations. (b) Analytic Crowd-Anticrowd theory. At high q, the two branches in (a) correspond to different dynamical attractor states, while the single branch in (b) represents an effective average (see text). The dotted line in (b) at $p = 0.25$, illustrates the modified analytical results for the upper branch if one assumes some knowledge of this branch's global output series. Parameters: $m = 1$, $S = 2$, and $N = 101$.

in the global outcome string is spontaneously broken in a manner reminiscent of a phase transition. Specifically, the global outcome series changes from the low-q phase where it resembles the period-4 Eulerian Trail $\ldots 00110011\ldots$, to a high-q phase where it comprises *two* distinct branches (see Fig. 5(a)).

Figure 6 shows individual runs near the critical noise threshold. The higher branch corresponds to a period-2 global outcome series $\ldots 1010\ldots$ which is *antiferromagnetic* if we denote 0 (1) as a *spatial* spin up (down) as opposed to a *temporal* outcome. The lower branch corresponds to the period-1 series of 'frozen' outcomes $\ldots 0000\ldots$ or $\ldots 1111\ldots$, i.e. *ferromagnetic*. In this high-q phase, the system will 'choose' one of these two global outcome branches spontaneously, as a result of the type and number of links each agent has. This symmetry-breaking of the global outcome series along the channel of minimum fluctuation in Fig. 5, $C_{\text{crit}}(q, p)$, originates in the internal coupling between the history dynamics, the strategy space and the individual agent networks. Because of the initial strategy allocation and connections, many agents will have an in-built bias towards one of the two possible actions and hence act in a deterministic or 'decided' way at a given timestep. However, there exist a few 'undecided' agents who need to toss an unbiased coin to decide between the equally balanced signals they gather from their local network. It is the

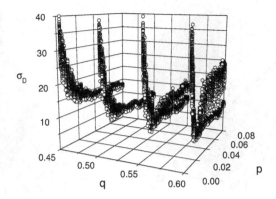

Fig. 6. Numerical results for individual runs, showing the fluctuation in excess demand, σ_D, around $q = 0.5$. Parameters as in Fig. 5.

fluctuations of these few 'undecided' agents who then push the system onto a particular branch.

We now discuss two technical details. First, there are many ties in strategy scores at very small m, and hence many tie-break coin-tosses. This means that the fluctuation terms $\{\epsilon_K(t)\}$ can no longer be ignored. The $m = 1$ surface in Fig. 5(b) was therefore produced by averaging over the $2 \cdot 2^m = 4$ timesteps in the Eulerian Trail. In other words, the double-average in Eq. (2) was evaluated over the $2 \cdot 2^m = 4$ timesteps in the Eulerian Trail. When a tie-break between the strategies $K = 1$ and $K = 2$ arises at one of the four timesteps, one replaces $n_{K=1}^{\text{mean}}$ and $n_{K=2}^{\text{mean}}$ by $\frac{1}{2}(n_{K=1}^{\text{mean}} + n_{K=2}^{\text{mean}})$ at that timestep. Likewise, for tie-breaks between any other K and K'. In this way, the average over the Eulerian Trail is easily evaluated analytically. As m increases, there are more timesteps over which one must average (i.e. $2P = 2 \cdot 2^m$ timesteps). However, since ties also become less frequent as m increases, one can simply ignore them without significant loss of accuracy (see Ref. [4] in which good agreement is obtained for the non-networked MG for a wide range of m values without considering ties). Second, the theory has assumed the non-networked MG result that the dynamics follow the Eulerian Trail. Only one branch therefore emerges in Fig. 5(b) at high q, appearing like some effective average over the global output series for all branches in Fig. 5(a). If instead one uses knowledge of the actual global output series for these separate branches, then results even closer to Fig. 5(a) can be obtained. This is illustrated at one particular p by the dotted line in fig. 5(b).

Figure 7 provides a contour plot of σ_D around the minimum. The black contour, centred around the critical curve $C_{\text{crit}}(q, p)$, effectively separates the two different regimes of behaviour. The low σ_D values around $C_{\text{crit}}(q, p)$ can be easily understood using the physical picture provided by the Crowd-Anticrowd

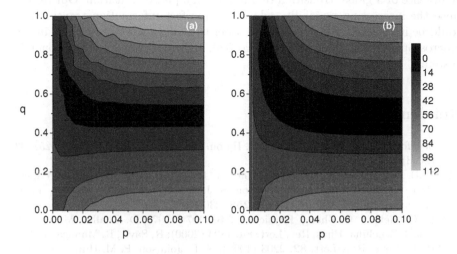

Fig. 7. Contour map version of fig. 5. Contours correspond to a constant value of σ_D as a function of the error-probability q and the network connectivity p. (a) Numerical results. (b) Analytic Crowd-Anticrowd theory. For clarity, only the upper branch of the numerical results is shown for the high-q phase.

theory: the stochasticity induced by q (i.e. noise) breaks up the size of the Crowds using a given strategy K, while simultaneously increasing the size of the Anticrowds using the opposite strategy $\overline{K}$. It is remarkable that a linear increase in the 'noise' q gives rise to such a non-linear variation in σ_D. Using the analytic expressions in this paper, an equation for $C_{\mathrm{crit}}(q,p)$ can be obtained – however we do not include it because it is cumbersome. As noted above, the theory neglects a full treatment of the dynamical fluctuations around n_K^{mean}. Hence, the theory overestimates the Crowd-Anticrowd cancellation arising in Eq. (2) and thus slightly underestimates σ_D in the neighbourhood of $C_{\mathrm{crit}}(q,p)$ (compare Figs. 7(a) and 7(b)). As p increases, $C_{\mathrm{crit}}(q,p)$ becomes less dependent on the connectivity p since more and more agents join the same network cluster. For $p \gtrsim 0.05$ the system passes the percolation threshold and hence is dominated by a giant, common cluster.

4 Conclusion

We have presented a general model of heterogeneous multi-agent systems competing for some limited resource, in which a communication network is used between agents. The numerical results can be explained quantitatively using the Crowd-Anticrowd theory. The network gives rise to a rich dynamical behaviour as a function of the inter-agent connectivity. We also found that less-than-perfect local information transmission between agents can lead to the emergence of a global transition reminiscent of a phase-transition. Our results raise the interesting possibility whereby imperfect information transmission could be induced at the local level in order to achieve a desired change in the macroscopic fluctuations within biological, informational or socio-economic systems.

References

1. See, for example, A. Soulier and T. Halpin-Healy, Phys. Rev. Lett. **90**, 258103 (2003).
2. W.B. Arthur, Science **284**, 107 (1999); W. B. Arthur, Am. Econ. Assoc. Papers Proc. **84**, 406 (1994); N. F. Johnson, S. Jarvis, R. Jonson, P. Cheung, Y. R. Kwong and P. M. Hui, Physica A **258**, 230 (1998).
3. D. Challet, Y. C. Zhang, Physica A **246**, 407 (1997); D. Challet, M. Marsili, and R. Zecchina, Phys. Rev. Lett. **84**, 1824 (2000); R. Savit, R. Manuca, and R. Riolo, Phys. Rev. Lett. **82**, 2203 (1999); N. F. Johnson, P. M. Hui, R. Jonson, and T. S. Lo, Phys. Rev. Lett. **82**, 3360 (1999); S. Hod and E. Nakar, Phys. Rev. Lett. **88**, 238702 (2002).
4. N.F. Johnson and P.M. Hui, preprint cond-mat/0306516; N.F. Johnson, S.C. Choe, S. Gourley, T. Jarrett, P.M. Hui, *Advances in Solid State Physics*, **44**, 427-438 (Springer Verlag, Germany, 2004); M. Hart, P. Jefferies, N. F. Johnson, P. M. Hui, Physica A **298**, 537 (2001); N.F.Johnson, P. Jefferies, P.M. Hui, *Financial Market Complexity* (Oxford University Press, Oxford, 2003).
5. A. Cavagna, J. P. Garrahan, I. Giardina, and D. Sherrington, Phys. Rev. Lett. **83**, 4429 (1999); M. Hart, P. Jefferies, N. F. Johnson, P. M. Hui, Phys. Rev. E **63**, 017102 (2000).
6. M. Anghel, Z. Toroczkai, K.E. Bassler, G. Korniss, Phys. Rev. Lett. **92**, 058701 (2004); I. Caridi, H. Ceva, cond-mat/0401372; M. Sysi-Aho, A. Chakraborti, and K. Kaski, Physica A **322**, 701 (2003); D. Challet, M. Marsili and G. Ottino, preprint cond-mat/0306445.
7. S. Gourley, S.C. Choe, N.F. Johnson and P.M. Hui, Europhys. Lett. **67**, 867 (2004).
8. S.C. Choe, N.F. Johnson and P.M. Hui, Phys. Rev. E (Rapid Communication), (2004) in press.
9. J.L. Casti, *Would-be Worlds* (Wiley, New York, 1997).
10. D. S. Callaway, M. E. J. Newman, S. H. Strogatz, and D. J. Watts, Phys. Rev. Lett. **85**, 5468 (2000); R. Albert and A.L. Barabási, Phys. Rev. Lett. **85**, 5234 (2000); M. E. J. Newman, C. Moore, and D. J. Watts, Phys. Rev. Lett. **84**, 3201 (2000); R. Albert and A.L. Barabási, Rev. Mod. Phys. **74**, 47 (2002).
11. T. Hogg, Nanotechnology **10**, 300 (1999); R.V. Sole, B. Goodwin, *Signs of Life* (Basic Books, New York, 2000); A.L. Barabási and Z.N. Oltvai, Nature (Genetics) **5**, 101 (2004).

Analysis of Complexity and Time Restriction in Resources Allocation Problems

Kiyoshi Izumi[1], Tomohisa Yamashita[2], and Koichi Kurumatani[3]

[1] Information Technology Research Institute (ITRI), AIST & CREST, JST
kiyoshi@ni.aist.go.jp
[2] ITRI, AIST & CREST, JST tomohisa@carc.aist.go.jp
[3] ITRI, AIST & CREST, JST k.kurumatani@aist.go.jp

Summary. In this paper, we constructed three types of agents, which are different in efficiency and accuracy of learning. They were compared using acquired payoff in a game-theoretic situation that is called Minority game. As a result, different types of learning methods got the highest payoff according to the complexity of environmental change and learning speed.

1 Introduction

1.1 Resource Allocation Problems and Minority Games

Resources allocation problems are seen in all the places in the real world. The allocation problem of bandwidth in the Internet or information communication, competition of the market share between companies. the route determination problem of traffic, consumers' purchase selection, and so on, are listed as examples. One of important features common to these problems is that the utility of the resources with less number to choose is higher. That is, the profit of players which choose alternatives different from other players are higher. Minority games, models of the game theory which abstract and simplify this feature, and have been studied by many researchers recently[3, 6]. The interesting result has been obtained especially about the phase transition of the relation between the frequency of the player which chooses each resource and the memory length of in resources selection. However, in the conventional research, all players determined selection using the same information, and have changed strategy by the same learning method. The heterogeneity among players was not taken into consideration. In this research, the heterogeneity of the information used for a player and the learning method was introduced into the minority game which is the model of a resources allocation problem. And two conditions, the complexity of the dynamics of the whole game and the time restrictions about learning were changed, and we investigated which type of agent becomes advantageous. Especially we were interested in the point whether the learning method which assumed others' intelligence at the time of what conditions becomes advantageous.

1.2 Complexity and Time Restriction

What conditions make one assume that the others have capability to process information, to determine action according to rules, and to update their behavioral rules? In this study, it is thought that the following two conditions are key points.

- The social systems that one belongs to have *complexity* about those dynamic behavior. The complexity originates mainly from the interaction among members in the social systems, not from external conditions.
- There is *time restriction*, when one tires to make decisions and to learn a new decision rules in social situations. He or she, therefore, must find not only accurate but also efficient decision rules and learning rules.

The point of view we take in this paper is that the above two conditions, complexity and time restriction, account for the advantage of the assumption that the others have intelligence.

- Because of the complexity of social systems, members must update their decision rules by learning, in order to adapt themselves to social situations and to acquire higher utility. When the members are heterogeneous, that is, they have various decision rules, they must use the others' information in learning.
- Because of the time restriction to learning, the members must find effective learning methods. They adopt the learning method with an assumption that the others have intelligence, because it is effective in social situations.

1.3 Hypothesis

The hypothesis of this research is as follows.

> *Assuming that the others have capability to process information, to update an internal state, and to determine action, have advantage when the two conditions hold; (a) the complexity of environmental change, and (b) the time restriction to learning.*

In this research, as a preliminary experiment for investigating this hypothesis, the computer simulation was performed on the topic of the game theory situation called a minority game.

2 Framework of Experiments

2.1 Minority Game as a Model of Resource Allocation Problems

The minority game is a repeated game in which N (odd number) players must choose one of two alternatives at each step. A payoff is given to minority group, players who chose the alternatives which fewer people chose between two alternatives. Arthur, who is one of the proposal persons of complexity systems economy, proposed the idea of a bar problem that people try to drink at the bar which fewer person chose between two bars[1]. Then, many researchers pointed out and analysis the feature as a nonlinear phenomenon and made various extensions[7].

Since the mechanism that a minority group win is seen also in resource allocation problems such as traffic flows, ecological systems, and trading in financial markets[4, 8]. In this research, standard minority game[3, 6] was developed and it considered as the framework of a preliminary experiment of the following framework interpreted as a model of resource selection problems. The agent of N (odd number) participates in a game, and time progresses discretely. Two resources exist in this game and all agents must choose which resource they use every time step. One time step in this game consists of the three stages; (1) determination of action, (2) calculation of payoffs, and (3) learning.

Determination of Action

Each agent i determines its own action $h^i(t) = \{+1, -1\}$. When an agent uses one resource, say, resource A, its action at time t, $h^i(t)$, is denoted by $+1$. When it uses the other, say, resource B, $h^i(t) = -1$. The action is decided based on information called *memory*, $\mathbf{I}^m(t-1)$, the time series data about which resource provided higher payoff at the past m time steps.

$$\mathbf{I}^m(t-1) = \{I(t-1), I(t-2), \cdots, I(t-m)\} \tag{1}$$

$I(\tau) = \{+1, -1\}$ expresses the higher payoff resource at τ. When the resource A provided the higher payoff at τ, $I(\tau) = +1$. When the resource B, $I(\tau) = -1$.

Each agent i has the rule that determines its resource selection behavior $h^i(t)$ according to each pattern of memory $\mathbf{I}^m(t-1)$. This rule is called strategy $S^i(t)$ (see Figure 1).

Calculation of Payoff

The resource selection behavior of all N agents are accumulated to calculate each agent's payoff at t. In this game, the resource that fewer agents chose can provide higher payoff to the agents that chose this resource. And it is assumed

$$
\begin{array}{ccc}
\text{Memory } \mathbf{I}^m(t-1) & & \begin{array}{c}\text{Action } h^i(t) \\ \text{A or B}\end{array} \\
\{-1,-1,-1,\cdots,-1,-1\} & \rightarrow & \left(\begin{array}{c} +1\text{ or } -1 \\ \{-1,-1,-1,\cdots,-1,+1\} \rightarrow \end{array}\right. \\
\end{array}
$$

Memory $\mathbf{I}^m(t-1)$ 　Action $h^i(t)$
　　　　　　　　　　　A or B
$\{-1,-1,-1,\cdots,-1,-1\} \rightarrow \begin{pmatrix} +1\text{ or } -1 \\ +1\text{ or } -1 \\ \vdots \\ +1\text{ or } -1 \end{pmatrix}$
$\{-1,-1,-1,\cdots,-1,+1\} \rightarrow$
　　　　$\vdots$
$\{+1,+1,+1,\cdots,+1,+1\} \rightarrow$
　　　　　　　　　　　$\uparrow$
　　　　　　　　Strategy $S^i(t)$

Fig. 1. Strategy

that the fewer agents select one resource, the larger payoff agents can acquire. Payoff$^i(t)$ is calculated to each agent i as follows.

$$\text{payoff}^i(t) = -h^i(t) \cdot \sum_{j=1}^{N} h^j(t) \tag{2}$$

When more agents choose resource A ($\sum_{j=1}^{N} h^j(t) > 0$), the agent that chooses resource B ($h^i(t) = -1$) belongs to a minority group, and the minority group's agents acquire positive payoffs. On the other hand, when more agents choose resource B ($\sum_{j=1}^{N} h^j(t) < 0$), the agent that chooses resource A ($h^i(t) = +1$) acquires a positive payoff.

The information $I(t)$ about which resource has the advantage at t is calculated similarly. When more agents choose resource B ($\sum_{j=1}^{N} h^j(t) < 0$), the resource A has the higher payoff, then, $I(t) = +1$. When more agents choose resource A ($\sum_{j=1}^{N} h^j(t) > 0$), then, $I(t) = -1$.

$$I(t) = \begin{cases} +1 & (\sum_{i=1}^{N} h^i(t) < 0) \\ -1 & (\sum_{i=1}^{N} h^i(t) > 0) \end{cases}$$

Learning

The learning in the standard minority game is the selection of the strategy $S_i(t)$ that is used in the determination of an action. Each agent has s strategies shown in Figure 1 and s is usually from 2 to 36. These strategies are generated randomly at the beginning of the game, and they are fixed throughout the game. Each strategy has a specific value called a virtual value. It is a number of times that the behavior derived from a strategy could acquire a positive payoff. Each agent chooses one strategy with the highest value from the s strategies, and uses it when it decides its own behavior at the next time step $t+1$.

In standard studies of the minority game, only the very simple learning was assumed and they analyzed in many cases about the relationship between the memory length m and the fluctuation of the number of agents that chose one resource. However, in a previous study[2], it is suggested that it has essential significance that all agents are homogeneous, that is, all agents share the same information and the same learning algorithm. It is shown that the same results was obtained in both cases that all agents used the real historical data $\mathbf{I}^m(t)$ and that all agents used the randomly generated data. In our study, agents are heterogeneous, that is, the information and the learning methods differ among agents.

2.2 Agents

We prepared 4 types of agents; one standard type and three extended types. First, **Pattern matcher (PM)** was a standard agent. This agent determines behavior based on the time series data $\mathbf{I}^m(t)$. It is extended from the player in the standard minority game, described in section 2.1.

In addition, we prepared three kinds of agent; **Hand imitators (HI)**, **Strategy imitators (SI)**, and **Perfect predictors (PP)**. They are different in terms of kinds of information that they use (see Table 1). Each agent took part in the minority game against Pattern matchers, and its payoff was compared with each other. Hand imitators perform simple learning of imi-

Table 1. Agent types

Efficiency	HI	>	SI	>	PP
Accuracy	HI	<	SI	<	PP
Information	Others' payoff Others' action		Others' payoff Others' action Memory about resources		Others' payoff Others' action Memory about resources Game structure

(HI: Hand imitator, SI: Strategy imitator, and PP: Perfect predictor)

tating the behavior of other agents with high payoff. Since it uses the least information, the speed of learning is quick and the efficiency is good, but the learning results will be inaccurate. On the other hand, Perfect predictor infers both other agents' strategy and the game structure (payoff matrix) using all kinds of information. Since it uses much information, the obtained result is exact, but the speed of learning is slow and the efficiency is low. Strategy imitators are in the middle of these two types. It performs only a inference of other agents' strategy, and imitates the strategy of other agents with high payoff. The efficiency and accuracy of its learning are in the middle of HIs and PPs.

Pattern Matcher (PM)

Pattern matchers' behavior rules are the same as stated in section 2.1. It is extended about learning. Since a agent in standard minority games continues to have its strategies given first without changing, as described in section2.1. Thus, it can not search for all of solution spaces. Then, we extended pattern matcher's learning method as follows, to enable it to search for all solution spaces.

Decision of behavior

Pattern matcher has one strategy described in figure 1. The pattern matching of the historical data of the past m steps, $\mathbf{I}^m(t-1)$, to the strategy $S^i(t)$ determines Pattern matchers' behavior $h^i(t)$.

Learning

When Pattern matchers acquires a positive payoff at t, it does not change its strategy. When it got a negative payoff, its strategy $S^i(t)$ is updated at a certain probability α (learning rate). The bit of the behavior rule (the resource A or B) to this historical pattern is reversed.

Hand Imitator (HI)

Hand imitator performs simple learning of imitating the behavior which other agents with a high payoff.

Decision of behavior

According to a certain probability p_A, Hand imitator uses the resource A. Probability to use the resource B p_B is $1 - p_A$.

Learning

The probability of dealing behavior of other agents with a high payoff is copied.

1. Inference of others' probability
   The estimated probability $\tilde{p}_A^j(t)$ that another agent j chooses the resource A is updated by the following equation.

   $$\tilde{p}_A^j(t) = (1 - \beta) \cdot \tilde{p}_A^j(t-1) + \beta \cdot \text{action}^j(t) \tag{3}$$

   $\text{action}^j(t)$ is probability which the agent j chooses the resource A at this term.

   $$\text{action}^j(t) = \begin{cases} 1 & (\text{Agent } j \text{ chooses the resource A at t}) \\ 0 & (\text{Agent } j \text{ chooses the resource B at t}) \end{cases}$$

   Parameter $0 \leq \beta \leq 1$ expresses the rate which updates the estimated value of the probability of behavior of other agents, and means the learning speed of others' models.

2. Accumulation of payoff

   About all agents j including itself, the accumulation value $R^j(t)$ of payoff is updated by the following equation.

$$R^j(t) = (1 - \gamma) \cdot R^j(t - 1) + \gamma \cdot \text{payoff}^j(t) \tag{4}$$

   $\text{payoff}^j(t)$ is a parameter showing the payoff of agent j this term. $0 \leq \gamma \leq 1$ expressed the update rate of the accumulation value of a payoff, and fixed it to 0.5 in this study.

3. Copy of the behavior according to the payoff

   Each agent copies the probability p_A from other agents with a certain probability α (learning rate). One agent j^* is chosen by the probability proportional to accumulation value $R^j(t)$ of each payoff from all the agents that he also contains first. And the probability $p_A^{j^*}(t)$ is copied to its own probability p_A.

Strategy Imitator (SI)

Strategy imitator performs only an estimation of other agents' strategy, and imitates the strategy of the high agent of a payoff.

Decision of behavior

The pattern matching of the historical data of the past m steps, $\mathbf{I}^m(t-1)$, to the strategy $S^i(t)$ determines behavior $h^i(t)$. It is the same as that of Pattern matcher.

Learning

Strategy imitator estimates others' strategies and imitates the other's strategy with higher payoff.

1. Estimation of others' strategies

   Strategy imitator estimates whether the other agent js to choose the resource A or B, using the pattern matching of the historical data $\mathbf{I}^m(t-1)$ to estimated others' strategies $\tilde{S}^j(t)$. If the estimated behavior is different from the actual behavior which agent j performed, by a certain probability $0 \leq \beta \leq 1$ (learning speed of others' models), the estimated strategy $\tilde{S}^j(t)$ will be updated. The bit of the agent j's behavior corresponding to the historical pattern in $\tilde{S}^j(t)$ will be inverted.

2. Accumulation of a payoff

   About all agents j including itself, the accumulation value $R^j(t)$ of payoff is updated by the equation 4. It is the same as that of Hand imitator.

3. Imitation of strategy according to the payoff

   A strategy is copied by a certain probability α (learning rate). One agent j^* is chosen by the probability proportional to accumulation value $R^j(t)$ of each payoff from all the agents. And the agent's strategy $\tilde{S}^{j^*}(t)$ is copied to its strategy.

Perfect Predictor (PP)

Perfect predictor estimates both other agents' strategies and the game structure (the payoff matrix) using all information.

Strategy

Perfect predictor estimates others' behavior and decide its own behavior according to the estimated payoff matrix.

1. Estimation of others' behavior
   Perfect predictor estimates whether the other agent js to buy or sell from the pattern matching of the historical data $\mathbf{I}^m(t-1)$ to estimated others' strategies $\tilde{S}^j(t)$.
2. Decision of behavior
   Perfect predictor has its own strategy that represents estimated game structure (the payoff matrix). That is, the strategy shows which behavior (selection of the resource A or B) can acquire a positive payoff corresponding to the others' behavior. According to the strategy and the other agent j's estimated behavior, Perfect predictor decides its own behavior.

Learning

Perfect predictor learns both others' strategies and the game structure.

1. Learning of others' model
   The estimated strategies $S^j(t)$ about other agent js are updated. Learning method is the same as that of Strategy imitator.
2. Learning of the game structure
   When its payoff is lower than the average of other agents' payoff, the bit of behavior corresponding to the estimated others' behavior is reversed by a certain probability α (learning rate). This means renewal of the knowledge about the payoff structure.

3 Simulation Results

3.1 Setting of the Simulation

Setting of the simulation of the preliminary experiment in this study is shown in Table 2.

25 agents participate in each trial of the minority game. 20 agents are Pattern matchers as standard agents in all trials. The remaining 5 agents are Hand imitators, Strategy imitators, or Perfect predictors.

The learning speed α of their strategy and the update rate of a payoff γ were fixed. The memory length m and the learning speed of others' models β of HI, SI, and PP were changed as shown in Table 2.

Table 2. Setting of the simulation

Number of agents	$N = 25$
Agents' combination	(20 PMs and 5 HIs), (20 PMs and 5 SIs), and (20 PMs and 5 PPs)
Memory length	$m = \{1, 2, \cdots, 15\}$
Learning speed	$\alpha = 0.8$ (fixed)
Learning speed of others' model	$\beta = \{0.1, 0.2, \cdots, 0.9\}$
Update rate of payoff	$\gamma = 0.5$ (fixed)
Number of simulation runs	10 times every parameter combination
Comparison method	Improvement rate of the payoff of 5 HIs, SIs, and PPs when the average of the payoff of 20 PMs is set to 100.

(PM: Pattern matcher, HI : Hand imitator, SI: Strategy imitator, and PP : Perfect predictor)

The simulation was performed 10 times every parameter set; {Agents' combination (PM vs. HI, SI, or PP) × Memory length m × Learning speed of others' modelsβ}. The agents (HI, SI, and PP) were compared by the improvement rate of payoff when the average of payoff of the 20 standard agents (Pattern matchers) wad set to 100.

3.2 Simulation Results

The simulation results are summarized in Figure 2 and examples of simulation results are shown in Figure 3(a)-(c).

Overall Results

Different types of learning methods got the highest payoff according to the memory length m and the learning speed of others' models β.

Our preceding study revealed that the memory length linked to the complexity of dynamics of the whole system[5]. When memory length is short, the whole system showed relatively simple patters that can be described by dynamic systems of finite dimensions. As the memory length got longer, the dynamics patters became more complex, and finally they can not be described by any finite dimensional dynamic systems. Therefore the memory length m can be considered as an index of complexity of the whole system.

The learning speed of others' models β is an index of the time restriction to learning. All agents continuously change their own strategies at the fixed learning speed α. When β is small, agents must track the others' strategies at low learning speed, so the time restriction to learning is strong. When β is large, agents can have relatively enough time to track the others' strategies, so the time restriction to learning is weak.

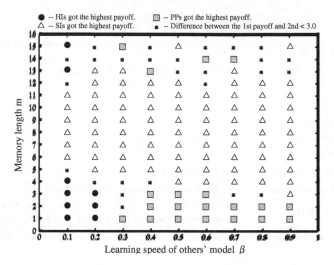

Fig. 2. Summary of results: Each symbol represents which agent type got the highest payoff. There are 4 distinct areas ($m \leq 3$ & $\beta < 0.3$, $m \leq 3$ & $\beta > 0.3$, $3 < m < 12$, and $m \geq 13$) according to the winners' types.

Figure 2 shows that there are 4 distinct areas according to the two conditions, the complexity and time restriction.

1. *Low complexity and strong time restriction area:* When the memory length is short ($m \leq 3$) and the learning speed of others' models is slow ($\beta < 0.3$), Hand imitators' payoff was high. This result was also shown in Figure 3a. This is because the Perfect predictors and Strategy imitators can not have enough time to learn the others' strategies because of the strong time restriction, and they showed the poor performance. Only Hand imitators could catch up with the fast change of Pattern matchers' strategies because of their simple and quick learning method. Moreover, Hand imitators could learn adequately accurate models because the whole system is comparatively simple.

2. *Low complexity and weak time restriction area:* When the memory length is short ($m \leq 3$) and the learning speed of others' models is fast ($\beta > 0.3$), Perfect predictor's payoff was high (Figure 3a) It is because the Perfect predictors can have enough time to learn the others' strategies and the game structure. Moreover the dynamics of whole systems is relatively simple in this area. They then could obtain exact learning results using all information and beat the other two types, the Hand imitators and Strategy imitators.

3. *Medium complexity area:* The Strategy imitators' payoff becomes high as memory length becomes longer ($3 < m < 13$). Figure 3b shows that the Strategy imitators could beat the other two type throughout all learning speed β. When other agents' strategy is complicated, simple learning

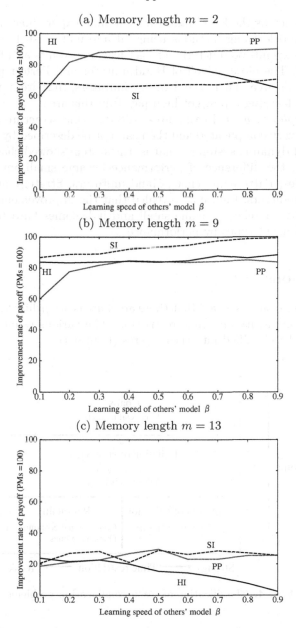

Fig. 3. Simulation results: (a) When memory length is short and the learning speed is slow, the payoff of HI (Hand imitator) is high. When the learning speed of an others model is fast, the payoff of PP (Perfect predictor) is high. (b) The payoff of SI (Strategy imitator) becomes high as memory length becomes long. (c) Finally the performance of all agent types have small difference.

methods such as the Hand imitators can only acquire inaccurate learning results. And learning methods using all information such as the Perfect predictors spend too much time to get learning results at a limited learning speed. Therefore, payoff of Hand imitator and Perfect predictor had fallen in this area. Then, the Strategy imitators, that are in the middle of these two learning types, got high payoff in this area.

4. *High complexity area:* In this area, the dynamic structure of the whole system is very complicated and then can not be described by any finite dimensional dynamic systems. That is, the system showed chaotic features. Therefore, the difference of performance became small among all the 4 agent types; Pattern matchers, Hand imitators, Strategy imitators, and Perfect predictors. Figure 3c revealed that the improvement rate against the Pattern matchers became small and it becomes hard to distinguish among all agent types by performance.

4 Conclusions

Our simulation results showed that there are 4 areas according to the two conditions, the complexity and time restriction. The various resource allocation problems can be classified into these 4 ares (Figure 4).

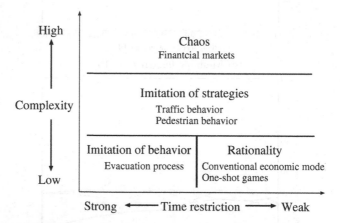

Fig. 4. Classification of resource allocation problems

Evacuation process is classified into the low complexity and strong time restriction area. When people try to determine evacuation routes, they may not use old memories but mainly use knowledge about the present situation. Therefore, the memory length may be short. Of course, the time restriction is strong in emergent situations. If one want to model evacuation process, he or she had better build a model with agents that do not consider the other agents' strategies or decision rules like the Hand imitators.

The situations that are described in conventional economic models, where the time restriction to learning is weak and environmental changes are not so complicated, rational agents have advantages. They try to find exact solution using all available information. This situation is so ideal that it is difficult to find examples in the real world, but the purchase of expensive goods may be an example.

The medium complexity area includes a wide range of the resource allocation problems such as the route selection of drivers or pedestrians. In this area, it is reasonable that people determine their actions assuming that the other also have decision rules or strategies. Then, the learning method based on the others' intelligence such as the Strategy imitators have the advantage.

Finally the high complexity area also contains various examples in the resource allocation problems such as financial markets. In this area, there is no universally correct answer of learning. Therefore, it is thought that the procedure of learning itself becomes a key factor not only the result of learning.

5 Discussions

In conventional economic models, when the participant of an economic system was not using all information, only some negative causality had been considered. For example, information cannot be acquired (in-accessibility). Information has many noises and it cannot be trusted (low reliability). Many costs are required to acquire information (high cost).

These models have an implicit assumption, "a forecast would become exact so much if many information is used." Although its attention had been directed to some extent by the cost which acquires information, probably, about the cost which learns using the acquired information, it had not almost been taken into consideration at all. It is because the condition that a participant's learning speed assumes infinitely the ideal condition of being fast, in the economic model with this classic, therefore the speed of change of the environment which is a learning target can be relatively regarded to a zero is considered. This is shown in figure 2 as an ultimate situation which is in the right at an infinite distant place. The algorithm which performs an exact learning in this condition using all information, without considering the efficiency of a learning is the most advantageous.

In classic economic models, the "true model" showing the behavior of the target economic phenomenon exists solemnly on a cloud as a given thing. And the participants who have an infinite learning speed using all information are learning toward a true model.

On the other hand, in the model by the agent simulation which was shown by this study, the behavior of an economic phenomenon is not being fixed and it has generated internally from the interaction between participants. Each participant thinks that he will forecast well and is performing various learnings. The learning behavior carries out an interaction and the behavior pattern of the whole economic system changes. While one mikoshi is shouldered all together and a motion of everybody collides, it is moving on the whole. The target to pursue also escapes or approaches according to a motion of a mikoshi.

As for a learning in the environment where the target of a learning changes behavior in response to reaction from the learning behavior of oneself, like a society and an economic system, learning using all information is not necessarily good. At the point, a classic economic model differs from the condition of having assumed implicitly. Then, what does actual human being do? Probably, paying attention to what information, heuristics is used about how it learns. Heuristics is the learning method discovered experientially. Although it cannot necessarily obtain an exact learning result, it can obtain efficiently the result which can be satisfied moderately. Strategy imitator is also a kind of heuristics of imitating the strategy of others who succeeded in fact.

It is also interesting to analyze the heuristics of a learning of actual human being found by the cognitive psychology and some learning methods in artificial intelligence in the viewpoint of the efficiency and accuracy of a learning, and to evaluate as a model of the learning in a society and an economic system. And finally it will become one approach to the elucidation of the function of an intelligence in social / economical situation.

References

1. W. B. Arthur. Inductive reasoning and bounded rationality (the el farol problem). *American Economic Review*, 84:406, 1994.
2. A. Cavagna. Irrelevance of memory in the minority game. *PHYSICAL REVIEW E*, 59(4):R3783–R3786, 1999.
3. D. Challet and Y.-C. Zhang. Emergence of cooperation and organization in an evolutionary game. *Physica A*, 246:407–418, 1997.
4. A. Galstyan, S. Kolar, and K .Lerman. Resource allocation games with changing resource capacities. In *Proceedings of the second international joint conference on Autonomous agents and multiagent systems*, pages 145–152. ACM Press, 2003.
5. K. Izumi. Complexity of agents and complexity of markets. In T. Terano, T. Nishida, A. Namatame, S. Tsumoto, Y. Osawa, and T.Washio, editors, *New Frontiers in Artificial Intelligence*, pages 110–120. Springer, 2001.
6. M. Marsili. Market mechanism and expectations in minority and majority games. *Physica A*, 299:93–103, 2001.
7. Minority Game's web page. http://www.unifr.ch/econophysics/.
8. Y.-C. Zhang. Modeling market mechanism with evolutionary games. *Europhys. News*, 29:51–54, 1998.

How Does Collective Intelligence Emerge in the Standard Minority Game?

Satoshi Kurihara[1], Kensuke Fukuda[2], Toshio Hirotsu[3], Osamu Akashi[2], Shinya Sato[2], Toshiharu Sugawara[4]

[1] I.S.I.R. Osaka University, 8-1, Mihogaoka, Ibaraki, Osaka, 567-0047 JAPAN
   kurihara@ist.osaka-u.ac.jp
   http://www.ai.sanken.osaka-u.ac.jp
[2] NTT Network Innovation Labs., Tokyo, JAPAN
[3] Toyohashi University of Technology, Aichi, JAPAN
[4] NTT Communication Science Labs., Nara, JAPAN

Summary. In this paper we analyze a simple adaptive model of competition called the Minority Game, which is used in analyzing competitive phenomena such as the operation of the market economy. The Minority Game is played by many simple autonomous agents, which develop collective self-organization as a result of simple behavioral rules. Many algorithms that produce the desired behavior in the game have been proposed. In all work to date, however, the focus has been on the macroscopic behavior of the agents as a whole. We focus on the behavior of individual agents, paying particular attention to the original form of the Minority Game. We suggest that the core elements responsible for the development of self-organization are (i) rules that place a good constraint on the behaviors of individual agents and (ii) the existence of rules that lead to effective indirect coordination. We also show that when efficient organization is formed, a power-law can be seen among behavior of individual agents.

1 Introduction

The individual animals in a school of fish or flock of flying birds seem to be organized in well-ordered overall behaviors by some form of intelligent coordination. The 'boid' was developed as a simple computer model that replicates such behavior [11]. Each boid runs the boids algorithm, and the boid collectively exhibit realistic flocking or schooling behavior. Such work has clearly shown that well-ordered overall behavior can be formed by merely local and very simple interaction among elements.

An environment consisting of many autonomous entities, every one of which can get a large profit even though the individuals are behaving in a selfish way governed only by simple rules that don't require knowledge of the overall system, is very attractive in term of modeling of the market economy and other socio-economic structures. Such an environment will be very useful as a general means for coordinating agents in multi-agent systems.

The Minority Game is a simulation program for analyzing models of adaptive competition. The game is like the market in that it is played by multiple autonomous agents. Each agent acts with the purpose of obtaining its own profit by using purely local information. The Minority Game has also been a popular object of study in 'econophysics' and has been looked at from various other angles since it was first proposed more than four years ago [1]. Many algorithms that produce the desired behavior in this game, i.e. large profits for nearly all agents, have been proposed [2-7, 9]. In nearly all of these studies, however, researchers were concentrating on the behavior of the agents as a whole; to date, we have not seen any papers on studies where the focus was on the behavior of the individual agents. Therefore, we investigated the behavior of the individual agents with a view to discovering the kind of individual behavior that leads to efficient overall behavior. We pay particular attention to the Minority Game in its initial form [1], because in this case the agents develop a well-ordered overall behavior through purely local and very simple interaction, even though each is behaving in a selfish way. We show that any set of rules which makes all of the agents behave in a well-ordered manner and get large profits produces fractal characteristics in the behaviors of the individual agents.

2 The Standard Minority Game

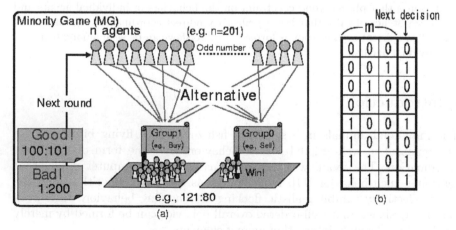

Fig. 1. Rules and strategy table

Firstly, we review the rules of the standard Minority Game (see Fig. 1 (a)). We have N agents, each of which is an autonomous entity that independently chooses between two alternatives (group 0 or group 1) according to its own behavioral rules. In each round of the game, all of the agents choose one alternative or the other, and the agents that then finish in the minority group are considered to be winners. Each winner is awarded one point, and the total numbers of points awarded to all agents is the profit in this round of the game. Therefore, the smaller the difference between the numbers of agents in the majority and minority groups, the better the result.

Each agent makes its selection based on one of multiple strategy tables that it holds. The entities in the table contain all combinations of m past winning-group choices along with next decisions that corresponds to each of the combinations (see Fig. 1 (b)). At the beginning of the game, each agent randomly prepares s strategy tables, and randomly set the next decision entries. In the first round of the game, each agent randomly selects one of the s strategy tables it holds. If the agent wins, that is, ends up in the minority group, one point is assigned as profit to the selected strategy table. If the agent loses, one point is deducted. In the second and subsequent rounds of the game, the strategy table that has the highest number of points is always selected. This cycle is repeated a predetermined number of times, and the final result of the game is the total number of points acquired by winning agents across all rounds.

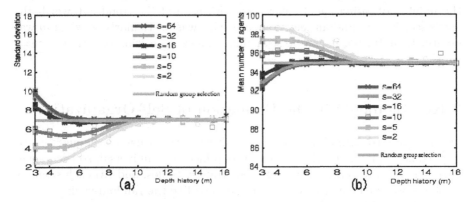

(a)

(b)

Fig. 2. (a) Standard deviations of the number of winning agents. (b) Mean numbers of winning agents

2.1 Coordinated Behaviors

We begin by verifying the collective behavior of the agents in the same way as was done in earlier studies [1-3]. Firstly, the standard deviation of the number of winning agents is shown in Fig. 2(a). The game was played for the number of rounds described below with the agents possessing various numbers of strategy tables, $s = \{2, 5, 10, 16, 32, 64\}$ and the strategy tables having various history depths, $m = 3$ to 16. One trial for each parameter pair, s, m, is 10,000 rounds of the game; ten trials were conducted for each pair. Fig. 2(b) shows the mean numbers of winning agents. The horizontal lines in Fig. 2(a) and (b) represent the standard deviation and mean value when all of the agents made random choices. These graphs show that, for the lower values of s, the standard deviation became lowest and the mean number of winning agents became highest when m was from three to five. This means that some kind of coordinated behavior among the agents was driving the winning-group ratio closer to 100:101 in these cases. The most interesting characteristic is that, although we would expect behavior based on longer histories to be more efficient, m values larger than 10 produce results that are the same as those of random behavior.

Even though the standard deviations for the lower s values are low in the range $m=3$ to 5, fixed sets of agents may always be ending up winners or losers. So, we investigated how many times each agent would be able to win in several situations. Figure 3 shows the rankings of the 201 agents by average score. In the case where each agent randomly selected "group 0" or "group 1", all of the agents were able to get approximately 4750 points. In contrast to this, the mean score was high when the standard deviation was small (m=3 to 5) and, although some differences between agents can be seen in the scores, all or almost all of the agents were able to achieve stable high scores. Interestingly, the number of losing agents increased with m, and the trend is towards the levels seen in the random selection case. This was particularly so when m was greater than 10. So, it can be seen that a phase transition occurred in $m=6$ to 9.

3 Key Elements in the Formation of Self-Organization

In the standard Minority Game, the following two rules are thought to be important elements: (i) the winning-group history is universal, so each agent must see the same entry in its selected strategy tables, and (ii) each strategy table's score is changed by every win and loss like the random-walk.

3.1 A Winning-Group History is Unnecessary

In the original algorithm [1], if the current winning-group history is "010," all agents see the entries for "010" in the strategy tables they hold, and each agent selects its currently highest-scoring strategy table. However, we hypothesize that the memory of winning-group history has no effect.

At this point, we raise the following questions.

(1) If each agent randomly selects the entry from its currently highest scoring strategy table, does self-organization occur?

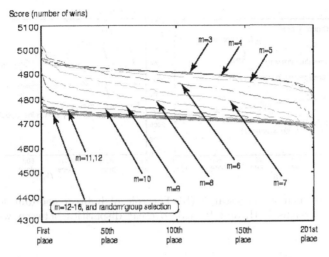

Fig. 3. Rankings of 201 agents for each value of m

(2) If only one agent, agent-A, selects an entry based on the winning-group history and the other agents use the same entry, does self-organization occur? In other words, agents do not obey the rule of winning-group history, but simply see the same entry as that selected by agent-A.

(3) If we intentionally generate a random winning-group history, does self-organization occur?

If self-organization develops in any of the above situations, our hypothesis that the memory of the winning-group history has no effect may be correct. Fig. 4(a) shows the result: the same self-organization as with the normal algorithm developed in cases (2) and (3). That is, even when we made a random winning-group history as in case (3), the agents were able to develop good overall organization. This result shows that the memory of the winning-group history may not be important in the development of self-organization. However, when each agent selects a random entry from its strategy tables, the behavior is the same as for random group selection.

We can thus infer an important point concerning the strategy tables. That is, in the normal algorithm, the rule that all agents depend on the winning-group history simply places a constraint on the agents. In other words, we may be able to use any rule that places a 'good' constraint on the agents, as in cases (2) and (3). Applying a good constraint to the agents essentially means decreasing their freedom. In earlier published work [8], we discussed how other frameworks lead to self-organization, and made the same suggestion as in this subsection.

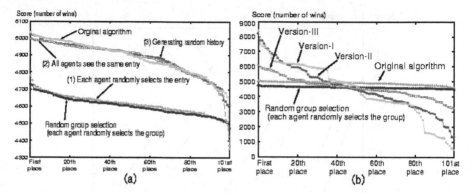

Fig. 4. (a) Is memory necessary? (b) The role of indirect coordination: despite having this, versions I, II, and III led to great disparities between winning and losing agents.

3.2 The Importance of Indirect Coordination

Indirect interaction between agents is also considered to be important element in the development of organization in the standard Minority Game. That is, each agent decides its behavior based on the results of each round, and each agent's decision indirectly affects the behaviors of the other agents. In the normal algorithm, one point is added to the selected strategy table when the agent wins, and one point is subtracted when it loses. As an example, consider an agent that has two strategy tables, table-A and table-B . If table-A has four points and table-B has one point, table-B will not selected until table-A has lost at least four times in a row. Will self-organization still develop if we change the rule for selecting the strategy tables? Fig. 4(b) shows the results of our investigation of this question. We implemented the following rules.

(Version I) Agents select strategy tables sequentially. The interval of exchange is randomly set.
(Version II) One point is added to the score for a selected strategy table that wins, but two points are subtracted if it loses.
(Version III) If the agent loses a single game, the strategy table is exchanged for the next-scoring table, even when the currently strategy table still has the most points.

Unfortunately, self-organization did not develop with any of these versions. The method of strategy-table selection must have a close relationship with the initial combinations of strategy tables held by each agent. Therefore, we investigated how the strategy tables are used by individual agents with the normal algorithm. In Fig. 5, (a) and (b) show the transitions of the scores for each strategy table held by the agents in 25th and 200th places among the 201 agents, in trials where all agents had $m=3$ and $m=14$, respectively. While

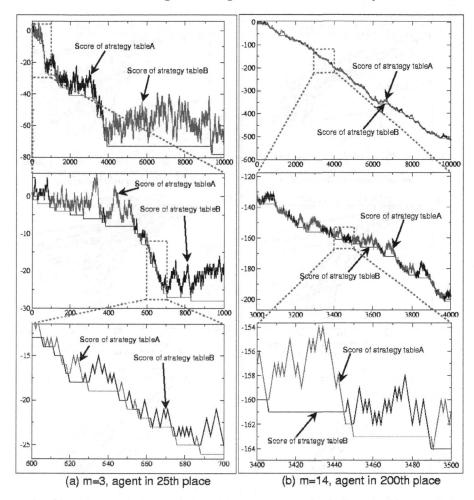

(a) m=3, agent in 25th place   (b) m=14, agent in 200th place

Fig. 5. Which strategy table was used?   Scores for (a) m=3, 25th place agent and (b) m=14, 200th place agent in (upper) 10000, (middle) 1000, and (lower) 100 games.

the agent in Fig. 5(a) used both strategy tables, there was no fixed period for the continuous use of one strategy table. A fractal characteristic is visible in these results; that is, the usage of strategy tables shows self-similarity. Fig. 6 shows, on log-log scales, per-agent histograms of the periods over which either of the strategy tables was continuously used. Results for $m=3$, 7, and 14 are given for (a) n = 101, and (b) n = 301. Power-law relations can be seen in both cases of $m=3$ (graphs were nearly straight lines) and this reflects the presence of a fractal characteristic.

Returning to Fig. 5, no fractal characteristic is visible in the results for the agent in 200th place (Fig. 5 (b)). Figure 2 showed that the performance

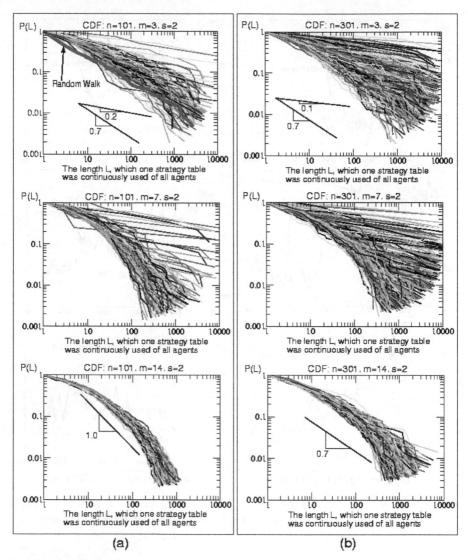

Fig. 6. No. of steps L over which either strategy table was in continuous use for (a) $n=101$ and (b) $n=301$ with (upper) For $m=3$ (all 101 agents and 301 agents), (middle) For $m=7$ (all 101 agents and 301 agents) (lower) For $m=14$.

of agents with m=14 is only as good as random selection. We thus do not see straight lines in the histograms for m=14 in Fig. 6; these results are similar to the result for random selection. This reflects a fixed period for the continuous use of each strategy table in these cases (as seen in the lower part of Fig. 5 (b)). The histograms for agents with m=7 show an interesting mix of the two types of results; some are similar to the graphs for m=3, while others are similar to those for m=14. It might be the case that the straight lines are for winning agents, while the curves similar to those for m=14 are for losing agents.

4 Discussion

At this point, why power-law relations can be seen in the cases of m=3? We think that the assigning and deducting rule of strategy table may be important. The rule of when the agent wins one point is assigned and when it loses one point is deducted is similar to "random-walk." Indeed, as the Fig. 4 (b)-VersionII shows, fractal characteristic cannot be seen by the rule of when the agent wins one point is assigned and when it loses two points are deducted, and as the Fig. 6 (a)-upper shows, each slope of agent and the slope of random-walk[5] is almost similar. In random-walk, probability that assigning-selection and deducting-selection are chosen is equal, so, when m value is small (like $m = 3$) it can be thought that probability of winning and losing may be equal. On the other hand, when m value is big (like $m = 14$), it can be thought that probability of losing may be higher than winning, so, fractal characteristic like random-walk cannot be seen.

As for power-law relations, Fig. 7 shows the slopes[6] of the histograms in Fig. 6 for all agents, from that in first place to that in last place ((a) n=101 and (b) n=301). As you can see, slopes are clearly greater for winning than for losing agents. This result is very interesting, because it indicates that individual agents have a way to find out whether they are behaving well or poorly without knowledge of the results for other agents.

Further detailed investigation will be necessary to clarify the essential mechanisms in the selection of strategy tables, and this is a topic for future work. However, we appear to have found a way to check whether or not agents are performing in a well-ordered manner.

5 Conclusion

In this paper we analyzed a simple adaptive-competition model called the Minority Game, which is used in the analysis of competitive phenomena in

---

[5] We ran random-walk 10000 steps, and calculated the CDF of histograms of the number of steps where a positive value or a negative value was consecutive.

[6] Slope is calculated by the least-squares method

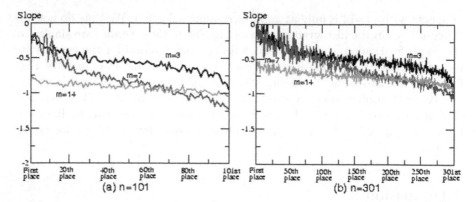

Fig. 7. Slopes of the histograms in Fig. 6.

markets. In particular, we focused on the standard Minority Game. We verified the validity of the following hypotheses: (i) the rules simply have to place a 'good' constraint on the behaviors of individual agents and (ii) the rules of the standard game lead to effective indirect coordination. A. Cavagana [10] has already demonstrated that the memory of the winning-group is irrelevant, and stated that the important point is the sharing of some data among the agents. However, we go further in suggesting that the important point is simply to place a 'good' constraint on the behaviors of individual agents, and that this constraint might take the form of the sharing of data which, in the normal algorithm, is the winning-group history. On point (ii), further detailed investigation will be necessary to clarify the essential mechanism in the selection of strategy tables that leads to effective indirect coordination. We will continue our analysis, with the aim of establishing a general algorithm that is applicable to other competition problems.

References

1. D. Challet and Y.-C. Zhang, Emergence of Cooperation and Organization in an Evolutionary Game, Physica A 246, 40 7 (1997).
2. Y.-C. Zhang, Modeling Market Mechanism with Evolutionary Games, Europhys. News 29, 51 (1998).
3. D. Challet, A. Chessa, M. Marsili, Y.C. Zhang, From Minority Games to Real Markets, Quantitative Finance (2001).
4. R. Savit et al., Adaptive Competition, Market Efficiency, Phase Transitions and Spin-Glasses (1997); on the major Minority Game web site:
   http: //www.unifr.ch/econophysics/minority/
5. D. Challet, M. Marsili and R. Zecchina, Phase Transition in a Toy market (1999); on the major Minority Game web site:
   http://www.unifr.ch/econophysics/minority/

6. Y. Li et al., Evolution in Minority Games I. Games with a Fixed Strategy Space; on the major Minority Game web site:
   http://www.unifr.ch/econophysics/minority/
7. Y. Li et al., Evolution in Minority Games II. Games with Variable Strategy Spaces; on the major Minority Game web site:
   http://www.unifr.ch/econophysics/minority/
8. S. Kurihara, R. Onai, and T. Sugawara, Self-organization based on Nonlinear Non-equilibrium dynamics of Autonomous Agents, Artificial Life and Robotics, Vol. 2, No. 3, 1998.
9. R. D'hulst and G. J. Rodgers, Three-sided Complex Adaptive Systems (1999); on the major Minority Game web site:
   http://www.unifr.ch/econophysics/minority/
10. A. Cavagana, Irrelevance of Memory in the Minority Game, Phys. Rev. E, 59, R3783 (1999).
11. C. W. Reynolds, Flocks, Herds, and Schools: A Distributed Behavior Model, Computer Graphics 21:4, pp. 25-34, (SIGGRAPH 87), 1987.

Part VII

Game-theoretic Approach

What Information Theory Says About Bounded Rational Best Response

David H. Wolpert[1]

NASA Ames Research Center, Moffett Field, CA, 94035, USA
dhw@email.arc.nasa.gov

Summary. Probability Collectives (PC) provides the information-theoretic extension of conventional full-rationality game theory to bounded rational games. Here an explicit solution to the equations giving the bounded rationality equilibrium of a game is presented. Then PC is used to investigate games in which the players use bounded rational best-response strategies. Next it is shown that in the continuum-time limit, bounded rational best response games result in a variant of the replicator dynamics of evolutionary game theory. It is then shown that for team (shared-payoff) games, this variant of replicator dynamics is identical to Newton-Raphson iterative optimization of the shared utility function.

1 Introduction

Recent work has used information theory [9, 12] to provide a principled extension of noncooperate conventional game theory to accommodate bounded rationality [25, 27]. Intuitively, this extension starts with the observation that in the real world ascertaining a game's equilibrium is an exercise in statistical inference: one is given (or assumes) partial information about the behavior of the players, and from that infers (!) what the joint mixed strategy is likely to be. There are many ways to do such statistical inference. The one investigated in [27] is based on information theory's version of Occam's razor: Predict the joint mixed strategy that has as little extra information as possible beyond the provided partial knowledge while being consistent with that knowledge. This version of Occam's razor is known as the Maximum entropy (Maxent) principle [9, 12]. It tells us that the mixed strategy of a game's equilibrium, $q(x \in X) = \prod_i q_i(x_i)$, is the solution to a coupled set of Lagrangian functions that are specified by the game structure and the provided partial knowledge.

Sec. 2 reviews how information theory can be used to derive bounded rational noncooperative game theory. Some simple examples of the bounded rational equilibrium solutions of games are then presented. Sec. 3 analyzes scenarios in which the players use bounded rational versions of best response strategies. Particular attention is paid to team games, in which the players share the same utility function. The analysis for this case provide insight into how to optimize the sequence of moves by the players, as far as their shared utility is concerned. This can be viewed as a formal way to optimize the organization chart of a corporation.

Best response strategies, even bounded rational ones, are poor models of real-world computational players that use Reinforcement Learning (RL) [20]. Sec. 4 considers iterated games in which players use a (bounded rational) variant of best response, a variant that is more realistic for computational players, and arguably for human players as well. In this variant the conditional expected utilities used by player i to update her strategy, expected payoff given move x_i, is a decaying average of recent conditional expected utilities. This decay biases the player to dampen large and sudden changes in her strategy. This variant is then explored for the case of team games. The continuum limit of the dynamics of such games is shown to be variant of the replicator dynamics. It is shown such continuum-limit bounded rational best response is identical to Newton-Raphson iterative optimization of the shared utility function of such games.

The formalism presented in this paper is a special case of the field of Probability Collectives (PC), a case in which the joint distribution over the variables of interest is a product distribution. This special case is known as Product Distribution (PD) theory [25, 27, 29, 28, 26, 7]. PC has many applications beyond those considered in this paper, e.g., distributed optimization and control [16, 15, 2, 29]. Finally, see [16] for relations to other work in game theory, optimization, statistical physics, and reinforcement learning.

2 Bounded Rational Noncooperative Game Theory

In this section we motivate PD theory as the information-theoretic formulation of bounded rational game theory. We use the integral sign ($\int$) with the associated measure implicit, i.e., it indicates sums if appropriate, Lebesgue integrals over $\mathbb{R}^n$ if appropriate, etc. In addition, the subscript (i) is used to indicate all index values other than i. Finally, we use $\mathcal{P}$ to indicate the set of all probability distributions over a vector space, and $\mathcal{Q}$ to indicate the subset of $\mathcal{P}$ consisting of all product distributions (i.e., the associated Cartesian product of unit simplices).

In noncooperative game theory one has a set of N **players**. Each player i has its own set of allowed **pure strategies**. A **mixed strategy** is a distribution $q_i(x_i)$ over player i's possible pure strategies. Each player i also has a **private utility** function g_i that maps the pure strategies adopted by all N of the players into the real numbers. So given mixed strategies of all the players, the expected utility of player i is $E(g_i) = \int dx \prod_j q_j(x_j) g_i(x)$.

In a **Nash equilibrium** every player adopts the mixed strategy that maximizes its expected utility, given the mixed strategies of the other players. More formally, $\forall i, q_i = \text{argmax}_{q_i'} \int dx \; q_i' \prod_{j \neq i} q_j(x_j) \; g_i(x)$. Perhaps the major objection that has been raised to the Nash equilibrium concept is its assumption of **full rationality** [10, 6, 18, 4]. This is the assumption that every player i can both calculate what the strategies $q_{j \neq i}$ will be and then calculate its associated optimal distribution. In other words, it is the assumption that every player will calculate the entire joint distribution $q(x) = \prod_j q_j(x_j)$.

In the real world, this assumption of full rationality almost never holds, whether the players are humans, animals, or computational agents [5, 17, 10, 3, 8, 1, 22, 14]. This is due to the cost of computation of that optimal distribution, if nothing else. This real-world **bounded rationality** is a major impediment to applying conventional game theory in the real world.

2.1 Review of the Minimum Information Principle

Shannon was the first person to realize that based on any of several separate sets of very simple desiderata, there is a unique real-valued quantification of the amount of syntactic information in a distribution $P(y)$. He showed that this amount of information is the negative of the Shannon entropy of that distribution, $S(P) = -\int dy \; P(y)\ln[\frac{P(y)}{\mu(y)}]$. So for example, the distribution with minimal information is the one that doesn't distinguish at all between the various y, i.e., the uniform distribution. Conversely, the most informative distribution is the one that specifies a single possible y. Note that for a product distribution, entropy is additive, i.e., $S(\prod_i q_i(y_i)) = \sum_i S(q_i)$.

Say we given some incomplete prior knowledge about a distribution $P(y)$. How should one estimate $P(y)$ based on that prior knowledge? Shannon's result tells us how to do that in the most conservative way: have your estimate of $P(y)$ contain the minimal amount of extra information beyond that already contained in the prior knowledge about $P(y)$. Intuitively, this can be viewed as a version of Occam's razor: introduce as little extra information beyond that you are provided in your inferring of P. This minimum information approach is called the maxent principle. It has proven extremely powerful in domains ranging from signal processing to supervised learning [12]. In particular, it is has been successfully used in many statistics applications, including econometrics[13]. It has even provided what many consider the cleanest derivation of the foundations of statistical physics [11].

2.2 Maxent Lagrangians

Much of the work on equilibrium concepts in game theory adopts the perspective of an external observer of a game. We are told something concerning the game, e.g., its cost functions, information sets, etc., and from that wish to predict what joint strategy will be followed by real-world players of the game. Say that in addition to such information, we are told the expected utilities of the players. What is our best estimate of the distribution q that generated those expected cost values? By the maxent principle, it is the distribution with maximal entropy, subject to those expectation values.

To formalize this, for simplicity assume a finite number of players and of possible strategies for each player. To agree with the convention in fields other than game theory (e.g., optimization, statistical physics, etc.), from now on we implicitly flip the sign of each g_i so that the associated player i wants to minimize that function rather than maximize it. Intuitively, this flipped $g_i(x)$ is the "cost" to player i when the joint-strategy is x.

With this convention, given prior knowledge that the expected utilities of the players are given by the set of values $\{\epsilon_i\}$, the maxent estimate of the associated q is given by the minimizer of the Lagrangian

$$\mathscr{L}(q) \equiv \sum_i \beta_i [E_q(g_i) - \epsilon_i] - S(q) \tag{1}$$

$$= \sum_i \beta_i [\int dx \prod_j q_j(x_j) g_i(x) - \epsilon_i] - S(q) \tag{2}$$

where the subscript on the expectation value indicates that it evaluated under distribution q. The $\{\beta_i\}$ are "inverse temperatures" implicitly set by the constraints on the expected utilities.

Solving, we get the coupled equations

$$q_i(x_i) \propto e^{-E_{q_{(i)}}(G|x_i)} \tag{3}$$

where the overall proportionality constant for each i is set by normalization, and $G \equiv \sum_i \beta_i g_i$ [1]. In Eq. 3 the probability of player i choosing pure strategy x_i depends on the effect of that choice on the utilities of the other players. This reflects the fact that our prior knowledge concerns all the players equally.

If we wish to focus only on the behavior of player i, it is appropriate to modify our prior knowledge. First consider the case of maximal prior knowledge, in which we know the actual joint-strategy of the players, and therefore all of their expected costs. For this case, trivially, the maxent principle says we should "estimate" q as that joint-strategy (it being the q with maximal entropy that is consistent with our prior knowledge). The same conclusion holds if our prior knowledge also includes the expected cost of player i.

Modify this maximal set of prior knowledge by removing from it specification of player i's strategy. So our prior knowledge is the mixed strategies of all players other than i, together with player i's expected cost. We can incorporate prior knowledge of the other players' mixed strategies directly, without introducing Lagrange parameters. The resultant **maxent Lagrangian** is

$$\mathscr{L}_i(q_i) \equiv \beta_i [\epsilon_i - E_q(g_i)] - S_i(q_i)$$

solved by a set of coupled **Boltzmann distributions**:

---

[1] The subscript $q_{(i)}$ on the expectation value indicates that it is evaluated according the distribution $\prod_{j \neq i} q_j$.

$$q_i(x_i) \propto e^{-\beta_i E_{q_{(i)}}(g_i|x_i)}. \tag{4}$$

Following Nash, we can use Brouwer's fixed point theorem to establish that for any non-negative values $\{\beta\}$, there must exist at least one product distribution given by the product of these Boltzmann distributions (one term in the product for each i).

The first term in $\mathscr{L}_i$ is minimized by a perfectly rational player. The second term is minimized by a perfectly *irrational* player, i.e., by a perfectly uniform mixed strategy q_i. So β_i in the maxent Lagrangian explicitly specifies the balance between the rational and irrational behavior of the player. In particular, for $\beta \to \infty$, by minimizing the Lagrangians we recover the Nash equilibria of the game. More formally, in that limit the set of q that simultaneously minimize the Lagrangians is the set of mixed strategy equilibria of the game, together with the set of delta functions about the pure Nash equilibria of the game. The same is true for Eq. 3.

Note also that independent of information-theoretic considerations, the Boltzmann distribution is a reasonable (highly abstracted) model of human behavior. Typically humans do some "exploration" as well as "exploitation", trying each move with probability that rises as the expected cost of the move falls. This is captured in the Boltzmann distribution mixed strategy.

One can formalize the concept of the rationality of a player in a way that applies to any distribution, not just a Boltzmann distribution. One does this with a **rationality operator** which maps a q and a g_i to a non-negative real value measuring the rationality of player i in adopting strategy q_i given private cost function g_i and strategies $q_{(i)}$ of the other players. For the solution in Eq. 4 and private cost g_i, the value of that operator is just β_i [27].

Eq. 3 is just a special case of Eq. 4, where all player's share the same private cost function, G. (Such games are known as **team games**.) This relationship reflects the fact that for this case, the difference between the maxent Lagrangian and the one in Eq. 2 is independent of q_i. Due to this relationship, our guarantee of the existence of a solution to the set of maxent Lagrangians implies the existence of a solution of the form Eq. 3. Typically players will be closer to minimizing their expected cost than maximizing it. For prior knowledge consistent with such a case, the β_i are all non-negative.

For each player i define $f_i(x, q_i(x_i)) \equiv \beta_i g_i(x) + \ln[q_i(x_i)]$. Then we can write the maxent Lagrangian for player i as

$$\mathscr{L}_i(q) = \int dx \, q(x) f_i(x, q_i(x_i)). \tag{5}$$

Now in a bounded rational game every player sets its strategy to minimize its Lagrangian, given the strategies of the other players. In light of Eq. 5, this means that we can interpret each player in a bounded rational game as being perfectly rational for a cost function that incorporates its computational cost. To do so we simply need to expand the domain of "cost functions" to include (logarithms of) probability values as well as joint moves.

2.3 Examples of Bounded rational Equilibria

It can be difficult to start with a set of cost functions and associated rationalities β_i and then solve for the associated bounded rational equilibrium q. Solving for q when prior knowledge consists of expected costs ϵ_i rather than rationalities can be even more tedious. (In that situation the β_i are not specified upfront but instead are Lagrange parameters that we must solve for.) However there is an alternative approach to constructing examples of games and their bounded rational equilibria that is quite simple. In this alternative one starts with a particular mixed strategy q and then solves for a game for which q is a bounded rational equilibrium, rather than the other way around.

To illustrate this, consider a 2-player single-stage game. Let each player have 3 possible moves, indicated by the numerals 0, 1, and 2. Say the (bounded rational) mixed strategy equilibrium is

$$q_1(0) = 1/2, \; q_1(1) = 1/4, \; q_1(2) = 1/4;$$
$$q_2(0) = 2/3, \; q_2(1) = 1/4, \; q_2(2) = 1/12 \,. \tag{6}$$

Now we know that at the equilibrium, $q_1(x_1) \propto e^{-\beta_1 E(g_1|x_1)}$, where β_1 is player 1's rationality, and g_1 is her cost function (the negative of her cost function). This means for example that

$$e^{-(\beta_1 [E(g_1|x_1=0)-E(g_1|x_1=1)])} \; = \; \frac{q_1(0)}{q_1(1)} \; = \; 2, \text{ i.e.,}$$

$$\beta_1 [E(g_1 \mid x_1 = 0) - E(g_1 \mid x_1 = 1)] = -\ln(2). \tag{7}$$

A similar equation governs the remaining independent difference in expectation values for player 1. The analogous two equations for player 2 also hold.

Now define the vectors $\mathbf{g}_{i;j}(.) \equiv g_i(x_i = j,.)$. So for example $\mathbf{g}_{1;0} = (g_1(x_1 = 0, x_2 = 0), g_1(x_1 = 0, x_2 = 1), g_1(x_1 = 0, x_2 = 2))$. Then we can express our equations compactly as four dot product equalities:

$$\beta_1(\mathbf{g}_{1;0} - \mathbf{g}_{1;1}) \cdot q_2 = -\ln(2) \; ; \quad \beta_1(\mathbf{g}_{1;0} - \mathbf{g}_{1;2}) \cdot q_2 = -\ln(2) \; ;$$
$$\beta_2(\mathbf{g}_{2;0} - \mathbf{g}_{2;1}) \cdot q_1 = -\ln(8/3) \; ; \quad \beta_2(\mathbf{g}_{2;0} - \mathbf{g}_{2;2}) \cdot q_1 = -\ln(8) \,. \tag{8}$$

We can absorb each β_i into its associated g_i; all that matters is their product.

We can now plug in for the vectors q_1 and q_2 from Eq. 6 and simply write down a set of solutions for the four three-dimensional vectors $\mathbf{g}_{i,j}$. For these $\{g_i\}$ the bounded ratinal equilibrium is given by the q of Eq. 6. If desired, we can evaluate the associated expected values of the cost functions for the two players; our q is the bounded ratinal equilibrium for those expected costs.

Note that the variables in the first pair of equalities in Eq. 8 are independent of those in the second pair. In other words, whereas the Boltzmann equations giving q for a specified set of g_i are a set of coupled equations, the equations giving the g_i for a specified q are not coupled. Note also that our equations for the $\mathbf{g}_{i;j}$ are (extremely) underconstrained. This illustrates how compressive the mapping from the g_i to the associated equilibrium q is. Bear in mind though that that mapping is also multi-valued in general; in general a single set of cost functions can have more than one equilibrium, just like it can have more than one Nash equilibrium.

The generalization of this example to arbitrary numbers of players with arbitrary move spaces is immediate. As before, indicate the moves of every player by an associated set of integer numerals starting at 0. Recall that the subscript (i) on a vector indicate all components but the i'th one. Also absorb the rationalities β_i into the associated g_i.

Now specify q and the vectors $g_i(x_i = 0, .)$ (one vector for each i) to be anything whatsoever. Then for all players i, the only associated constraint on the i'th cost function concerns certain projections of the vectors $g_i(x_i > 0, .)$ (one projection for each value $x_i > 0$). Concretely, $\forall i, x_i > 0$,

$$\int dx'_{(i)} \, g_i(x_i, x'_{(i)}) \prod_{j \neq i} q_j(x'_j) = -\ln\left(\frac{q_i(0)}{q_i(x_i)}\right) + \int dx'_{(i)} g_i(0, x'_{(i)}) \prod_{j \neq i} q_j(x'_j),$$

i.e., $\forall i, x_i > 0, \mathbf{g}_i(x_i, .) \cdot q_{(i)} = -\ln\left(\frac{q_i(0)}{q_i(x_i)}\right) + \mathbf{g}_i(0, .) \cdot q_{(i)}.$ \hfill (9)

All the terms on the right-hand side are specified, as well as the $q_{(i)}$ term on the left-hand side. Any $\mathbf{g}_i(x_i, .)$ that obeys the associated equation has the specified q as a bounded rational equilibrium.

See [27] for discussion of alternative interpretations of this information-theoretic formulation of bounded rationality. That reference also discusses kinds of prior knowledge that do not result in the Maxent Lagrangian, in particular knowledge based on finite data sets (Bayesian inference). A scalar-valued quantification of the rationality of a player is also presented there.

3 Bounded Rational Versions of Best Response

One crude way to try to find the q given by Eq. 4 would be an iterative process akin to the best-response scheme of game theory [10]. Given any current distribution q, in this scheme all agents i simultaneously replace their current distributions. In this replacement each agent i replaces q_i with the distribution given in Eq. 4 *based on the current* $q_{(i)}$. This scheme is the basis of the use of Brouwer's fixed point theorem to prove that a solution to Eq. 4 exists. Accordingly, it is called **parallel Brouwer updating**. (This scheme goes by many names in the literature, from Boltzmann learning in the RL community to block relaxation in the optimization community.)

Sometimes conditional expected costs for each agent can be calculated explicitly at each iteration. More generally, they must be estimated. This can be done via Monte Carlo sampling, iterated across a block of time. During that block the agents all repeatedly and jointly IID sample their (unchanging) probability distributions to generate joint moves, and the associated cost values are recorded. These are then use to estimate all the conditional expected costs, which then determine the parallel Brouwer update [2].

This is exactly what is done in RL-based schemes in which each agent maintains a data-based estimate of its cost for each of its possible moves, and then chooses its actual move stochastically, by sampling a Boltzmann distribution of those estimates. (See [25] for ways to get accurate MC estimates more efficiently than in this simple scheme, e.g., by exploiting the bias-variance tradeoff of statistics.)

One alternative to parallel Brouwer updating is **serial** Brouwer updating, where we only update one q_i at a time. This is analogous to a Stackelberg game, in that one agent makes its move and then the other(s) respond [4, 6]. In a team game, any serial Brouwer updating must reduce the common Lagrangian, in contrast to the case with parallel Brouwer updating.

There are many versions of serial updating. In **cyclic** serial Brouwer updating, one cycles through the i in order. In **random** serial Brouwer updating, one cycles through them in a random fashion.

In **greedy** serial Brouwer updating, instead of cycling through all i, at each iteration we choose what single player to update based on the associated drop in the common Lagrangian. Those drops can be evaluated without calculating the associated Boltzmann distributions. To see how, use N_i to indicate the normalization constant of Eq. 4. Then define the **Lagrangian gap** at q for player i as $\ln[N_i] + \int dx_i q_i(x_i) E_{q_{(i)}}(g_i \mid x_i) + \int dx_i q_i(x_i)\ln[q_i(x_i)]$. This is how much $\mathscr{L}$ is reduced if only q_i undergoes the Brouwer update [3].

Another obvious variant of these schemes is mixed serial/parallel Brouwer updating, in which one subset of the players moves in synchrony, followed by another subset, and so on. Such updating in a team game can be viewed as a simple model of the organization chart of the players. For example, this is the case when the players are a corporation, with G being a common cost function based on the corporation's performance.

---

[2] Parallel Brouwer updating has minimal memory requirements on the agents. Say agent i has just made a particular move, getting cost r, and that the most recent previous time it made that time was T iterations ago. Then the new estimated cost for that move, E', is related to the previous one, E, by $E' = \frac{r + k^T E a}{1 + k^T a}$, where k is a constant less than 1, and a is initially set to 1, while itself also being updated according to $a+ = k^T$. So agent i only needs to keep a running tally of E, a, and T for each of its possible moves to use data-aging, rather than a tally of all historical time-cost pairs

[3] Proof outline: Write the entropy after the update as a sum of non-i entropies (which are unchanged by the update) plus i's new entropy. Then expand i's new entropy. This gives the value of the new Lagrangian as -ln[N_i]. Then do the subtraction.

Say we observe the functioning of such an organization over time, and view those observations as Monte Carlo sampling of its behavior. Then we can use those samples to statistically estimate how best to do serial/parallel Brouwer updating, for the purpose of minimizing the shared cost function G. This can be viewed as a way to optimize the organization chart coupling the players.

4 Parallel Brouwer with Data-aging is Nearest Newton

This section considers a variant of best-response that is more realistic (more accurately modeling RL-based computational players that are actually used in machine learning, and arguably more accurately modeling human players as well). In this variant the expected cost used by each player to update her strategy is a decaying average of recent expected utilities; this decay reflects a conservative preference for dampening large changes in strategy.

Such a bias is used (implicitly or otherwise) in most multi-player RL algorithms. For example, in the COIN framework each agent i collects a data set of pairs of what value its private cost function has at timestep t together with the move it made then. It then estimates its cost for move x_i as a weighted average of all the cost values in its data set for that move. The weights are exponentially decaying functions of how long ago the associated observation was made. This **data-aging** is crucial to reflect the non-stationarity of agent i's environment, i.e., that the other agents are changing their strategies with time. Arguably, humans use similar modifications to best response. Indeed, in idealized learning rules like ficticious play, such dampening is crucial.

4.1 The Dynamics of Brouwer Updating

Consider a multi-stage game where at the end of iteration t, each player i updates her distribution $q_i(.,t)$ to

$$q_i(x_i, t) = \frac{e^{-\Phi_i(x_i,t)}}{\int dx_i' e^{-\Phi_i(x_i',t)}}. \tag{10}$$

This is a generalization of parallel Brouwer updating, where the function being exponentiated can be Q values (as in Q-learning[24]), single-instant reward values, distorted versions of these (e.g., to incorporate data-aging), etc.

As an example, for single-instant rewards (i.e., conventional parallel Brouwer), $\Phi_i(x_i, t)$ is player i's estimate of (β_i times) her conditional expected cost for taking move x_i at time $t - 1$. If that estimate were exact, this would mean

$$\Phi_i(x_i, t) = \beta E(g_i \mid x_i) = \beta \int dx_{(i)} q_{(i)}(x_{(i)}, t-1) g_i(x_i, x_{(i)}). \tag{11}$$

As another example, for Q-learning, one player is Nature and her distribution is always a delta function. In this case $\Phi_i(x_i, t)$ is the Q-value for player i

taking action x_i, when the state of Nature is as specified by the associated delta function in $q(., t-1)$.

Note that there's no Monte Carlo sampling being done here, as there is in most real-world RL; this is a somewhat abstracted version of such RL. Alternatively, the analysis here becomes exact when Φ_i is evaluated closed form, or (as when Φ_i is an empirical expectation value) there's enough samples in a Monte Carlo block so that empirical averages effectively give us exact values of expected quantities.

At this point we have to say something about how Φ_i evolves with time. Consider the case where Φ_i is an estimate of some function ϕ_i, formed by exponential aging of the previous ϕ values. In our case (since everything is evaluated closed form) assuming there have been an infinite number of preceding timesteps, this is the same as geometric data-aging:

$$\Phi_i(x_i, t) = \alpha \phi_i(x_i, q(t-1)) + (1-\alpha)\Phi_i(x_i, t-1) \tag{12}$$

for some appropriate function ϕ_i [4]. For example, in parallel Brouwer updating, $\phi_i(x_i, t) = \beta E(g_i \mid x_i, q_{(i)}(t))$, while $\Phi_i(x_i, t)$ is a geometric average of the previous values of $\phi(x_i)$.

4.2 The Continuum-time Limit

To go to the continuum-time limit, let t be a real variable, and replace the temporal delay value of 1 in Eq. 12 with δ and α with $\alpha\delta$ (we'll eventually take $\delta \to 0$). In addition differentiate Eq. 10 with respect to t to get

$$\frac{dq(x_i, t)}{dt} = -q_i(x_i, t)[\frac{d\Phi_i(x_i, t)}{dt} - \int dx_i' q_i(x_i', t)\frac{d\Phi_i(x_i', t)}{dt}]. \tag{13}$$

In the $\delta \to 0$ limit, assuming q is a continuous function of t, Eq. 12 becomes

$$\frac{d\Phi_i(x_i, q)}{dt} = \alpha[\phi_i(x_i, q) - \Phi_i(x_i, q)]. \tag{14}$$

where from now on the t variable is being suppressed for clarity.

If we knew the dynamics of ϕ_i, we could solve Eq. 14 via integrating factors, in the usual way. Instead, here we'll plug that equation for $\frac{d\Phi_i}{dt}$ into Eq. 13. Then use Eq. 10 to write $\Phi_i(x_i, q) = \text{constant} -\ln(q_i(x_i))$. The result is

$$\frac{dq_i(x_i)}{dt} = \alpha q_i(x_i) \left[\phi_i(x_i, q) + \ln(q_i(x_i))\right]$$
$$\int dx_i' \, \alpha q_i(x_i')[\phi_i(x_i') + \ln(q_i(x_i'))]. \tag{15}$$

---

[4] To see this is exponential data-aging with exponent γ set $\gamma = -\ln(1-\alpha)$.

4.3 Relation with Nearest Newton Descent and Replicator Dynamics

As mentioned previously, there are many ways to find equilibria, and in particular many distributed algorithms for doing so. This is especially so in team games, where finding such equilibria reduces to descending a single overarching Lagrangian. One natural idea for descent in such games is to use the Newton-Raphson descent algorithm. However that algorithm cannot be applied directly to search across q in a distributed fashion, due to the need to invert matrices coupling the agents. As an alternative, one can consider what new distribution p the Newton algorithm would step to if there was no restriction that p be a product distribution. One can then ask what product distribution is closest to p, according to Kullback-Leibler distance[9]. It turns out that one *can* solve for that optimal product distribution. The associated update rule is called the **Nearest Newton** algorithm[29].

It turns out that when one writes down the Nearest Newton update rule, it says to replace each component $q_i(x_i)$ with the exact quantity appearing on the right-hand side of Eq. 15, where α is the stepsize of the update, and $\phi_i(x_i, t) = \beta E(G \mid x_i, q_{(i)}(t))$, as in parallel Brouwer updating for a team game [5]. In other words, in team games, the continuum limit of having each player using (bounded rational) best response is identical to the continuum limit of the Newton-Raphson algorithm for descending the Lagrangian, with the data-aging parameter α giving the stepsize.

Eq. 15 arises in other yet other contexts as well. In particular, say Φ_i is conditional expected rewards (i.e., $\phi_i(x_i, t-1) = E(g_i \mid q(., t-1)))$. Then the $\beta \to \infty$ limit of Eq. 15 reduces to a simplified form of the replicator dynamics equation of evolutionary game theory[21, 23]. (If the stepsize α is an appropriately increasing function of $E(G)$ other versions of that dynamics arise.) This is because in that limit the ln term disappears, and the righthand side of Eq. 15 involves only the difference between player i's expected cost and the average expected cost of all players. This 3-way connection suggests using some of the techniques for solving replicator dynamics to expedite either parallel Brouwer or Nearest Newton.

4.4 Convergence and Equilibria

By Eq. 15, at equilibrium, for each i, $q_i(x_i)[\phi_i(x_i, q) + \ln(q_i(x_i))]$ must be independent of i. One way this can occur is if it equals 0. However $q_i(x_i)$ can never be 0, by Eq. 10. This means we have an equilibrium at $q_i(x_i) \propto e^{-\phi_i(x_i, q)}$. Intuitively, this is exactly what we want, according to Eq. 10 and our interpretation of $\phi_i(x_i, q)$ as an estimate of $\phi_i(x_i, q)$. Note also that this solution means that $\phi_i(x_i, q) = \Phi_i(x_i, q)$, so that (according to Eq. 14) $\Phi_i(x_i, q)$ has also reached an equilibrium.

---

[5] More generally, Nearest Newton uses this update rule with $\phi_i(x_i, t) = \beta E(g_i \mid x_i, q_{(i)}(t))$ where each $g_i(x) = G(x) - D(x_{(i)})$ for some function D. See [29].

When our equilibrium has $q_i(x_i)[\phi_i(x_i,q) + \ln(q_i(x_i))] = A \neq 0$, we have

$$q_i(x_i) \propto e^{-q_i(x_i)\phi_i(x_i,q)}. \tag{16}$$

In light of Eq. 10, this means that $\Phi_i(x_i,q) \neq \phi_i(x_i,q)$. So by Eq. 14, $\Phi_i(x_i,q)$ hasn't reached an equilibrium in this case:

$$\frac{d\Phi_i(x_i,q)}{dt} = \alpha\phi_i(x_i,q)[1 - q_i(x_i)]. \tag{17}$$

If both $q_i(x_i)$ and $\phi_i(x_i,q)$ were frozen at this point, this solution for $\Phi_i(x_i,q)$ would not obey Eq. 12. So either $q_i(x_i)$ and/or $\phi_i(x_i,q)$ cannot be frozen. In fact, if $\phi_i(x_i,q)$ varies with time, then we know by Eq. 15 that $q_i(x_i)$ varies as well. So in either case $q_i(x_i)$ must vary, i.e., this equilibrium is not stable.

Although the dynamics has the desired fixed point, it may take a long time to converge there. There are several ways to analyze that: One is to examine the second derivatives (with respect to time) of the q_i and/or the Φ_i. Another is to examine the time-dependence of the residual error,

$$r_i^{ge}(x_i,t) \equiv \frac{e^{-\Phi_i(x_i,t)}}{\int dx_i' e^{-\Phi_i(x_i',t)}} - \frac{e^{-\phi_i(x_i,t)}}{\int dx_i' e^{-\phi_i(x_i',t)}}. \tag{18}$$

The next subsection includes a convergence analysis involving residual errors, but for a different variant of Brouwer from the ones considered so far.

4.5 Other Variants of Brouwer Updating

Data-aging can be viewed as moving only part-way from the current Φ_i to what it should be (i.e. to ϕ_i). An alternative is to dispense with the Φ_i and ϕ_i altogether, and instead step part-way from the current q to what it should be, i.e., partially move to the (bounded rational) best response mixed strategy. Formally, this means replacing Eq. 10 so that the update is not implicit, in how $\Phi_i(x_i,t)$ depends on the past value of $q(t-1)$ (Eq. 12), but explicit:

$$q_i(x_i,t) = q_i(x_i,t-1) + \alpha[h_i(x_i,q_{(i)}(t-1)) - q_i(x_i,t-1)] \tag{19}$$

where $h_i(x_i,q_{(i)}(t))$ is the Boltzmann distribution of what $q_i(x_i,t)$ would be, under ideal circumstances, and we implicitly have small stepsize α.

The only fixed point of this updating rule is where $q_i = h_i$ $\forall i$. So just like with continuum-limit parallel Brouwer, we have the correct equilibrium. To investigate how fast the update rule of Eq. 19 arrives at that equilibrium, write its error at time t as the residual

$$\begin{aligned}
r_i^{st}(x_i,t) &= q_i(x_i,t) - h_i(x_i,q_{(i)}(t)) \\
&= q_i(x_i,t-1)[1-\alpha] + \alpha h_i(x_i,q_{(i)}(t-1)) - h_i(x_i,q_{(i)}(t)) \\
&= q_i(x_i,t-1)[1-\alpha] + \alpha h_i(x_i,q_{(i)}(t-1)) \\
&\quad - h_i[x_i,q_{(i)}(t-1) + \alpha[h_{(i)}(q(t-1)) - q_{(i)}(t-1)]]
\end{aligned} \tag{20}$$

where we have assumed that all all players other than i are updating themselves in the same that i does (i.e., via Eq. 19), and $h_{(i)}(q(t-1))$ means the vector of the values of all $h_{j\neq i}(x_j)$ evaluated for $q(t-1)$.

With obvious notation, rewrite Eq. 20 as

$$r_i^{st}(x_i, t) = q_i(x_i, t-1)[1 - \alpha]$$
$$+ \alpha h_i(x_i, q_{(i)}(t-1))$$
$$- h_i[x_i, \ q_{(i)}(t-1) - \alpha r_{(i)}(t-1)]. \tag{21}$$

Now use the fact that α is small to expand the last h_i term on the righthand side to first order in its second (vector-valued) argument, getting the result

$$r_i^{st}(x_i, t) \approx r_i(x_i, t)[1 - \alpha] + \alpha \nabla h_i \cdot r_{(i)}(t-1) \tag{22}$$

where the gradient of h_i is with respect to the vector components of its second argument. Accordingly, if $r_i^{st}(x_i)$ starts much larger than the other residuals, it will be pushed down to their values. Conversely, if it starts much smaller than them, it will rise.

There are other ways one can reduce a stochastic game to a deterministic continuum-time process. In particular, this can be done in closed form for ficticious play games and some simple variants of it [19, 10].

Acknowledgements: I would like to thank Stefan Bieniawski, Bill Macready, George Judge, Chris Henze, and Ilan Kroo for helpful discussion.

References

1. Al-Najjar, N. I., and R. Smorodinsky, "Large nonanonymous repeated games", Game and Economic Behavior 37, 26-39 (2001).
2. Antoine, N., S. Bieniawski, I. Kroo, and D. H. Wolpert, "Fleet assignment using collective intelligence", Proceedings of 42nd Aerospace Sciences Meeting, (2004), AIAA-2004-0622.
3. Arthur, W. B., "Complexity in economic theory: Inductive reasoning and bounded rationality", American Economic Review 84, 2 (May 1994), 406-411.
4. Aumann, R.J., and S. Hart, Handbook of Game Theory with Economic Applications, North-Holland Press (1992).
5. Axelrod, R., The Evolution of Cooperation, Basic Books NY (1984).
6. Basar, T., and G.J. Olsder, Dynamic Noncooperative Game Theory, Siam Philadelphia, PA (1999), Second Edition.
7. Bieniawski, S., and D. H. Wolpert, "Adaptive, distributed control of constrained multi-agent systems", Proceedings of AAMAS 04 , (2004).
8. Boutilier, C., Y. Shoham, and M. P. Wellman, "Editorial: Economic principles of multi-agent systems", Articial Intelligence Journal 94 (1997), 1-6.
9. Cover, T., and J. Thomas, Elements of Information Theory, Wiley-Interscience New York (1991).
10. Fudenberg, D., and D. K. Levine, The Theory of Learning in Games, MIT Press Cambridge, MA (1998).

11. Jaynes, E. T., "Information theory and statistical mechanics", Physical Review 106 (1957), 620.
12. Jaynes, E. T., and G. Larry Bretthorst, Probability Theory : The Logic of Science, Cambridge University Press (2003).
13. Judge, G., D. Miller, and W. Cho, "An information theoretic approach to ecological estimation and inference", Ecological Inference:New methodological Strategies (King, Rosen, and Tanner eds.), Cambridge University Press (2004).
14. Kahneman, D., "A psychological perspective on economics", American Economic Review (Proceedings) 93:2 (2003), 162-168.
15. Lee, C. Fan, and D. H. Wolpert, "Product distribution theory for control of multi-agent systems", Proceedings of AAMAS 04 , (2004).
16. Macready, William, and David H. Wolpert, "Distributed constrained optimization with semi-coordinate transformations", submitted (2004).
17. Neyman, A., "Bounded complexity justifies cooperation in the finitely repeated prisoner's dilemma", Economics Letters 19 (1985), 227-230.
18. Osborne, M., and A. Rubenstein, A Course in Game Theory, MIT Press Cambridge, MA (1994).
19. Shamma, J.S., and G. Arslan, "Dynamic fictitious play, dynamic gradient play, and distributed convergence to nash equilibria", submitted (2004).
20. Sutton, R. S., and A. G. Barto, Reinforcement Learning: An Introduction, MIT Press Cambridge, MA (1998).
21. Tuyls, K., D. Heytens, A. Nowe, and B. Manderick, "Extended replicator dynamics as a key to reinforcement learning in multi-agent systems", Lecture Notes in Artificial Intelligence, LNAI, (ECML 2003), (2003).
22. Tversky, A., and D. Kahneman, "Advances in prospect theory: Cumulative representation of uncertainty", Journal of Risk and Uncertainty 5 (1992), 297-323.
23. Verbeeck, K., A. Nowe, and K. Tuyls, "Coordinated exploration in stochastic common interest games", Proceedings of AAMAS-3. University of Wales, Aberystwyth, (2003).
24. Watkins, C., and P. Dayan, "Q-learning", Machine Learning 8, 3/4 (1992), 279-292.
25. Wolpert, D. H., "Factoring a canonical ensemble", cond-mat/0307630.
26. Wolpert, David H., "Finding bounded rational equilibria part 1: Iterative focusing", Proceedings of the International Society of Dynamic Games Conference, 2004 , (2004), in press.
27. Wolpert, D. H., "Information theory — the bridge connecting bounded rational game theory and statistical physics", Complex Engineering Systems (A. M. D. Braha and Y. Bar-Yam eds.), (2004).
28. Wolpert, D. H., and S. Bieniawski, "Adaptive distributed control: beyond single-instant categorical variables", Proceedings of MSRAS04 (A. S. et al ed.), Springer Verlag (2004).
29. Wolpert, D. H., and S. Bieniawski, "Distributed control by lagrangian steepest descent", Proceedings of CDC 04 , (2004).

Evolution of Reciprocal Cooperation in the Avatamsaka Game

Eizo Akiyama[1] and Yuji Aruka[2]

[1] Institute of Policy and Planning Sciences, University of Tsukuba, Tel
  +81-298-54-9455, Fax +81-298-53-3848 eizo@sk.tsukuba.ac.jp
[2] Faculty of Commerce, Chuo University, aruka@tamacc.chuo-u.ac.jp

Summary. The Avatamsaka game is investigated both analytically and using computer simulations. The Avatamsaka game is a *dependent game* in which each agent's payoff depends completely not on her own decision but on the other players'. Consequently, any combination of mixed strategies is a Nash equilibrium.

Analysis and evolutionary simulations show that the socially optimal state becomes evolutionarily stable by a Pavlovian strategy in the repeated Avatamsaka game, and also in any kind of dependent game. The mechanism of the evolutionary process is investigated from the viewpoint of the agent's memory and mutation of strategies.

1 The Avatamsaka Game as a Dependent Game

1.1 Robinson Crusoe Economy and Multiple-person Games

In our daily life, we must sometimes make decisions to improve our current situation. If utilities for the results depend only upon our own decision, the problem is usually called an optimization problem, which a solitary person, such as Robinson Crusoe (or a person whose payoff is unrelated to the decisions of other people) meets ("Robinson Crusoe economy" [3]). What matters first in this kind of situation is the possibility of constructing the function itself. Once the function is constructed, solving for the optimal behavior is a computational problem to determine the strategy that optimizes the function. As von Neumann and Morgenstern pointed out in their book "Theory of Games and Economic Behavior," the difficulty in solving for the optimal strategy in one-person games, such as the Traveling Salesman Problem, is a *technical* rather than *conceptual* one.

In multiple person problems, however, the result for each one will depend in general not merely upon his decision but on those of the others as well. Moreover, all participants desire maxima simultaneously. Consequently, it is usually not possible for a player to decide his best behavior without considering others' decision-making processes. Such a situation in general is called a *game*[1].

1.2 Independent / Dependent Game

Let us refer to the Robinson Crusoe problem, in which decisions of others do not have any effect on his utility, as *an independent game*; Robinson Crusoe's effort can be termed *independent optimization*. (In the context of the evolutionary model, it will be called an *independent adaptation*.) In this paper, we consider a class of games that is opposite of the independent game: *dependent games*.

In a dependent game, the result for each agent will depend merely on the decisions of others, but not at all on his own decision. Every player's effort to increase his profit in a dependent game is called a *dependent optimization*. (In the context of the evolutionary model, it will be called a *dependent adaptation*.) Any multiple-person problem is representable as a formulation that lies between the independent game and the dependent game, in principle.

By definition, any change of a player's action in a dependent game does not directly improve her utility. However, if the game is repeated there is a chance her current action might serve as a signal to induce beneficial behaviors from others in the future.

For an example of dependent games, let us consider internet auction sites. Many internet auction sites adopt a "user rating system," with which each buyer rates his experiences with sellers, in order to prevent dishonesty among sellers. In this case, rating sellers high or low does not affect the current utility of the buyers at all, which exemplifies the dependent game situation. Note that not a few members of auction sites *repeatedly* participate in auctions there. As a result, the rating behaviors are repeated for the future, which usually motivates the buyers to give honest ratings. If their participation were not repeated, it would not matter for a player, for example, to rate a good seller "low" without any consideration (and forever leave the auction site).

1.3 The Avatamsaka Game

As the simplest example of dependent games, let us consider the *Avatamsaka Game*[2][1]. Its payoff matrix is shown in Table 1. In this game, a player's

---

[1] In the theory of games, von Neumann and Morgenstern had an insight into the fact that this problem is not an ordinary maximization problem and that the formalization of this type of problem should take the form of a *matrix*. As a result, they succeeded in formalizing the problem as one in which every individual *"can determine the variables which describe his own actions, but not those of the others."* In addition, *"those alien variables cannot, from his point of view, be described by statistical assumptions."*

Table 1. Payoff matrix for the Avatamsaka game: Two matrices below show the points gained by Player 1 (left) and Player 2 (right). The columns indicate the actions that Player 1 would choose and the rows indicate those for Player 2. Because we only deal with symmetric games (i.e. the payoff structures from the viewpoints of Player 1 and of Player 2 are identical), such as the game in this table, we will show the payoff matrices only for Player 1 in the remainder of this paper.

Payoff for player 1

	D	C
D	1	2
C	1	2

Payoff for player 2

	D	C
D	1	1
C	2	2

points depend only on the other agent's behavior. If player 2 chooses behavior D (defect), the point player 1 can gain is only 1.0, no matter what behavior player 1 chooses. If player 2 chooses behavior C (cooperation), player 1 is bound to receive 2.0.

In this game, any combination of both players' mixed strategies is a *Nash equilibrium* because neither player has an incentive to change their strategy. In addition, any mixed-strategy is a *neutrally stable strategy* from the viewpoint of evolutionary game theory.

1.4 Repeated Interactions in a Society, Players' Cognitive Abilities, and Mistakes

Let us imagine that we are actually under circumstances similar to the Avatamsaka game. In this case, various thoughts might come to mind such as "Should I be satisfied with his defection, even though we are in a Nash equilibrium state?" "Can I somehow induce his cooperation?" "Can I somehow use my behavior as a signal to him?" etc. The reason we may not be satisfied with our current game play is that, if we actually encounter a situation like

---

[2] Avatamsaka is a well-known Mahayana Buddhist Sutra. The late Shigeo Kamata, Professor Emeritus, University of Tokyo, who was working under the field of Buddhist philosophy, skillfully illustrated the situation of Heaven and Hell in view of Avatamsaka. (See Kamata [6], pp.167-168.)

Suppose that two men are seated across from each other at a table. They are bound with ropes except for one arm. Each is given an overly long spoon. Thereby, they cannot serve themselves because of the awkwardness of the long spoon, but there is sufficient food for both of them on the table. If they cooperate in concert to provide each other with a meal, they can both be happy. This defines the Heaven, or Paradise. Alternatively, one will be kind enough to provide the other with a meal, but the other might not have the same feeling of cooperation. This case pays the other, which must give rise to a feeling of hate for the inconsiderate opponent. This describes a situation denoted as Hell. (See [1] for the philosophical and sociological significance of this game.)

this game, something other than the activity of "analyzing the Table 1 in the way of traditional game theory" appears in our mind, which might cause mutual cooperation under Avatamsaka game interactions.

For example, we usually interact not only with a certain individual, but also with other community members in the real world. Therefore, it is necessary for us to consider, comprehensively, relationships with various types of individuals (*various interactions in a society*). In addition, we sometimes, or often, consider not only current, but also future relationships with the person (*repeated interactions*). In this case, although my current action has no meaning for me now, but it might have some meaning as a "signal" to derive cooperation of the opponent in the future. Whether or not repeated interaction works as a signal to derive mutual cooperation depends on the perceptions and cognition of both players. Note that, in the course of repeated interactions, we must also consider the fact that we are beings who cannot be free from mistakes in the real world. And so, cognitive abilities to deal with players' mistakes are crucial for formation of cooperation (*players' cognitive abilities and mistakes*). Furthermore, we do not fix our strategy. We sometimes change our strategy to adapt to our current situation (*mutation / modification of the strategies*).

This paper investigates the following points to study the evolution of cooperation under a dependent game: the Avatamsaka game.

1. How can we derive cooperative behavior of the opponents in Avatamsaka situations, when we live in a society in which we interact repeatedly with various players who might sometimes make mistakes? Players surely need cognitive abilities to induce another's cooperation. To what extent are such abilities required for deriving cooperation?
2. How do social phenomena change with players' cognitive abilities? For example, an agent who remembers only the previous actions of the opponents, such as so-called "Tit-For-Tat strategy" in the Prisoner's Dilemma game, may bring about completely different social phenomena from agents who remember the previous behaviors of themselves and of their opponents.
3. Do capricious changes of strategies (mutation) have a fundamental effect on the evolutionary phenomena?

2 The Model

This study addresses a *multi-stage (or repeated, iterated)* Avatamsaka game; we refer to one iteration in the game as a *stage game*. Furthermore, we refer to C or D in a stage game as an *action*, while a complete plan in a multi-stage game for a player to decide his action based on the past information as a *strategy*. We investigate evolutionary phenomena in the Avatamsaka game using an evolutionary model whose formulation was presented by [4] for the study of the Prisoner's Dilemma game.

2.1 Evolutionary Dynamics with Mutational Effect

Suppose that there are infinitely many agents in a population (a game world). All agents in the population play a round-robin tournament with each other in a generation. In each match, an infinitely iterated Avatamsaka game is played between two agents. Suppose that the total population of the agents in the game world is 1.0 and that the fraction of the agents having strategy i $(i = 0, 1, \ldots, n-1)$ in a generation t is x_i, where $\sum_i x_i = 1.0$.

The fitness f_i of strategy i is defined as the expected value of the payoffs in all games in which strategy i is involved in the population. Because the expected payoff of strategy i depends on the probability of meeting with strategies against whom he can gain high (or low) payoffs, f_i depends on the strategy distribution $\boldsymbol{x} = (x_0, x_1, \ldots, x_{n-1}) \in \Delta$ over the population, where $\Delta = \{\boldsymbol{x} : \forall i\, x_i \geq 0 \text{ and } \sum_i x_i = 1.0\}$

Assume that the population share of a strategy either increases or decreases according to the fitness of the strategy and that a small fraction of newborn agents appear as mutants of other strategies. Considering the fact that the average fitness of all agents is $\frac{x_i f_i}{\sum_i x_i f_i}$, the population share x_i' of strategy i in the next generation $t+1$ would be the following.

$$x_i' = \left(\frac{f_i}{\sum_i x_i f_i} x_i + u \right) / (1 + nu), \; i = 0, \ldots, n-1,$$

where u is the constant term representing the effect of the uniform and time-independent mutation: the mutation rate. For computer simulations, we choose two cases $u = 0$ (no mutation) and $u = 0.0004$ (with mutation) to illustrate the effects of mutation. The change of u does not substantially affect the results if $0 < u \ll 1$.

2.2 Iterated Game with Action-noise and the Strategies with Past Memory

In the repeated Avatamsaka game, each agent uses a strategy to determine her action in the next stage game based on the memory of actions that she and the opponent made in previous rounds. We investigate the following three cases in this paper: (1) Each agent continues the initial action without any consideration of past information. This condition is denoted herein as the "$m = 0$ (no memory)" case. (2) Each agent chooses her action referring to the previous action of the opponent ("$m = 1$") (3) Each agent determines her action referring to the previous actions of herself and of the opponent ("$m = 2$"). As shown below, the number of possible pure strategies in this repeated game depends on the agents' memory size. Let us denote the nth strategy with memory size of m by S_n^m.

In the case of no memory ($m = 0$), only two possible pure strategies exist: "Always D (which we denote as AllD or S_0^0)" and "Always C (AllC or S_1^0)."

In the case of $m = 1$, we denote a strategy of an agent as $p_0 p_1$ ($p_0, p_1 \in [0, 1]$), where $p_0(p_1)$ is the probability to play "C" when the opponent's last move was "C" ("D"). For example, the strategy "00" (i.e. $p_0 = 0$, $p_1 = 0$) always plays "D." Four possible pure strategies exist for $m = 1$. They are 00, 01, 10, 11, which we denote respectively as S_0^1, S_1^1, S_2^1, and S_3^1. Note that 00=S_0^1 and 11=S_3^1 are equivalent to AllD and AllC. 10=S_2^1 means that the agent acts in the same manner as the opponent did previously: the so-called Tit-For-Tat (TFT) strategy. 01=S_1^1 means that the agent chooses the opposite action of the opponent's previous action: the Anti-TFT (ATFT) strategy.

The strategy in the case of $m = 2$ can also be described as $p_0 p_1 p_2 p_3$ ($p_i \in [0, 1]$), where p_0, p_1, p_2 and p_3 represent the probabilities to play "C" when, in the previous round, (the agent's move, the opponent's move) = (C, C), (C, D), (D, C) and (D, D), respectively. There are 16 possible pure strategies, such as 0000, ..., 1111 = S_0^2 S_{15}^2. Similarly, the total number of possible pure strategies is 2^{2^m} for the memory size of m.

For simplicity, we assume that each game would be repeated *ad infinitum* between two agents chosen from the population. Moreover, agents are assumed to make *mistakes* with probability p during the repetition. (e.g. The agent with the strategy to play "C" may sometimes play "D" by mistake against his will.) This means that uncertainty (*action-noise*) exists in agents' cognitive functioning. Taking into account the fact that the transition of the probability distribution for the 2^m states above is a Markov process, the transition matrix can be defined uniquely depending on the two agents' strategies. The probability distribution in the steady state corresponds to the normalized eigenvector with eigenvalue 1 of the transition matrix as long as $p > 0$. The average payoff in an infinitely iterated game can be given from the probability distribution of the steady state and the stage game payoff matrix. (See [4].) A slight change of p does not substantially affect the analysis result introduced in the next section as long as $p << 1$. (Basically we assume $0 < p << 1$. Theoretical values found in the next section such as the equilibria of game dynamics are the values gained at $p \to 0$ limit. For computer simulations, we use $p = 0.01$.)

As a short note for strategies with $m = 2$, S_0^2=0000, and S_{15}^2=1111 mean AllD, AllC and S_{10}^2=1010 correspond to TFT, and S_5^2=0101 to ATFT. Moreover, S_8^2=1000 is "GRIM" (that plays "C" only when the previous choices of the both players were "C"), S_9^2=1001 is so-called "PAVLOV" [5]. The last two strategies can be formed only if $m = 2$. Note that the nomenclature used in the above is the one usually used in "Prisoner's Dilemma" studies. For that reason, it might be sometimes misleading to use the above strategy names in *other* games, but we use those names in this paper for simplicity.

3 Analysis and Simulation Results

In this section, we show both the analytical results and the results of computer simulations of the evolutionary Avatamsaka game. To capture the nature of evolutionary phenomena, we especially investigate (i) the dynamics of the population share of the strategies, and (ii) the dynamics of agents' actions in each iterated Avatamsaka games.

3.1 Agents Without any Memory of Past Actions ($m = 0$)

When agents have no memory of past actions ($m = 0$), only two possible pure strategies exist: AllD (S_0^0) and AllC (S_1^0). The payoff matrix for $m = 0$ case is given in Table 2. Obviously, any combination of both players' mixed strategies is a Nash equilibrium. Therefore, the subset of strategies $\mathbf{x} \in \Delta$ that are in Nash equilibrium with themselves, $\Delta^{NE} = \{\mathbf{x} \in \Delta : (x, x)$ is a Nash equilibrium$\}$, is equivalent with the whole mixed-strategy space Δ.

Furthermore, any mixed strategy can be a Neutrally Stable Strategy (NSS), but there is no Evolutionary Stable Strategy (ESS) in this game. (The set of NSS, $\Delta^{NSS} = \Delta$, and the set of ESS, $\Delta^{ESS} = \emptyset$.)

Table 2. Payoff matrix of the iterated Avatamsaka game for $m = 0$ strategies: The procedure to derive this matrix is given in Section 2.2. Because $m = 0$ matrix is simple, we can understand its intuitive meaning. If the opponent is AllD, a player's payoff is $(1 - p) + 2p = 1 + p$ because the opponent's action is "D" with probability $1 - p$ and "C" with probability p, where p is the probability of agent's making mistake.

	AllD	AllC
AllD (S_0^0)	1 + p	2 - p
AllC (S_1^0)	1 + p	2 - p

An example of population dynamics in the case where the agents have no memory ($m = 0$) and no mutation occurs ($u = 0$) is shown in Fig. 1(a). Frequencies of the strategies remain exactly the same as in the initial generation. In the Avatamsaka game, an agent always obtains an identical payoff whether he chooses C or D unless the other acts differently. Because we cannot change the future actions of the strategies AllD and AllC (S_0^0 and S_1^0) that play "D" and "C" forever by their definition, the fitness of either AllD or AllC is determined only by the distribution of AllD and AllC over the population. Therefore, the respective fitness values of AllD and AllC are identical: $f_0 = f_1$. Consequently, the population share of AllC or of AllD does not change with generations. This population dynamics corresponds to the fact that Δ^{NSS} of this game is identical to Δ.

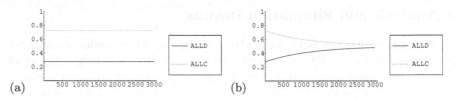

Fig. 1. Population dynamics in the Avatamsaka game world in which agents have no memory of past actions, $m = 0$: The fraction of the strategies in the population is shown as a function of generation. (a) When the mutation rate $u = 0$, the population share of each of the two strategies does not change with generations. (b) When $u = 0.0004$, both converge to 0.5.

If the mutation process is introduced ($u = 0.0004$) in this $m = 0$ Avatamsaka game world, the observed phenomenon is altered completely. In this case, the frequencies (x_0, x_1) of the strategies (AllD, AllC) converge to the equal population share, $(1/2, 1/2)$, with generations (Fig. 1(b)), and the entropy of the configuration of the population is maximized.

3.2 Considering the Opponent's Previous Action ($m = 1$)

Payoff Structure

When agents have a memory of the opponent's previous action ($m = 1$), they can implement four kinds of pure strategies, AllD, ATFT, TFT and AllC, which are denoted respectively as S_0^1, S_1^1, S_2^1, and S_3^1. The payoff matrix for $m = 1$ case is given in Table 3.

In this case, $\Delta^{NE} = \{(u(1-v), v/2, v/2, (1-u)(1-v)) : u, v \in [0,1]\}$ because the expected payoff for any mixed strategy $x \in \Delta$ against $\mathbf{x}^* \in \Delta^{NE}$ is always $3v/2 + (1-v)(1-u-p)$, which is independent of the value of x. Furthermore, the set of NSS, $\Delta^{NSS} = \{((1-v)/2, v/2, v/2, (1-v)/2) : v \in [0,1]\}$. However, there is no ESS.

Table 3. Payoff matrix of the iterated Avatamsaka game for $m = 1$ strategies: The procedure to derive this matrix is given in Section 2.2.

	AllD	ATFT	TFT	AllC
AllD (S_0^1)	$1+p$	$2 - 2p - 2p^2$	$1 + 2p - 2p^2$	$2 - p$
ATFT (S_1^1)	$1+p$	$3/2$	$3/2$	$2 - p$
TFT (S_2^1)	$1+p$	$3/2$	$3/2$	$2 - p$
AllC (S_3^1)	$1+p$	$1 + 2p - 2p^2$	$2 - 2p - 2p^2$	$2 - p$

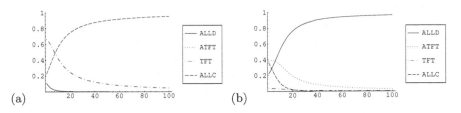

Fig. 2. Avatamsaka game: $m = 1$, mutation rate $= 0$

Population Dynamics

The initial strategy distribution has no influence on the future population dynamics in the case of $m = 0$, but it does in the case of $m = 1$ with no strategic mutation, in most cases.

In this game world, where each agent may have a memory of the previous move of the other, the population share $x = (x_0, x_1, x_2, x_3)$ converges either to $(1, 0, 0, 0)$ or $(0, 0, 0, 1)$ (the dominance by AllD (S_0^1) or by AllC (S_3^1)) with generations in most cases other than the knife-edge cases, for which x is exactly in Δ^{NE} in the initial generation. ($x \in \Delta^{NE}$, of course, does not change with generations.) Convergence depends on the initial distributions of ATFT(S_1^1) and TFT(S_2^1) (Fig. 2). AllC continues to increase when the frequency of TFT is greater than that of ATFT ($x_2 > x_1$); the AllD population continues to increase when $x_1 > x_2$.

Because AllD always chooses D regardless of past actions, the average payoff of *any* opponent of AllD is always 1 in the Avatamsaka game, while that of the opponent of AllC is always 2. Consequently, the effect of the strategy distribution of AllD/AllC on the fitness of any strategy is always the same: it does not affect the population dynamics.

It is ATFT or TFT that may affect the population dynamics. When meeting with AllC, TFT always chooses "C" in response to AllC's previous "C," while ATFT always chooses "D" (except for the change of the action by noise.) That is, TFT benefits AllC and, in the same way, *ATFT benefits AllC*.

As shown above, the appearance of strategies with $m = 1$ gives rise to population dynamics that are dependent on the strategy distribution in the initial generation. Note that the relation $x_2 > x_1$ or $x_1 > x_2$ will not change with generations because the expected payoff of TFT and that of $ATFT$ are identical, which can be known by Table 3. Consequently, if $x_2 > x_1$ in the initial generation, the fraction of AllC continues to increase while that of AllD continues to decrease.

Effect of Mutation

The power of the convergence to the AllD/AllC society observed in the above is not so strong because it results solely from the difference between the relative fraction of TFT to ATFT, not on the interaction *within* AllD (AllC)

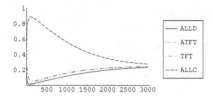

Fig. 3. Avatamsaka game: $m = 1$, mutation rate $u = 0.0004$

agents. When the mutation is introduced to the evolutionary process, **x** converges to $(1/4, 1/4, 1/4, 1/4)$ with generations (Fig. 3). What happens here is the triumph of the mutational effect that increases the entropy of the populational configuration over the adaptation power, which is also observed in the $m = 0$ case.

The world of $m = 1$ shares the same nature with that of $m = 0$ in the sense that all existing strategies cannot conduct independent optimization because the increase of the fitness of AllC (AllD) depends completely on TFT(ATFT). Consequently, the fitness of AllC also approaches that of AllD with generations because the fraction of TFT, whose fitness is always identical to ATFT's, approaches that of ATFT by mutational effect.

The mutation mechanism in principle has the capability of increasing the entropy of the population structure. On the other hand, there usually exists in a game the power that comes from players' adaptation to the payoff matrix. We can say that the Avatamsaka game at $m = 0$ or $m = 1$ is one of the games where the effect of the adaptation is overshadowed by the mutational effect that increases the entropy. In other words, dependent adaptation (See section 1.2) by itself is insufficient to overcome the mutational effect.

3.3 Considering the Player's and the Opponent's Actions (m=2)

Let us consider the case of $m = 2$, in which an agent can refer to both her own and the opponent's previous action to decide the next action. The salient difference is that, while the fitness of each of $m = 0, 1$ strategies depends completely on the population shares of the other strategies, several $m = 2$ strategies exist whose fitness depends on the shares of all strategies including that of the strategies themselves.

The payoff matrix for $m = 2$ strategies is a little complicated function of p; because we have insufficient space for describing the entire matrix (and doing it does not seem so informative), we visualize the payoff structure in Fig. 4 by computing payoffs for $p = 0.01$ and by drawing them as a figure and a graph. It can be confirmed analytically that the "PAVLOV[3] ($1001=S_9^2$)" is the ESS of this game. That fact can also be confirmed with Fig. 4.

---

[3] As in the study of the iterated Prisoner's Dilemma game [5], PAVLOV is a good strategy to form and maintain mutual cooperation under the circumstances with

Population dynamics of the $m = 2$ Avatamsaka game without mutation are shown in Fig. 5. In this setting, the game world converges to the exclusive winning of PAVLOV. Note that the result does not substantially differ even if there is mutation.

Dominance by PAVLOV is realized for any initial condition as long as the initial population share corresponds to the interior of the mixed strategy space Δ. In other words, "independent adaptation" through the use of a $m = 2$ strategy, PAVLOV, overshadows the effect of mutation.

Let us consider why PAVLOV is strong in this game. The PAVLOV policies are as follows: (1) If I feel good (the opponent's action is C), keep doing the current action, whether C or D. (2) Never comply with the situation if I feel uneasy (the opponent's action is D) by sending a signal to the opponent by changing my action[4]. With these two features, PAVLOV can maintain a reciprocal relationship once it is formed because a Pavlovian player continues C as long as he feels good.

Furthermore, two PAVLOVs interacting with each other can quickly recover from betrayal relationships caused by action noise. Assume that two PAVLOVs play a repeated Avatamsaka game, as shown below. PAVLOV's action should be C if both cooperated (CC) in the previous round.

```
PAVLOV | C C D* D C C C ...
PAVLOV | C C C  D C C C ...
```

In this figure, the mark "*" denotes the betrayal of the upper player by mistake (noise) that is supposed to happen with probability p in this study. While the upper player continues D, the lower Pavlovian player, who feels uneasy, sends a signal to the upper player by changing her action into D in the next round (DD), thereby engendering the recovery of mutual cooperation CC two rounds later.

Analysis of the $m = 2$ payoff matrix shows that PAVLOV is the ESS for $m \leq 2$. Furthermore, some additional consideration allows us to know that PAVLOV is also *evolutionarily stable in the whole pure-strategy space* including the longer-memory case if $w < 1$, in which w is a constant probability to another round of repetition[5].

---

action noise. However, PAVLOV is not an evolutionarily stable strategy in the Prisoner's Dilemma with noise because the best response to PAVLOV is AllD in the Prisoner's Dilemma.

[4] Choosing C or D functions only as a "signal" in the Avatamsaka game because the choice does not affect the player's own payoff.

[5] The only thing players in the Avatamsaka game should do is to derive the other's C as frequently as possible. In so doing, the best response to PAVLOV's D that happens by mistake should be D in the next step, which will stop the PAVLOV's continuing D *as quickly as possible*. Therefore, the best response should have a code ((a) $CD \rightarrow D$). The best-response strategy that has fallen into mutual defection with PAVLOV would recover from the tragic state toward the mutual cooperation in the next step ((b) $DD \rightarrow C$). The best response strategy, of course,

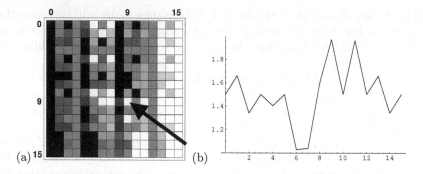

Fig. 4. (a) An illustration of the payoff matrix of the $m = 2$ Avatamsaka game: The vertical axis shows the pure strategy of player 1 and the horizontal axis shows that of player 2. The figures 0, 9, 15 on the vertical and horizontal axes correspond to pure strategies S_0^2, S_9^2, S_{15}^2, respectively. The shade on the tiles represents the payoff for the iterated Avatamsaka game with $p = 0.01$. The higher the payoff, the whiter the color on the correspondent tile. We can visually know the best-response relations among $m = 2$ pure strategies in this figure. (b) The payoff when facing PAVLOV (S_9^2); $p = 0.01$ is plotted as a function of the strategy number. Both (b) and the arrow in (a) indicate that the only best response for PAVLOV is PAVLOV itself. (The payoff for PAVLOV against PAVLOV is $2 - 3p + 6p^2 - 4p^3$.)

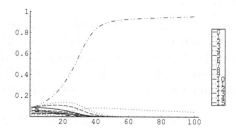

Fig. 5. Population dynamics with $m = 2$ and $u = 0$.

would retain mutual cooperation ((c) $CC \rightarrow C$) If the best-response strategy has produced a mistake, it would defect again ((d) $DC \rightarrow D$) to revert quickly to mutual cooperation with PAVLOV. (Otherwise, PAVLOV would continue D.) (a)–(d) as a whole shows that the only strategy that can derive most frequent cooperation from PAVLOV is PAVLOV itself, provided $w < 1$. (If $w = 1$ (infinite repetition), a strategy that differs only in his first, say, 100 rounds from PAVLOV can have exactly the same payoff as PAVLOV, on average.) Consequently, if all players' strategies are PAVLOV, they are in a strong perfect equilibrium [2], and so PAVLOV is the ESS.

4 Discussion

This paper shows evolutionary phenomena in a dependent game, the Avatamsaka game. Let us sum up the results.

4.1 Memory Size, Mutation and Social Phenomena

First, the memory size generally has a large effect on evolutionary dynamics. For a memory size of $m = 0$, strategic interactions do not have any effect on the population dynamics. Consequently, the population shares of the strategies do not change from the initial generation. When $m = 1$, a strategic interaction becomes more influential than in the case of $m = 0$. As a result, the population converges to either AllC or AllD state, but the convergence depends strongly on the initial configuration of population. If the population consists of agents who consider the previous actions of both themselves and the opponent ($m = 2$), the cooperative society is self-organized in the population by PAVLOV for any initial distribution of strategies that corresponds to a mixed strategy in the interior of Δ. However, further extension of memory size ($m \geq 2$) has no substantial effect on evolutionary phenomena: PAVLOV or its analog always dominates the population. Particularly, if $w < 1$, PAVLOV alone dominates the population. In this sense, an *effective memory-size* of 2 exists in the Avatamsaka game.

Second, the effect of mutation is remarkable in cases of the Avatamsaka game with $m = 0, 1$, in which only dependent adaptation is possible. However, in the case of $m = 2$, the mutational effect is overshadowed by the independent adaptation by PAVLOV, so that the introduction of mutation cannot have a strong effect on evolutionary phenomena. Summing up, investigation of the Avatamsaka game shows clearly that the entropy of a populational configuration is maximized by mutation if there is only dependent adaptation, while the entropy can be decreased when there is an independent adaptation for which a certain size of memory is required for agents in the Avatamsaka game.

4.2 Efficiency in the Dependent Game, and the Origin of Inefficiency

Third, PAVLOV is the only ESS in the Avatamsaka game with $m \leq 2$. In addition, PAVLOV is the only ESS in the Avatamsaka game for any m if $w < 1$, where w is a constant probability of another repetition.

Note that, in *any kind of dependent game* with any number of players and with any number of strategies, a *socially optimal state* in which everybody mutually cooperates is achievable as an evolutionarily stable state through PAVLOV-like strategies. Let us consider a Pavlovian strategy used by an agent who continually changes his action until all others each chooses the most beneficial action for him; that is, the Pavlovian strategy keeps sending signals by changing his action to the others until he feels comfortable. It can be easily understood that all players' mutual cooperation is the only stable state. Furthermore, it is the Pavlovian strategy that derive most frequent cooperation of opponent Pavlovians, which means that the best response to Pavlovian strategies is the Pavlovian strategy itself, and that Pavlovian strategy is ESS (See footnote 5 in Section 3.3)

Summing up, the world of the *dependent game* will eventually engender a stable state that is socially optimal. The world of the *independent game*, in which players each independently confront an optimization problem, will also lead to a socially optimal state wherein the only thing that agents should do is to choose their own optimal strategies, if we disregard computational complexity of the problem. Although players in dependent games should have memories of previous actions and players in independent games do not, both the dependent and the independent games should engender socially optimal states through adaptation.

Seen from another perspective, the origin of social inefficiency that we observe in reality arises from both technical and conceptual difficulties: Technical difficulty lies in computational problems for players to calculate optima, while conceptual difficulty comes from the existence of the *entanglement* of dependent and independent game structures (e.g. the Prisoner's Dilemma) but neither from the structure of completely dependent games nor from that of completely independent games.

Acknowledgment

This research was partially supported by the Ministry of Education, Science, Sports and Culture, Grant-in-Aid for Scientific Research (B), 2002–2003, 14580486. The first author (EA) is partially supported by a Ministry of Education, Science, Sports and Culture, Grant-in-Aid for Young Scientists (B), 2002–2003, 14780342.

References

1. Aruka, Y. (2001), "Avatamsaka Game Structure and Experiment on the Web," in Aruka, Y. (ed.), *Evolutionary controversies in Economics*, Springer, Tokyo, 115–132.
2. Boyd, R. (1989), "Mistakes allow evolutionary stability in the repeated prisoner's dilemma game," *Journal of Theoretical Biology* **136**,47–56.
3. von Neumann, J. and O. Morgenstern (1944), *Theory of Games and Economic Behavior*, Princeton University Press, NJ, USA
4. Nowak, M. A. and Sigmund, K. (1993), "Chaos and the evolution of cooperation." *Proc. Natl. Acad. Sci. USA* **90**, 5091–5094.
5. Nowak, M. A. and Sigmund, K. (1993), "A strategy of win-stay lose-shift that outperforms tit-for-tat in the Prisoner's Dilemma game." *Nature* **364**, 56–58.
6. Kamata, S. (1988), *Kegon no Shiso (The Thought of Avatamsaka)*, in Japanese, Kodan-sha, Tokyo.

Game Representation - *Code Form*

Maria Cristina Peixoto Matos[1] and Manuel Alberto M. Ferreira[2]

[1] Instituto Politécnico de Viseu - Escola Superior de Tecnologia de Viseu
Departamento de Matemática
Campus Politécnico 3504-510 Viseu PORTUGAL
`cristinapeixoto@mat.estv.ipv.pt`
[2] Instituto Superior de Ciências do Trabalho e da Empresa Departamento de
Métodos Quantitativos
Av. das Forças Armadas 1649-026 Lisboa PORTUGAL
`vice.presidente@iscte.pt`

Summary. When we analyze a game we want a full representation, i.e., we want
to be able to look at a game and perceive all of its details. If it allows us to visualize
the respective solution all the better. This aim was the reason for researching a
new game representation which is described in this paper. The *code form* of a game
consists mainly in a table containing all of the information with no suppression or
"adulteration" with regards to the game. Initially presented at Fifth Spanish Meeting
on Game Theory & European Voting Games (Seville 2002), based on examples,
code form is put forth in this paper thoroughly. We begin by presenting *code form*,
exemplifying it with four games. We finish proving the existence of Nash Equilibrium
in *code form* games.
KEY WORDS: Game theory, *Code Form*.

1 Introduction

The idea of a new game representation occurred when we analyzed a game
with 3 players in the normal form. When we model a game we want to capture
the most of the relevant details. A game may have a complex temporal and
information structure. This structure could be very significant to understand
the way the game will be played. Some of these details are not considered
either in the normal form or in the extensive form of a game. It seemed
that a representation that gave us a global picture of the game would be
the ideal. As a consequence of this idea, we tried a game representation that
agglutinated the information of the extensive form and of the normal form
and add some additional information. The fundamental of the *code form* is a
table where the strategies available to any player are codified. We imagined
the *code form* representation and we verified that this representation has great
advantages when the number of players is greater or equal than , when the

game has imperfect information, when the game is a repeated one and mainly when it is sequential. We begin for presenting the mathematical *code form* formalization and then we present four examples, a voting game, a market entry game, a bargaining game and the matching pennies. In the voting game we have three players that have to approve a pay rise for themselves by voting. This is a sequential game that is played only once. In the market entry we have two firms. One of them has a monopoly market and the other decides to enter or not. This is a game with sequential and simultaneous moves that is played only once. The bargaining game is a repeated game with sequential moves where two players are negotiating how they split a given benefit. Each one hoping to retain the bigger share. Finally in the matching pennies we have two players. Each player has a coin hidden in a hand. Then they show simultaneously the coins. One of the players gets the coin of the other if the both coins show the same face. Otherwise is the other player that gains his opponent coin. This is a game with simultaneous moves that is played only once. We finish proving that the *code form* verifies the Nash Theorem once the model definition verifies the conditions of this theorem.

2 The Representation of a Game in *Code Form*

Game theory is a highly developed science, which allows vast and interesting outcomes in the classification, formulation, and resolution of situations both in business and in human relations. This scientific area is also useful in the management of day-to-day conflicts in all domains, whether human or organizational, and even in all situations that involve strategic interaction. As in other sciences, game theory is based on models. The models of this discipline are rigorous abstract presentations of classes of situations taken from real life. Its abstraction allows a wide range of phenomena to be studied. Obviously the main objective of the models is to present solutions. "Solutions" refer to a systematic description of the consequences resulting from one type of game. In following section, we present the mathematical formalization of a new game model.

3 *Code Form* Game - Mathematical Formalization

Definition 1. *A* code form *game consists of a finite table, with evident extension in the case of infinite moves and infinite players, where only some cells are filled. We fill the cells in the same order that the game is played. For that we need:*

- $R = \{1, 2, ..., R\}$, *a set of rounds.*

- $J = \{1, 2, ..., J\}$, *a set of moves.*

- $C = \{1, 2, ..., C\}$, *a set of columns.*

- $L = \{1, 2, ..., L\}$, *a set of rows.*

- $N = \{1, 2, ..., N\}$, *a set of players.*

- $E_n = \{e_1, e_2, ..., e_E\}$, *a set of strategies available to each player.*

- $E = E_1 \times E_2 \times ... \times E_N$, *a set of all such strategy profiles - space of strategies profiles.*

- $\begin{aligned} RN : L \times \{1\} &\to R \\ a_{i1} &\to r \end{aligned}$, *a function that indicate de round number.*

- $\begin{aligned} PN : L \times \{2\} &\to J \\ a_{i2} &\to j \end{aligned}$, *a function that indicate de move number.*

- $\begin{aligned} JE : L \times \{C\} &\to N \times E_N \\ a_{ic} &\to (n, e_n) \end{aligned}$, $c \neq 1, 2, C,$ *a function that indicates who moves and what action is played.*

- $\begin{aligned} PJ : L \times \{C\} &\to IR^n \\ a_{ic} &\to (u_1, u_2, ...u_N) \end{aligned}$, *a function that gives the payoff of every players* *where* $u_n : E \to IR$ *is a von Neumann-Morgenstern utility function.*

We denote when $RN(a_{i1}) = RN(a_{i-1,1})$, $PN(a_{i2}) = PN(a_{i-1,1})$ and $JE(a_{ic}) = PN(a_{i-1,c})$, the cells are not filled. The row is changed when the play change. The column is changed when the player change.

4 Examples

4.1 Voting for a Pay Rise

Let us consider a game where three company board members have to approve a pay rise for themselves by voting. If two of the members vote in favour of the rise it will pass, otherwise the salary will remain the same. Whatever the decision taken, it will be conveyed to the company employees. For this reason, while each of the board members wants to vote against the measure, he wants it to be approved. This position will guarantee a political advantage in so far as, at the eyes of the employees, the administrator will be recognized as a conscientious and unambitious businessman. In this way each player has two possible strategies: approve the rise or to disapprove the rise. The payoffs are established taking into account each player's preference: approval of the rise without his vote.
Thus:

- Rise approved /Yes: 1
- Rise approved /No: 2
- Rise disapproved /Yes:-1
- Rise disapproved/No: 0

Suppose, as this is a sequential game, that player 1 votes first, player 2 votes second and finally player 3 votes.

The following table illustrates the *code form* game representation.

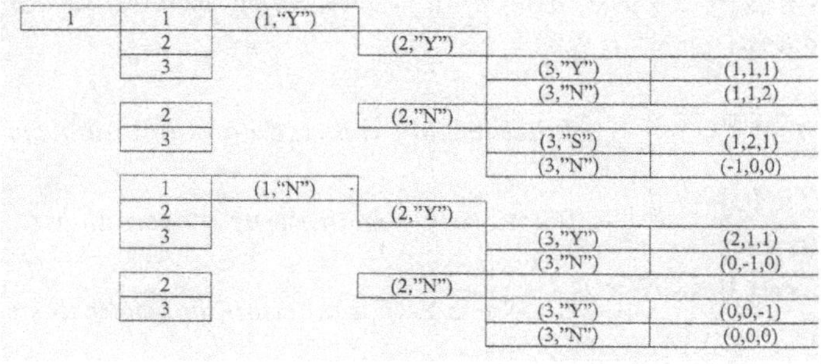

1	1	(1,"Y")			
	2		(2,"Y")		
	3			(3,"Y")	(1,1,1)
				(3,"N")	(1,1,2)
	2		(2,"N")		
	3			(3,"S")	(1,2,1)
				(3,"N")	(-1,0,0)
	1	(1,"N")			
	2		(2,"Y")		
	3			(3,"Y")	(2,1,1)
				(3,"N")	(0,-1,0)
	2		(2,"N")		
	3			(3,"Y")	(0,0,-1)
				(3,"N")	(0,0,0)

Fig. 1. Voting for a Pay Rise

Reading from left to the right, the first table column indicates the period number and the second column indicates the move number. The following columns mention who moves when and in what circumstances and what action is played when somebody is called upon to move. Last column indicates the payoffs vector in accordance with the strategies chosen by the players. As we can see, the *code form* led us to an equilibrium strategy: Player 1 does not approve the pay rise, Player 2 approves the pay rise and Player 3 approves the pay rise. This is the game solution.

4.2 Matching Pennies

Let us consider a game with two players, player 1 and player 2. Both players place simultaneously the coin on the table.
The strategies are:

- If they are both heads or both tails, player 1 takes 1.
- Otherwise player 2 takes 1.

The game is represented by the following table:
What changed in this representation compared to the previous game representation? We can verify, in the simultaneous moves, that the options of the players are at the same level. In fact, we can see the players acting at same time. The remaining keeps in the same way.

1	1	(1,"H")	(2,"H")	(1,-1)
		(1,"H")	(2,"T")	(-1,1)
		(1,"T")	(2,"H")	(-1,1)
		(1,"T")	(2,"T")	(1,-1)

Fig. 2. Matching pennies

4.3 Market Entry

Let us now imagine that we want to analyze games which simultaneously have sequential moves and simultaneous moves. How to translate this situation into *code form*? Knowing that *code form* is supposed to show the whole game information, and also to present to the reader the simultaneous moves.
Let us consider the following example:
Suppose that firm 1 has a monopoly in a given market. Firm 2 is considering to enter this market. If firm 2 decides to enter, then both companies simultaneously decide whether to attack or to accept this situation.
Firstly, let us identify the strategies available to each firm:

- Firm 2 can choose to enter - E - or to stay out of the market - O.
- If it decides to enter then it can choose to attack - T - or to accept - A.
- On the other hand, firm 1 has the possibility, if firm 2 enters the market, to attack - T or to accept - A.

This situation leads us to a game with imperfect information. The following table illustrates the codified representation of the game where the first number of the payoff belongs to the company that wants to penetrate the market and the second number corresponds to the established company.

1	1	(2,"O")			(0,5)
		(2,"E")			
	2		(1,"T")	(2,"T")	(-2,-1)
			(1,"T")	(2,"A")	(-3,1)
	2		(1,"A")	(2,"T")	(0,-3)
			(1,"A")	(2,"A")	(1,2)

Fig. 3. Market Entry

Like the above example, we can verify, in the simultaneous moves, the players acting at same time. The remaining keeps in the same way.

4.4 Sequential Bargaining

The above examples are about one round games. What happens to the *code form* when we have a repeated game? How *code form* translates this situation?

Let us consider a game with two players, player 1 and player 2, which are negotiating how they will split a given benefit, each one hoping to retain the bigger share. Suppose that the offers and counter-offers take time, T periods, and that each player applies a discount factor δ on future benefits, $0 < \delta < 1$ with . The next table represents this situation considering $T = 3$. Let be $x_i(t)$ the sum that player i asks for at moment t.

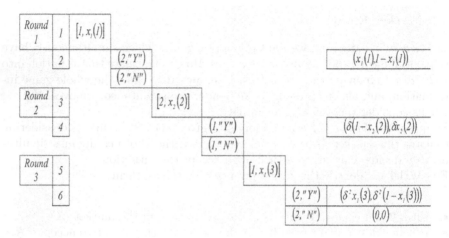

Fig. 4. Sequential bargaining

Again the *code form* allows visualizing all the game information. For, through the *code form*, to identify if is a repeated game, it is enough to reorganize the information. All columns keep the meaning presented in the previous examples.

5 Mixed Strategies

In the above model we defined E_n as the finite set of available strategies for the player. The cartesian product of these sets was denoted by E . So a typical element of this set is $e = (e_1, e_2, ..., e_N)$ where each e_n is a pure strategy [3] for player n, that is, it is an element of E_n . We call $e = (e_1, e_2, ..., e_N)$ a pure strategy profile. When a player plays a pure strategy, his actions are deterministic, that is he does not include uncertainty in his opponent. If he wants to he must act in a random way and then he will have to evaluate probabilistically what the other players will do. In other words, when a player chooses one of his strategies, every other player might be uncertain about which pure strategy player will choose.

---

[3] Pure strategy in the sense of that when the player chooses one of these strategies he does not act in a random way.

Definition 2. *A mixed strategy of player n is a lottery over the pure strategies of player n. One of player $n's$ mixed strategies is denoted σ_n and the set of all player $n's$ mixed strategies is denoted Σ_n.*

Thus $\sigma_n = \left(\sigma_n(e_n{}^1), \sigma_n(e_n{}^2), ..., \sigma_n(e_n{}^{k_n})\right)$ where k_n is the number of pure strategies of player n and $\sigma_n(e_n{}^i) \geq 0, i = 1, 2, .., k_n$ and $\sum \sigma_n(e_n{}^i) = 1$. The cartesian product $\sum = \sum_1 \times \sum_2 \times... \times \sum_N$ is the set of all mixed strategy profiles.

Formally, the mixed strategy set for each player is the probability distribution set over his pure strategy set.

Definition 3. *A $n-dimensional$ simplex defined by the $n+1$ points $x_0, x_1, ...$ x_n in $IR^p, p \geq n$, is denoted $\langle x_0, x_1, ..., x_n \rangle$ and is defined to be the set $\{IR^p \mid$*

$$x = \sum_{j=0}^{n} \theta_j x_j, \sum_{j=0}^{n} \theta_j = 1, \theta_j \geq 0\}$$ *. The simplex is nondegenerate if the n*

vectors $x_1 - x_0, ..., x_n - x_0$ are linearly independent. If $x = \sum_{j=0}^{n} \theta_j x_j$ the numbers

$\theta_0, \theta_1, ..., \theta_n$ are called the barycenter coordinates of x. The barycenter of the simplex $\langle x_0, x_1, ..., x_n \rangle$ is the point having barycenter coordinates $\theta_0 = \theta_1 = ... = \theta_n = \dfrac{1}{n+1}$.

Taking the above definition, since E_n is a finite set and k_n is its cardinal, we can identify Σ_n with $(k_n - 1) - dimensional$ simplex, Δ_{k_n-1} [4]. With mixed strategies the payoff function extends to the mixed strategy profiles set $\Sigma = \Sigma_1 \times \Sigma_2 \times ... \times \Sigma_N$ in expected payoffs terms. To simplify we keep the same notation for the function who represents effective payoffs function and expected payoffs function, that is, u_n represents the expected payoff function of player n as a function of the mixed strategy profile $\sigma = (\sigma_1, \sigma_2, ..., \sigma_N)$. Thus we have:

$$u_n(\sigma) = \sum_{e_1 \in E_1} \sum_{e_2 \in E_2} ... \sum_{e_n \in E_n} \sigma(e_1)\sigma(e_2)...\sigma(e_N)u_n(e_1, e_2, ..., e_N)$$

6 Nash Equilibrium

6.1 Definition

As it was said above, the formulation and resolution of a game is very important to game theory. There are several game solution concepts. Unhappily many of these concepts are restricted to a certain kind of games. The most important solution concept was defined by John Nash (1950) and is known by Nash equilibrium. As we will see the Nash equilibrium existence is guaranteed for a large set of games.

---

[4] $k_n - dimensional$ vectors non negative whose components sum equal to the unit.

Definition 4. *A Nash equilibrium of a game is a profile of mixed strategies* $\sigma = (\sigma_1, \sigma_2, ..., \sigma_N)$ *such that for each* $n = 1, 2, ..., N$, *for each* e_n *and* $e_n{}'$ *in* E_n, *if* $\sigma_n(e_n) > 0$ *then:*
$$u_n(\sigma_1, \sigma_2, ..., \sigma_{n-1}, e_n, \sigma_{n+1}, ..., \sigma_N) \geq u_n(\sigma_1, \sigma_2, ..., \sigma_{n-1}, e_n{}', \sigma_{n+1}, ..., \sigma_N).$$

¿From the above definition we can see that an equilibrium is a profile of mixed strategies such that a player knows what strategies the other players will go to choose, and no player has incentive to deviate from the equilibrium since that he cannot improve his payoff through an unilateral change of his strategy. In fact, the Nash equilibrium induces a necessary condition of strategical stability. however does not induce an enough condition for the analysis of all the games. However, as Ratliff [5] argues "Current attempts to satisfactorily justify the application of Nash equilibrium to a wider class of games many ultimately prove successful, in which the equilibria of these games will be relevant. Furthermore, Nash equilibrium would be the relevant solution concept if there were some pregame dynamic which assured that players' beliefs were in agreement prior to their strategy selection".

Let's go back to our two examples and calculate the equilibria.

As we can see, in "voting for a pay rise" the Nash equilibrium leads us to an equilibrium strategy: Player 1 does not approve the pay rise, Player 2 approves the pay rise and Player 3 approves the pay rise. This is the game solution.

In "matching pennies" we have that there is no equilibrium in pure strategies. The solution is to randomize and play a mixed strategy. Suppose that player 1 estimates that the player 2 will play H (head). Obviously player 1 prefer to play H. But then, from the definition of Nash equilibrium, player 2 has to set positive weight in his estimation of that player 1 would play only on H. Thus, player 2 prefers to play T (tail), contradicting the supposition that player 1 assessed him as playing H. In fact this belief supposing that player 1's estimation of player 2 set weight strictly greater than a half on the fact that player 2 would play H. Similarly we can argue by supposing that player 1's estimation of player 2 set weight strictly less than a half on the fact that player 2 would play H. Thus player 1's estimation of player 2's choices must be $(\frac{1}{2}, \frac{1}{2})$. And similarly player 2's assessment of player 1's choices must also be $(\frac{1}{2}, \frac{1}{2})$.

In "market entry" we are led us to an equilibrium strategy:

$$((Enter, Accept), (Accept)).$$

---

[5] Lecture notes from a PhD game-theory course at the University of Arizona during 1992-1997.

Note that we will only look at the Nash equilibrium, the game presents two credible equilibria:

$((Stayout, Attack), Attack) and ((Enter, Accept), (Accept))$

and a non-credible one:

$((Stayout, Accept), Attack).$

To the "sequential bargaining" we can conclude:

- Round 3:The player 2 should say Yes iff $\delta^2(1 - x_1(3)) \geq 0$, that is $x_1(3) \leq 1$. The player 1 should choose $x_1(3)$ to maximize $\delta^2 x_1(3)$ such that the player 2 says Yes, that is, choose $x_1(3) = 1$.
- Round 2: The player 1 should say Yes iff $x_2(2) \geq \delta x_1(3) = \delta \cdot 1 = \delta$, that is $x_2(2) \geq \delta$. Player 2 offer $x_2(2)$ to maximize $\delta(1 - x_2(2))$ subject to player 1 saying Yes. To get the player 1 to say Yes, $x_2(2)$ must satisfy $x_2(2) \geq \delta$. But it should be clear that the player's 2 maximizes $\delta(1 - x_2(2))$ by making $x_2(2)$ as small as possible. Thus, the solution is for player 2 to set $x_2(2) = \delta$.
- Round 1: The Player 2 should say Yes iff $1 - x_1(1) \geq \delta(1 - x_1(1)) = \delta \cdot (1 - \delta) = \delta - \delta^2$, that is $x_1(1) \leq 1 - \delta + \delta^2$. The player 1 should offer $x_1(1)$ to maximize $x_1(1)$ subject to the player 2 saying Yes. It should be clear that the solution is for the player 2 to set $x_1(1) = 1 - \delta + \delta^2$. Note that, if Player 1 demands a higher price, the player 2 will say no, counter-offer $x_2(2) = \delta$, which the Player 1 will do best to say Yes too, thus getting the smaller payoff of $\delta x_2(2) = \delta^2$.

6.2 Existence

As we see, a particular selection of mixed strategy choices of the players forms Nash equilibrium if and only if each player is using a best response to the strategy choices of the other players, that is, is using a strategy which maximizes the player's payoff given the choices of all of the other players. But in what circumstances is there Nash equilibrium? John Nash answered this question in 1950-1951. Before starting to discuss the answer witch John Nash gave we state for reference an important fixed point theorem.

Theorem 1. [6] *Let $M \subset IR^n$ be a compact convex set. Let $F \to\to M$ an upper hemi-continuous convex valued correspondence. Then the correspondence F has a fixed point.*

Proof:
First suppose the case that M is a nondegenerate simplex in IR^n.

---

[6] Kakutani 1941

Let $M = \langle x_0, x_1, ..., x_n \rangle$, that is, the simplex defined by the $n + 1$ points $x_0, x_1, ..., x_n$.

Now form the mth barycentric subdivision of M [7]. We define the continuous function $f^m : M \rightarrow M$ such that:

> If x is the vertex of any cell of the subdivision let $f^m(x)$ for some $y \inf(x)$. For any other x we define $f^m(x) = y$ by extending the function in a linear manner inside each cell. That is, if x is in the cell $\langle x_0, x_1, ..., x_n \rangle$ and $x = \sum_{j=0}^{n} \theta_j x_j$ with $x = \sum_{j=0}^{n} \theta_j = 1, \theta_j \geq 0$ then $f^m(x)$ is defined to be $x = \sum_{j=0}^{n} \theta_j f^m(x_j)$.

Applying Brouwer's Theorem [8] to obtain a fixed point of the map f^m, say x^m. If is a vertex of one of the cells in the subdivision then we are done since $x^m = f^m(x^m) \in F(x^m)$. If x^m is not a vertex of one of the cells then let the cell in which it does lie be $\langle x_0{}^m, x_1{}^m, ..., x_n{}^m \rangle$, and let $\theta_0{}^m, \theta_1{}^m, ..., \theta_n{}^m \rangle$ be the barycentric coordinates of x^m relative to that cell. Thus:

$$x^m = \sum_{j=0}^{n} \theta_j{}^m x_j{}^m;$$

and

$$x^m = f^m(x^m) = \sum_{j=0}^{n} \theta_j{}^m y_j{}^m \qquad (1)$$

where

$$y_j{}^m = f^m(x_j{}^m) \in F(x_j{}^m \text{ for all } j = 0, 1, ..., n).$$

Now choose a subsequence of $m \rightarrow \infty$ say $m_k \rightarrow \infty$ such that:

$$x^{m_k} \rightarrow x^*, \theta^{m_k} \rightarrow \theta^j, \text{ and } y_j{}^{m_k} \rightarrow = y_j.$$

Also, since the cells shrink to points as $m_k \rightarrow \infty$ each of the vertices of the cell containing x^{m_k} also converges to x^*, i.e., $x^{m_k} \rightarrow x^*$.

Thus from (1)

---

[7] A $0 - dimensional$ simplex is itself its subdivided simplex. A $i - dimensional$ simplex $\langle x_0, x_1 \rangle$ is subdivided into two simplexes of the same dimension $\langle x_0, x_1 \rangle$ and $\langle x_0, x_1 \rangle$ where y is the barycentre of $\langle x_0, x_1 \rangle$.

[8] Brouwer's Theorem (1910): Let $f : B^n \rightarrow B^n$ be a continuous function from the $n - ball$ to itself. Then there is some $x^* \in B^n$ such that $x^* = f(x^*)$, i.e., there is fixed point.

$$x^* = \sum_{j=0}^{n} \theta_j y_j.$$

Also since F is upper hemi-continuous $y_j \in F(x^*)$.

Since $F(x^*)$is convex and x^* is a convex combination of the $y_j's$ this implies that $x *^\in F(x^*)$as required.

Now suppose M is not a simplex. We take some simplex M' containing M and a retraction $\varphi : M' \to M$ [9].

Then $F' : M' \to\to M'$defined by $F'(x) = F(\varphi(x))$ is clearly an upper hemi-continuous correspondence and a fixed point of F' clearly lies in M and so is also a fixed point of F.

Theorem 2. [10] *The mixed extension of every finite game has, at least, one strategic equilibrium.*

Proof [11]:

Consider the set-valued mapping (or correspondence) that maps each strategy profile, x, to all strategy profiles in which each player's component strategy is a best response to x(that is, maximizes the player's payoff given that the others are adopting their components of x). If a strategy profile is contained in the set to which it is mapped (is a fixed point) then it is an equilibrium. This is so because a strategic equilibrium is, in effect, defined as a profile that is a best response to itself.

Thus the proof of existence of equilibrium amounts to a demonstration that the best response correspondence has a fixed point. The fixed-point theorem of Kakutani asserts the existence of a fixed point for every correspondence from a convex and compact subset of Euclidean space into itself, provided two conditions hold. One, the image of every point must be convex. And two, the graph of the correspondence (the set of pairs (x, y) where is in the image of x) must be closed.

Now, in the mixed extension of a finite game, the strategy set of each player consists of all vectors (with as many components as there are pure strategies) of non-negative numbers that sum to 1; that is, it is a simplex. Thus the set of all strategy profiles is a product of simplexes. In particular, it is a convex and compact subset of Euclidean space. Given a particular choice of strategies by the other players, a player's best responses consist of all (mixed) strategies that put positive weight only on those pure strategies that yield the highest expected payoff among all the pure strategies. Thus the set of best responses is a sub simplex. In particular, it is convex. Finally, note that the conditions that must be met for a given strategy to be a best response to a given profile are all weak polynomial inequalities, so the graph of the

---

[9] Continuous function taking M' to M that leaves all points of M fixed.

[10] Nash 1950, 1951

[11] This proof was given by John Hillas for the intensive Course on the Game Theory of Strategic Equilibrium, given at SUNY at Sony Brook in the summer of 1990.

best response correspondence is closed. Thus all the conditions of Kakutani's theorem hold, and this completes the proof of Nash's theorem.

7 A Process to Find the Game Solution

As we see, in the above examples, there are many games where players' moves are sequential, that is players move in accordance with a given order. These games have a great applicability in economics, business and political science. We present an algorithm that helps us to solve these games in the simplest manner. The process consists of considering the rationality of the players the way the game should be played and to find its equilibrium. This process has a key idea which is the fact that if a player can foresee what the other players will do, he is in fact in a position to determine the strategy that will allow him to get the best payoff, and have the ability to evaluate the credibility of his opponents' strategies.

In this way we can see how selecting the optimum payoff, at each respective sets of payoffs of the deciding player we will easily be able to find the game equilibrium. This selection is made in an order contrary to the development of the game, that is, the process starts off by selecting the payoffs of the player who has the last word and ends with the selection of the payoffs of whom initially plays.

7.1 Best Payoff Method

Algorithm:
1. Let N be the player who takes the last decision in the game. Consider any move that comes just before terminal move. Select the payoff sets associated with each move. In each of those sets select the player $N's$ optimum payoff;
2. Consider the subgame that is obtained by excluding the column which refers to player N;
3. Let N' be the player which takes the decision immediately prior to player N;
4. Let $N = N'$. Repeat the process from step 1 until the player who makes the first decision is found;
5. Select the strategy profile induced by the payoff vector obtained.

8 Conclusion

We presented a new game model which provides a global view of the game more clear and provides more information about the game than the other game representations. Considering that the importance of a representation rests on its ease of interpretation, ease of solving a game and truthful information about the game, *code form* representation is the only one, which fulfils all these requirements. We also observed that a Nash equilibrium exists if and only if there exists a fixed point of a particular best-response correspondence. Kakutani's fixed-point theorem guarantees the existence of this fixed point when the correspondence satisfies certain conditions. We verify that the best-response correspondence satisfies the Kakutani conditions and hence prove the existence of a Nash equilibrium. Looking at *code form* we can see that game in *code form* satisfies the Nash Theorem.

References

1. Aubin, J. P.; "Applied Functional Analysis"; John Wiley & Sons, Inc; New York, 1979
2. Bicchieri, Cristina; Jeffrey, Richard, Skyrms, Brian, "The Logic of Strategy". Oxford University Press, Inc.. 1999
3. Davis, Morton D., "Introducción a la Teora de Juegos", Spanish translation by José Carlos Gómez Borrero. Ciencia e Tecnología, Alianza Editorial. Madrid. 1986
4. Brezis, H.; "Analyse Fonctionelle"; Mac Millan Publishing Co. Inc.; New York, 1979
5. Dresher, Melvin; "Jeux de Stratgie- Théorie et Applications""; Springer-Verlag, Dunod, Paris; 1965
6. Eberwein, Curtis, "The Sensitivity of Repeated Bargaining Outcomes to the Choice of First Proposer". June. 2000
7. Fudenberg, Drew; Tirole, Jean, "Game Theory". Cambridge. Mass: Mit. Press. 1991
8. Gardner, Roy, "Juegos para Empresarios e Economistas". Spanish translation by Paloma Calvo e Xavier Vila. Antoni Bosch, Editor, S.A. Barcelona. March. 1996
9. Gibbons, Robert, "Game Theory for Applied Economist". Princeton University Press. 1992
10. Gibbons, Robert, "Um Primer Curso de Teora de Juegos". Spanish translation by Paloma Calvo and Xavier Vila. Antoni Bosch, Editor, S.A. Barcelona. October. 1993
11. Kakutani, S: "A Generalization of Brouwer's Fixed Point Theorem", Duke Mathematics Journal., 8, 1941
12. Kennan, John; Wilson, Robert, "Bargaining with Private Information". Journal of Economic Literature 31. 1993
13. Matos, M.C.; Ferreira, M.A.M., "Games in Code Form". Presented at the Fifth Spanish Meeting on Game Theory. Seville. 1-3 July. 2002
14. Matos, M. C.; Ferreira, M. A. M., "Games in Code Form versus games in Extensive Form". Presented at Quantitative Methods in Economics (Multiple Criteria Decision Making XI). The Slovak Society for Operations Research. Faculty of Economics and Management /Slovak Agricultural University in Nitra). Nitra Slovakia. 5-7 December. 2002

15. Matos, M. C.; Ferreira, M.A.M., "Jogos na Forma Codificada". Temas em Métodos Quantitativos 3. Editores: Elizabeth Reis e Maria Manuela Hill. ISCTE. Edições Sílabo. Lisbon. 2003
16. Matos, M. C.; Ferreira, M. A. M., "Best Payoff Method". Presented at XV IMGTA. Italian Meeting on Game Theory and Applications. Urbino. Italy. 9-12 July. 2003
17. Matos, M. C.; Ferreira, M. A. M., "Impaciência nos Jogos de Negociação". Revista Portuguesa e Brasileira de Gestão. N2/2003. Volume VIII. September 2003. FGV. EBAPE. INDEG. ISCTE
18. Morris, Peter; "Introduction to Game Theory" Springer-Verlag, New York Inc.; 1977
19. Nash. J, "Non-Cooperative Games". Annals of Mathematics, 54, 1951
20. Neumann, J. von; Morgenstern, O., "Theory of Games and Economic Behaviour". John Wiley & Sons, Inc. New York. 1967
21. Osborne, Martin J., "An introduction to game theory". Oxford University Press. 2000

Effect of Mutual Choice Metanorm in Group Dynamics for Solving Social Dilemmas

Tomohisa Yamashita, Kiyoshi Izumi, Koichi Kurumatani

Investigation of Improvement of Traffic Efficiency by Route Information Sharing
Information Technology Research Institute
National Institute of Advanced Industrial Science and Technology
2-41-6, Aomi, Koto-ku Tokyo 135-0064, Japan
tomohisa.yamashita@aist.go.jp, kiyoshi@ni.aist.go.jp,
k.kurumatani@aist.go.jp

Summary. In this research, our aim is the increase of both each individual utility and social welfare in road transportation systems by decreasing traffic congestion. For the attainment of our purpose, we propose a simple route choice mechanism based on the concept of mass user support. Through multiagent simulation, we analyze the effect of our proposed mechanism on the improvement of traffic efficiency of individual drivers and the whole systems.

1 Introduction

Recently, ubiquitous computing environment in road transportation systems has been developed rapidly; car navigation systems have been spread widely, accuracy of GPS (Global Positioning System) are advanced increasingly, a Vehicle Information and Communication System (VICS) started and then the range which VICS can provide congestion information has been extended [11], in-vehicle communication devices, sensors, processors have been developed.

In road transportation systems, one of most important information service is navigation from origin to destination. Many researchers have been trying to develop navigation systems based various kinds of traffic information [3, 7]. The purpose of navigation systems is to maximize the efficiency of individual users by providing the routes that satisfying the intension of user, e.g., the route with shortest travel distance or travel time. In previous research, it has been revealed optimization of the performance of road transportation systems is difficult to improve by providing current congestion information [6, 9, 10]. The reason is that congestion is often caused by route choice behaviors for congestion avoidance. Based on the current state of congestion, navigation system provides the shortest time route that is not congested currently. If many drivers using such systems also choose one route, they concentrate one route simultaneously. As a result, traffic congestion is caused in this route where navigation systems provided as the shortest route in distance. To make matters worse, traffic congestion sometimes becomes heavier in spite of the increase of drivers using traffic information. Based on current state that car navigation systems have been diffused rapidly, this is a serious problem. Furthermore, because this kind of behavior is observed not only in road transportation systems but also in large scale theme park and event hall [2, 8], general solutions for congestion have been expected.

Here, we introduce the concept of mass user support [4, 5] in order to avoid a cause of unintentional congestion. Mass user support provides information services to users, or groups, or society as mass beyond conventional personal services. Although conventional service considers only utilities of individual users, mass user support considers not only the summation of personal services to individual users but also the interaction among users. The purpose of mass user support is increase of both each individual utility but totally social welfare, and definitely not the increase of totally social welfare at the sacrifice of utilities of certain users. That is to say, mass user support requires some kind of social coordination, which leads mutual concession among users [4] . Here, the best solution in whole road transportation systems isn't necessarily required. The reason is that, if the performance of road transportation systems can be improved slightly by mass user support, the profit on the society is immeasurable because of large-scale of road transportation systems. As a result of improvement, there are following merits of the improvement of the performance; i) reduction of directly economic loss, and ii) decrease of the harmful effects of pollutant included in exhaust gas of vehicles, iii) provision of more comfortable driving environment to drivers.

In this research we challenge to construct a navigation system that improves the traffic efficiency of individual drivers and the whole systems. For this purpose, based on mass user support, we propose a simple route choice mechanism that makes a mutual concession among users autonomously. By our proposed mechanism, it is expected that the increase of both each individual utility and totally social welfare can be realize. With multiagent simulation, we examine the effect of our proposed mechanism on the following two points. i) As the drivers using our proposed mechanism increases, the efficiency of them, e.g., travel time from an origin to a destination, becomes better monotonously. ii) The efficiency of the drivers using our proposed mechanism is always better than the drivers using other kind of mechanisms.

2 Multiagent Modeling

2.1 Traffic Model

In this paper, in order to examine interdependency between traffic congestion and route choice behaviors in road transportation systems, we construct traffic flow model as simple as possible. Therefore, we don't consider following

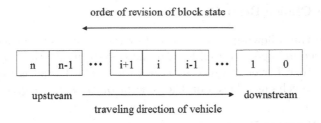

Fig. 1. Direction of movement of vehicles and revision of block

factors; traffic signal (i.e., stop at a red light), waiting for right turn of on-coming cars in intersections, right turn lane, multi-lane, overtaking, a blind alley, U-turn on lane except intersections.

In our traffic flow model, the road between intersections is called a link, which is divided into several blocks. The length of a block is equal to the distance that a vehicle runs at free flow speed of the link during one simulation step. And then, an order is assign to each block from downstream to upstream. Concerning the block assigned i-th, we define K_i as the density of block i, L_i as the length of block i, N_i as the number of the vehicles in block i, and V_i as the velocity of vehicle in block i. Here, K_i is the division of N_i by L_i. In block i, V_i is revised based on Greenshield's V-K relationship. This relationship represents that speed and density on block i are linearly related, and is described as

$$V_i = b - aK_i, \qquad (1)$$

where a and b are coefficients determined from field observations. In this simulation, we set these coefficients as $a = 99.20$ and $b = 13.89$.

The process of the flow calculation between the two neighboring blocks i and $i + 1$ is as follows. At every step, the speed of vehicles in each block is revised according to the V-K relationship, and then vehicles move forward based the speed. The move of vehicles is processed from downstream to up-stream as shown in Figure 1. Based on V_i, vehicle j can move forward. When vehicle j moves into block i from block $i + 1$, its speed changes to V_i from V_{i+1}. If the density of block i is over jam density K_{jam}, no vehicles can move into block i from block $i + 1$. After the process of the move of vehicle j_1 in front of vehicle j_2, if vehicle j_1 is within the distance that vehicle j_2 can move forward at the speed of vehicles V_i, vehicle j_2 approaches vehicle j_1 to the minimum distance between two cars. Although vehicle j_2 has enough speed to move forward more, it has to stay at the back of vehicle j_1. At the next step in block i, when V_i is revised based on K_i, vehicles can accelerate or slowdown to V_i immediately regardless of the speed in the last step.

2.2 Route Choice Behavior

We prepare the following three types of drivers in a route choice behavior. First and second types are well-known and easy-to-understand because they simply minimize travel distance or travel time. Third type is our proposed route choice mechanism based on the concept of mass user support.

Shortest Distance Route

The drivers using the Shortest Distance Route (SD) decide a route based on the knowledge of the map without information of traffic congestion. The drivers using the SD a route that has the shortest distance from its origin to destination. They use the only distance of a link and search the shortest route in distance and don't consider traffic congestion at all.

Shortest Time Route

The drivers using the Shortest Time Route (ST) decide a route based on the knowledge of the map with information of current traffic congestion. The drivers of this type represent the dynamic route choice based on not only map information but also current congestion information of entire network at anytime from traffic information center (e.g., VICS Center) through vehicle equipments. They search the route that has the shortest travel time from their origins to destinations, and revise their route whenever they face an intersection based on current congestion information.

The driver of using the ST is assumed to receive current density of all blocks through vehicle equipments from traffic information center. They considers the expected travel time of a link, which is calculated based on the current density of each block as follows. At first, the speed on block i is calculated based on the V-K relationship with density K_i. And then, the passage time of block i is calculated based on length L_i of block i and speed V_i on block i. Finally, the passage time of a link is calculated by summing the passage time of all block on the link. We define expected travel time ETT_l as the summation of the passage time of all blocks on link l. The driver using the ST searches the shortest route in the expected travel time from its current position to destination.

Shortest Time Route with Route Information Sharing

The drivers using the Shortest Time Route with Route Information Sharing (ST-RIS) decide a route based on the knowledge of the map, information of current traffic congestion, and information of the routes of other drivers using ST-RIS. The drivers using the ST-RIS also search the route and revise their route whenever they face the intersections. The difference of the ST is that the drivers using the ST-RIS share their routes from thier origin to destination.

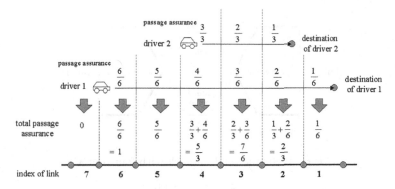

Fig. 2. Example of calculation of total passage assurance: Driver 1 has the route through six links 6, 5, 4, 3, 2, 1 from current position on link 6 to its destination on link 1. Driver 2 has the route through three links 4, 3, 2 from current position on link 4to its destination on link 2.

This sharing can be realized easily with communication devices, e.g., cellular phone, through route information server.

At first, the driver using the ST-RIS searches the shortest route from its origin to destination in the expected travel time. And then, the drivers notify their routes to route information server. Route information server collects the routes of all drivers using the ST-RIS, and assigns the passage assurance of a driver to a link based on their routes, current positions, and destinations as follows. The passage assurance means the degree of the assurance that a driver will passes through a link in the future. We define passage assurance $PA_{l,j}$ of driver j to link l as follows. If a route passes through p links from its current position to destination, route information server assigns a order to each link from a destination to current position of the route. And then, route information server divides the order by p and regard it as a weight of a link. For example, $1/p$ is assigned a link including a destination, 1 $(=p/p)$ is assigned a link including a current position.

Furthermore, we define total passage assurance TPA_l as the summation of the passage assurance of all drivers to link l. Route information server calculates TPA_l as follows,

$$TPA_l = \sum_{k \in ST-RIS}^{k} PA_{l,k}, \qquad (2)$$

where $ST\text{-}RIS$ is the set of the drivers using the ST-RIS. Figure 2 shows an example of calculating total passage assurance of links.

Route information server provides the total passage assurance of all links for the drivers using the ST-RIS. Finally, we define prospective traffic volume PTV_l of link l as the product of ETT_l and $(TPA_l + 1.0)$. The driver using the ST-RIS searches the shortest route in the prospective traffic volume from its current position to destination.

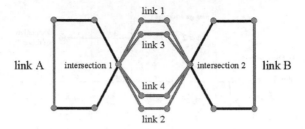

Fig. 3. Simple road network with four routes

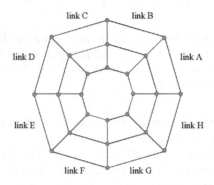

Fig. 4. Radial and ring network

3 Computer Simulation

3.1 Simulation Settings

In order to evaluate effects of our proposed route choice mechanism in various conditions, i.e., in the different components of route choice behaviors, we change the share of three route choice types, the Shortest Distance Route (SD), the Shortest Time Route (ST), and the Shortest Time Route with Route Information Sharing (ST-RIS). Concretely, we prepare three cases of the share of three types. In Case 1 and 2, we treat simple situation as follows. In Case 1, we omit the ST-RIS, and change the share of the SD and the ST. On the other hand, in Case 2, we omit the ST, and change the share of the SD and the ST-RIS. In the future, traffic information is easier to access for many drivers because navigation systems have spread widely. Therefore, in Case 3, we fixed the share of the SD to 20%, and change the share of the ST and the ST-RIS.

Furthermore, in order to evaluate route choice behaviors in various configurations of road networks, we use two kinds of the configurations of road networks. One network configuration is a simple road network that all drivers choose one route out of four routes to reach destination. These configurations are shown in Figure 3 and Table 1. In the case of this simple road network,

Table 1. Conditions of the simple network and the radial and ring network.

	simple network	radial and ring
Number of links	16	40
Number of nodes	14	24
Number of vehicles	1200	1650
Total length (km)	125.9	173.1

links 1 and 3 have the same length, and links 2 and 4 also have the same length. The length of links 1and 3 is slightly longer than that of links 2 and 4. An origin of a driver is assigned randomly to any block on link A or B in Fig. 3. If an origin of a driver is on link A, a destination of it is assigned to any block on link B. Otherwise, a destination of it is assigned on link A. All links has the same capacity except links in links 1 to 4. In the case of four routes, the capacity of links 1, 2, 3, 4 is 1/4 of other links. The other is a radial and ring network shown in Figure 4 and Table 1. In this road network, all links have the same capacity. An origin of a driver is assigned randomly to any block on link A to H in Fig. 4. And then, its destination is assigned to any block on a symmetric link. For example, if an origin is assigned on link B, its destination is assigned on link F.

In order to make the vehicle density in the simple road network and the radial and ring network all the same value, we arrange the number of vehicles in the network based on each total length. At every simulation step, twenty vehicles are generated randomly on links A or B in the case of the simple road network, and on links A to H in the case of the radial and ring networks. After reaching its destination, a driver returns to its origin. And then, a driver goes to its destination again. A driver repeats to go to its destination and return to its origin throughout simulation.

In order to compare our simulation results under the different shares of route choice type in the different road networks, we define ideal travel distance, ideal travel time, and travel time efficiency. The ideal travel distance of a driver is defined as the distance from an origin to destination when a driver passes through the shortest route in distance. The ideal travel time is defined as the time required from an origin to destination when a driver passes through the shortest route in distance at free flow speed. The travel time efficiency is defined as the ratio of the time required actually from an origin to destination and the ideal travel time.

3.2 Simulation Results

As a result of simulations under the conditions about the share of drivers' types and the configurations of road networks, especially we focus on the average travel time efficiency of each type and all drivers. The results of our simulation are shown in Figure 5 to 10. The travel time efficiency is calculated

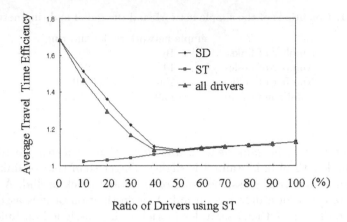

Fig. 5. Average travel time efficiency of Case 1 in simple road network with four routes

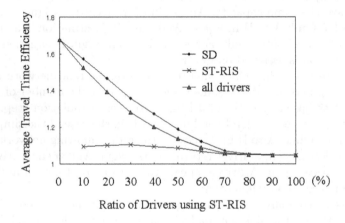

Fig. 6. Average travel time efficiency of Case 2 in simple road network with four routes

based on the average of 10 trials. In these graphs, the horizontal axis is the share of the ST or ST-RIS, and the vertical axis is the travel time efficiency.

Figure 5, 6 and 7 show the result in the simple network with four routes. Fig. 5 shows the average travel time efficiency in Case 1. The ST continued rising monotonously from beginning to end. Until 50% of the share of the ST, the SD and all drivers fell steadily, and then increased slightly. After 50%, all of three are almost the same. Fig. 6 shows the average travel time efficiency in Case 2. Until 30% of the share of the ST-RIS, the ST-RIS went up slightly. After that, the ST-RIS went down slightly. Until 50% of the share of the ST, the SD and all drivers fell steadily from beginning to end. Fig. 7 shows the

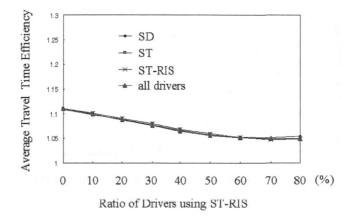

Fig. 7. Average travel time efficiency of Case 3 in simple road network with four routes

average travel time efficiency in Case 3. Up to 60% of the share of the ST-RIS, the SD, ST, ST-RIS, and all drivers decreased steadily. After 60%, however the average of the SD went up slightly, the rest remained almost constant. Furthermore, the ST-RIS is worse than others until 60%, and after that, the ST-RIS is better than others. Throughout simulation, their values were very close.

Figure 8, 9 and 10 show the result in the radial and ring network. In the graphs in Fig. 8 and 9, there are the parts of dashed line. This means that the travel time efficiency of these parts was not a valid value for comparing with the other parts. In the conditions that 0 or 10% of all drivers used the ST and the rest used the SD, 80% of all vehicles were caught in a deadlock. In our traffic simulation, the deadlock is defined as follows; vehicles in the top of link could not move into next links because these links are filled with other vehicles completely. Furthermore, the vehicles in the top of such a link also could not move into next links on the same reason.

Fig. 8 shows the average travel time efficiency in Case 1. After 20% of the share of the ST, the ST increased slightly and steadily. At first, the SD and all drivers dropped dramatically. After 80%, all of them declined marginally. Fig. 9 shows the average travel time efficiency in Case 2. Although the shapes of the SD and all drivers were similar to these in Fig. 8, the ST-RIS decreased slightly and steadily. Fig. 10 shows the average travel time efficiency in Case 3. From beginning to end, all of them continued decreasing steadily. Until 50% of the share of the ST-RIS, all are ranked in descending order as the SD, ST-RIS, all drivers, and ST. After that, the rank was changed to the ST-RIS, all drivers, SD, and ST.

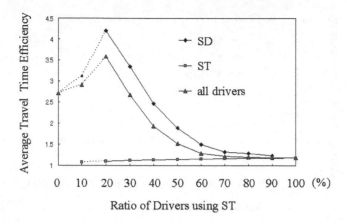

Fig. 8. Average travel time efficiency of Case 1 in the radial and ring network

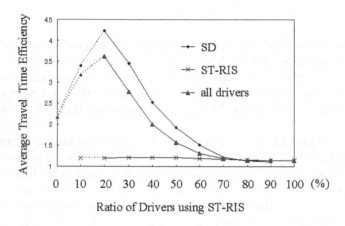

Fig. 9. Average travel time efficiency of Case 2 iin the radial and ring network

4 Discussion

In this section, we discuss the effect of our proposed mechanism to increase both each individual utility and social welfare.

At first, concerning the results in the simple network with four routes, there was the improvement of traffic congestion with the increase of the ST-RIS. In Fig. 5, as the ST increased, the travel time efficiency of the ST became worse. The reason is that the drivers using the ST often concentrated simultaneously one route that traffic information system told that there was no traffic congestion . In Fig. 6, as the ST-RIS increased, the travel time efficiency of the ST-RIS and all drivers became better. In Fig. 7, as the share of the ST-RIS rose, the travel time efficiency of all types was improved monotonously.

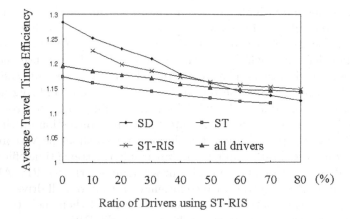

Fig. 10. Average travel time efficiency of Case 3 in the radial and ring network

The drivers using the ST-RIS were able to avoid concentrating simultaneously one certain route where there is no traffic congestion currently. Our proposed mechanism realized a mutual concession among users autonomously. Furthermore, although the efficiency of the ST-RIS was not always better than the drivers of the SD and ST, there were little differences between the ST-RIS and the other types. Therefore, the ST-RIS was roughly satisfied with the two features that we require. In the simple network with four routes, our proposed mechanism works efficiently in the improvement of the travel time efficiency.

Next, concerning the results in the radial and ring network, traffic congestion was improved as the ST-RIS increased. In Fig. 8, as the ST increases, the travel time efficiency of the ST becomes worse. On the other hand, in Fig. 9, as the ST-RIS increased, that of the ST-RIS became better In Fig. 10, as the share of the ST-RIS rose, the travel time efficiency of all types was improved monotonously. Similarly to the simple network with four routes, the drivers using the ST-RIS were able to avoid concentrating simultaneously one certain route. The difference is that the travel time efficiency of the ST is always better than that of the ST-RIS, and that of SD was better than that of the ST-RIS after 50% share of the ST-RIS. These relationships between the travel time efficiency of the ST and ST-RIS, i.e., i) as the ST-RIS increased, the travel time efficiency of the ST and the ST-RIS was improved monotonously, and ii) the travel time efficiency of the ST was always better than that of the ST-RIS, are satisfied the conditions of the social dilemmas [1].

According to Dawes the social dilemmas are caused in the situation where individuals may contribute to a group voluntarily but all members share the benefits. Based on game theoretical analysis, it is evident that the decision to do nothing is always the best strategy regardless of what the rest does. The reason is that the individuals can acquire more benefit whether it contributes or not. However, all individuals would acquire more if all contribute.

In our traffic model, it can be considered that cooperative behavior is the ST-RIS, and defective behavior is the SD (and the ST) because the travel time efficiency of all drivers increases as the ST-RIS increases, and the drivers using the ST always acquire better efficiency. In the radial and ring network, a mutual concession among the ST-RIS, i.e., avoiding congested routes and taking the longer route each other, improves traffic congestion on the whole system. The ST can pass through the shorter route without heavy congestion in distance than the ST-RIS and can improve only the travel time efficiency of only them at the expense of a mutual concession among the ST-RIS. Although the ST-RIS improves the travel time efficiency of itself and all drivers, the ST and SD free-ride the contribution of the ST-RIS, and then their travel time efficiency is improved more significantly than the ST-RIS.

Concerning the construction of the effective mechanism based on route information sharing, there is room of improvement of our proposed mechanism. Currently, our proposed mechanism is very simple because information server receives route from individual drivers and return the accumulative information to them regardless of current traffic congestion, the ratio of drivers using our proposed mechanism, and the intention and status of drivers. If our mechanism can adapt dynamic traffic status, it will be able to improve the efficiency of the driver using it and all drivers. However, we observed that the route choice behaviors face the problem of the social dilemmas. The drivers using other types like the SD and ST free-ride the mechanism based on mass user support making a mutual concession. It is not difficult relatively to develop the mechanisms that can continue improving the travel time efficiency monotonously with the increase of their share. However, it is difficult significantly to develop the mechanisms that can realize the better travel time efficiency than other types. In order to solve the social dilemmas in route choice behaviors, we need to construct new types of route navigation mechanisms against the social dilemmas. Because the social dilemmas have been taken up in various research fields earnestly, we apply the knowledge to our mechanism and then will be able to acquire new solutions suited to the features of traffic flow.

5 Conclusion

In this paper, in order to decrease traffic congestion, we proposed the navigation mechanism with route information sharing based on mass user support. With multiagent modeling, we construct a simple traffic flow model based on the V-K relationships. As the route choice behavior, three types, the Shortest Distance Route (SD), the Shortest Time Route (ST), and the Shortest Time Route with Route Information Sharing (ST-RIS), are prepared. For examine the effect of this mechanism in various conditions, we used three kinds of the shares of the SD, ST, and ST-RIS, and two kinds of road networks, one is the simple networks with four routes from an origin to a destination, and the other is the radial and ring network. As a result of simulations, we conformed that i) in the simple network with four routes, the ST-RIS works efficiently, ii) in the radial and ring network, the ST-RIS can make a mutual concession, and can improve the efficiency of the ST-RIS and all drivers. However, the social dilemma situation is caused, and then the ST and SD free-ride the ST-RIS. Therefore, it is difficult for the ST-RIS to realize the best travel time efficiency. Based on such results, we pointed that it is necessary to modify our proposed mechanism and construct new types of mechanisms to conquer the social dilemmas in route choice behaviors.

References

1. Dawes, R. M. 1981. Social Dilemmas. Annual Review of Psychology 31: 169-193.
2. Kawamura, H.; Kurumatani, K.; and A. Ohuchi. 2003. Modeling of Theme Park Problem with Multiagent for Mass User Support, In Working Note of The International Joint Conference of Artificial Intelligence 2003, Workshop on Multiagent for Mass User Support, 1-7.
3. Klugl, F.; Bazzan, A.L.C.; and Wahle, J. 2003. Selection of information types based on personal utility: a testbed for traffic information markets. In Proceedings of the second International Joint Conference on Autonomous Agents and Multiagent systems, 377- 384.
4. Kurumatani, K. 2003. Mass User Support by Social Coordination among Users. In Proceedings of the International Joint Conference of Artificial Intelligence 2003, Workshop on Multiagent for Mass User Support MAMUS-03, 58-59.
5. Kurumatani, K. 2003. Social Coordination with Architecture for Ubiquitous Agents: CONSORTS. In Proceedings of International Conference on Intelligent Agents, Web Technologies and Internet Commerce 2003 (CD-ROM).
6. Mahmassani, H. S.; Jayakrishnan, R. 1991. System Performance and User Response under Real-Time Information in a Congested Traffic Corridor. Transportation Research 25A(5): 293-307.
7. Shiose, T.; Onitsuka, T.; and Taura, T. 2001. Effective Information Provision for Relieving Traffic Congestion. In Proceedings of 4th International Conference on Intelligence and Multimedia Applications,138-142.
8. Suzuki, R.; Arita T. 2003. Effects of Information Sharing on Collective Behaviors in Competitive Populations. In Proceedings of the Eight International Symposium on Artificial Life and Robotics, 36-39.
9. Tanahashi, I.; Kitaoka, H.; Baba, M.; H. Mori, H.; Terada, S.; and Teramoto, E. 2002. NETSTREAM, a Traffic Simulator for Large-scale Road Networks, R & D Review of Toyota CRDL, 37(2):47-53 (in Japanese).
10. Yoshii, T.; Akahane, H.; and Kuwahara, M. 1996. Impacts of the Accuracy of Traffic Information in Dynamic Route Guidance Systems. In Proceedings of The 3rd Annual World Congress on Intelligent Transport Systems (CD-ROM).
11. http://www.vics.or.jp

Lecture Notes in Economics and Mathematical Systems

For information about Vols. 1–475
please contact your bookseller or Springer-Verlag

Vol. 476: R. Demel, Fiscal Policy, Public Debt and the Term Structure of Interest Rates. X, 279 pages. 1999.

Vol. 477: M. Théra, R. Tichatschke (Eds.), Ill-posed Variational Problems and Regularization Techniques. VIII, 274 pages. 1999.

Vol. 478: S. Hartmann, Project Scheduling under Limited Resources. XII, 221 pages. 1999.

Vol. 479: L. v. Thadden, Money, Inflation, and Capital Formation. IX, 192 pages. 1999.

Vol. 480: M. Grazia Speranza, P. Stähly (Eds.), New Trends in Distribution Logistics. X, 336 pages. 1999.

Vol. 481: V. H. Nguyen, J. J. Strodiot, P. Tossings (Eds.). Optimization. IX, 498 pages. 2000.

Vol. 482: W. B. Zhang, A Theory of International Trade. XI, 192 pages. 2000.

Vol. 483: M. Königstein, Equity, Efficiency and Evolutionary Stability in Bargaining Games with Joint Production. XII, 197 pages. 2000.

Vol. 484: D. D. Gatti, M. Gallegati, A. Kirman, Interaction and Market Structure. VI, 298 pages. 2000.

Vol. 485: A. Garnaev, Search Games and Other Applications of Game Theory. VIII, 145 pages. 2000.

Vol. 486: M. Neugart, Nonlinear Labor Market Dynamics. X, 175 pages. 2000.

Vol. 487: Y. Y. Haimes, R. E. Steuer (Eds.), Research and Practice in Multiple Criteria Decision Making. XVII, 553 pages. 2000.

Vol. 488: B. Schmolck, Ommitted Variable Tests and Dynamic Specification. X, 144 pages. 2000.

Vol. 489: T. Steger, Transitional Dynamics and Economic Growth in Developing Countries. VIII, 151 pages. 2000.

Vol. 490: S. Minner, Strategic Safety Stocks in Supply Chains. XI, 214 pages. 2000.

Vol. 491: M. Ehrgott, Multicriteria Optimization. VIII, 242 pages. 2000.

Vol. 492: T. Phan Huy, Constraint Propagation in Flexible Manufacturing. IX, 258 pages. 2000.

Vol. 493: J. Zhu, Modular Pricing of Options. X, 170 pages. 2000.

Vol. 494: D. Franzen, Design of Master Agreements for OTC Derivatives. VIII, 175 pages. 2001.

Vol. 495: I. Konnov, Combined Relaxation Methods for Variational Inequalities. XI, 181 pages. 2001.

Vol. 496: P. Weiß, Unemployment in Open Economies. XII, 226 pages. 2001.

Vol. 497: J. Inkmann, Conditional Moment Estimation of Nonlinear Equation Systems. VIII, 214 pages. 2001.

Vol. 498: M. Reutter, A Macroeconomic Model of West German Unemployment. X, 125 pages. 2001.

Vol. 499: A. Casajus, Focal Points in Framed Games. XI, 131 pages. 2001.

Vol. 500: F. Nardini, Technical Progress and Economic Growth. XVII, 191 pages. 2001.

Vol. 501: M. Fleischmann, Quantitative Models for Reverse Logistics. XI, 181 pages. 2001.

Vol. 502: N. Hadjisavvas, J. E. Martínez-Legaz, J.-P. Penot (Eds.), Generalized Convexity and Generalized Monotonicity. IX, 410 pages. 2001.

Vol. 503: A. Kirman, J.-B. Zimmermann (Eds.), Economics with Heterogenous Interacting Agents. VII, 343 pages. 2001.

Vol. 504: P.-Y. Moix (Ed.), The Measurement of Market Risk. XI, 272 pages. 2001.

Vol. 505: S. Voß, J. R. Daduna (Eds.), Computer-Aided Scheduling of Public Transport. XI, 466 pages. 2001.

Vol. 506: B. P. Kellerhals, Financial Pricing Models in Con-tinuous Time and Kalman Filtering. XIV, 247 pages. 2001.

Vol. 507: M. Koksalan, S. Zionts, Multiple Criteria Decision Making in the New Millenium. XII, 481 pages. 2001.

Vol. 508: K. Neumann, C. Schwindt, J. Zimmermann, Project Scheduling with Time Windows and Scarce Resources. XI, 335 pages. 2002.

Vol. 509: D. Hornung, Investment, R&D, and Long-Run Growth. XVI, 194 pages. 2002.

Vol. 510: A. S. Tangian, Constructing and Applying Objective Functions. XII, 582 pages. 2002.

Vol. 511: M. Külpmann, Stock Market Overreaction and Fundamental Valuation. IX, 198 pages. 2002.

Vol. 512: W.-B. Zhang, An Economic Theory of Cities. XI, 220 pages. 2002.

Vol. 513: K. Marti, Stochastic Optimization Techniques. VIII, 364 pages. 2002.

Vol. 514: S. Wang, Y. Xia, Portfolio and Asset Pricing. XII, 200 pages. 2002.

Vol. 515: G. Heisig, Planning Stability in Material Requirements Planning System. XII, 264 pages. 2002.

Vol. 516: B. Schmid, Pricing Credit Linked Financial Instruments. X, 246 pages. 2002.

Vol. 517: H. I. Meinhardt, Cooperative Decision Making in Common Pool Situations. VIII, 205 pages. 2002.

Vol. 518: S. Napel, Bilateral Bargaining. VIII, 188 pages. 2002.

Vol. 519: A. Klose, G. Speranza, L. N. Van Wassenhove (Eds.), Quantitative Approaches to Distribution Logistics and Supply Chain Management. XIII, 421 pages. 2002.

Vol. 520: B. Glaser, Efficiency versus Sustainability in Dynamic Decision Making. IX, 252 pages. 2002.

Vol. 521: R. Cowan, N. Jonard (Eds.), Heterogenous Agents, Interactions and Economic Performance. XIV, 339 pages. 2003.

Vol. 522: C. Neff, Corporate Finance, Innovation, and Strategic Competition. IX, 218 pages. 2003.

Vol. 523: W.-B. Zhang, A Theory of Interregional Dynamics. XI, 231 pages. 2003.

Vol. 524: M. Frölich, Programme Evaluation and Treatment Choise. VIII, 191 pages. 2003.

Vol. 525: S. Spinler, Capacity Reservation for Capital-Intensive Technologies. XVI, 139 pages. 2003.

Vol. 526: C. F. Daganzo, A Theory of Supply Chains. VIII, 123 pages. 2003.

Vol. 527: C. E. Metz, Information Dissemination in Currency Crises. XI, 231 pages. 2003.

Vol. 528: R. Stolletz, Performance Analysis and Optimization of Inbound Call Centers. X, 219 pages. 2003.

Vol. 529: W. Krabs, S. W. Pickl, Analysis, Controllability and Optimization of Time-Discrete Systems and Dynamical Games. XII, 187 pages. 2003.

Vol. 530: R. Wapler, Unemployment, Market Structure and Growth. XXVII, 207 pages. 2003.

Vol. 531: M. Gallegati, A. Kirman, M. Marsili (Eds.), The Complex Dynamics of Economic Interaction. XV, 402 pages, 2004.

Vol. 532: K. Marti, Y. Ermoliev, G. Pflug (Eds.), Dynamic Stochastic Optimization. VIII, 336 pages. 2004.

Vol. 533: G. Dudek, Collaborative Planning in Supply Chains. X, 234 pages. 2004.

Vol. 534: M. Runkel, Environmental and Resource Policy for Consumer Durables. X, 197 pages. 2004.

Vol. 535: X. Gandibleux, M. Sevaux, K. Sörensen, V. T'kindt (Eds.), Metaheuristics for Multiobjective Optimisation. IX, 249 pages. 2004.

Vol. 536: R. Brüggemann, Model Reduction Methods for Vector Autoregressive Processes. X, 218 pages. 2004.

Vol. 537: A. Esser, Pricing in (In)Complete Markets. XI, 122 pages, 2004.

Vol. 538: S. Kokot, The Econometrics of Sequential Trade Models. XI, 193 pages. 2004.

Vol. 539: N. Hautsch, Modelling Irregularly Spaced Financial Data. XII, 291 pages. 2004.

Vol. 540: H. Kraft, Optimal Portfolios with Stochastic Interest Rates and Defaultable Assets. X, 173 pages. 2004.

Vol. 541: G.-y. Chen, X. Huang, X. Yang, Vector Optimization. X, 306 pages. 2005.

Vol. 542: J. Lingens, Union Wage Bargaining and Economic Growth. XIII, 199 pages. 2004.

Vol. 543: C. Benkert, Default Risk in Bond and Credit Derivatives Markets. IX, 135 pages. 2004.

Vol. 544: B. Fleischmann, A. Klose, Distribution Logistics. X, 284 pages. 2004.

Vol. 545: R. Hafner, Stochastic Implied Volatility. XI, 229 pages. 2004.

Vol. 546: D. Quadt, Lot-Sizing and Scheduling for Flexible Flow Lines. XVIII, 227 pages. 2004.

Vol. 547: M. Wildi, Signal Extraction. XI, 279 pages. 2005.

Vol. 548: D. Kuhn, Generalized Bounds for Convex Multistage Stochastic Programs. XI, 190 pages. 2005.

Vol. 549: G. N. Krieg, Kanban-Controlled Manufacturing Systems. IX, 236 pages. 2005.

Vol. 550: T. Lux, S. Reitz, E. Samanidou, Nonlinear Dynamics and Heterogeneous Interacting Agents. XIII, 327 pages. 2005.

Vol. 551: J. Leskow, M. Puchet Anyul, L. F. Punzo, New Tools of Economic Dynamics. XIX, 392 pages. 2005.

Vol. 552: C. Suerie, Time Continuity in Discrete Time Models. XVIII, 229 pages. 2005.

Vol. 553: B. Mönch, Strategic Trading in Illiquid Markets. XIII, 116 pages. 2005.

Vol. 554: R. Foellmi, Consumption Structure and Macroeconomics. IX, 152 pages. 2005.

Vol. 555: J. Wenzelburger, Learning in Economic Systems with Expectations Feedback (planned) 2005.

Vol. 556: R. Branzei, D. Dimitrov, S. Tijs, Models in Cooperative Game Theory. VIII, 135 pages. 2005.

Vol. 557: S. Barbaro, Equity and Efficiency Considerations of Public Higer Education. XII, 128 pages. 2005.

Vol. 558: M. Faliva, M. G. Zoia, Topics in Dynamic Model Analysis. X, 144 pages. 2005.

Vol. 559: M. Schulmerich, Real Options Valuation. XVI, 357 pages. 2005.

Vol. 560: A. von Schemde, Index and Stability in Bimatrix Games. X, 151 pages. 2005.

Vol. 561: H. Bobzin, Principles of Network Economics. XX, 390 pages. 2006.

Vol. 562: T. Langenberg, Standardization and Expectations. IX, 132 pages. 2006.

Vol. 563: A. Seeger (Ed.), Recent Advances in Optimization. XI, 455 pages. 2006.

Vol. 564: P. Mathieu, B. Beaufils, O. Brandouy (Eds.), Artificial Economics. XIII, 237 pages. 2005.

Vol. 565: W. Lemke, Term Structure Modeling and Estimation in a State Space Framework. IX, 224 pages. 2006.

Vol. 566: M. Genser, A Structural Framework for the Pricing of Corporate Securities. XIX, 176 pages. 2006.

Vol. 567: A. Namatame, T. Kaizouji, Y. Aruga (Eds.), The Complex Networks of Economic Interactions. XI, 343 pages. 2006.

Vol. 568: M. Caliendo, Microeconometric Evaluation of Labour Market Policies. XVII, 258 pages. 2006.

Vol. 569: L. Neubecker, Strategic Competition in Oligopolies with Fluctuating Demand. IX, 233 pages. 2006.

Vol. 570: J. Woo, The Political Economy of Fiscal Policy. X, 169 pages. 2006.

Vol. 571: T. Herwig, Market Conform Valuation of Options. VIII, 104 pages. 2006.